This book series aims to provide details of blockchain implementation in technology and interdisciplinary fields such as Medical Science, Applied Mathematics, Environmental Science, Business Management, and Computer Science. It covers an in-depth knowledge of blockchain technology for advance and emerging future technologies. It focuses on the Magnitude: scope, scale & frequency, Risk: security, reliability trust, and accuracy, Time: latency & timelines, utilization and implementation details of blockchain technologies. While Bitcoin and cryptocurrency might have been the first widely known uses of blockchain technology, but today, it has far many applications. In fact, blockchain is revolutionizing almost every industry. Blockchain has emerged as a disruptive technology, which has not only laid the foundation for all crypto-currencies, but also provides beneficial solutions in other fields of technologies. The features of blockchain technology include decentralized and distributed secure ledgers, recording transactions across a peer-to-peer network, creating the potential to remove unintended errors by providing transparency as well as accountability. This could affect not only the finance technology (crypto-currencies) sector, but also other fields such as:

Crypto-economics Blockchain
Enterprise Blockchain
Blockchain Travel Industry
Embedded Privacy Blockchain
Blockchain Industry 4.0
Blockchain Smart Cities,
Blockchain Future technologies,
Blockchain Fake news Detection,
Blockchain Technology and It's Future Applications
Implications of Blockchain technology
Blockchain Privacy
Blockchain Mining and Use cases
Blockchain Network Applications
Blockchain Smart Contract
Blockchain Architecture
Blockchain Business Models
Blockchain Consensus
Bitcoin and Crypto currencies, and related fields

The initiatives in which the technology is used to distribute and trace the communication start point, provide and manage privacy, and create trustworthy environment, are just a few examples of the utility of blockchain technology, which also highlight the risks, such as privacy protection. Opinion on the utility of blockchain technology has a mixed conception. Some are enthusiastic; others believe that it is merely hyped. Blockchain has also entered the sphere of humanitarian and development aids e.g. supply chain management, digital identity, smart contracts and many more. This book series provides clear concepts and applications of Blockchain technology and invites experts from research centers, academia, industry and government to contribute to it.

If you are interested in contributing to this series, please contact msingh@endicott.ac.kr OR loyola.dsilva@springer.com

Wendy Charles
Editor

Blockchain in Life Sciences

 Springer

Editor
Wendy Charles ⓘ
University of Colorado
Centennial, CO, USA

ISSN 2661-8338   ISSN 2661-8346 (electronic)
Blockchain Technologies
ISBN 978-981-19-2978-6   ISBN 978-981-19-2976-2 (eBook)
https://doi.org/10.1007/978-981-19-2976-2

This Springer imprint is published by the registered company Springer Nature Singapore Pte Ltd.
The registered company address is: 152 Beach Road, #21-01/04 Gateway East, Singapore 189721,
Singapore

To Chris and Katie for your tremendous patience and understanding.

Foreword

"Science will be blockchained by 2025."

This was the bold claim I made way back in 2017 after 20 years of scientific research and a year of investigating this nascent technology that was transforming other industries from finance to supply chain. As someone who had been involved across the entire spectrum of medical research from bench to bedside, this future incorporation of blockchain to science seemed inevitable even if my title and timeline were meant to be provocative. Some were skeptical of the idea, others were hostile (my federal bosses at the time ordered me to stop talking about blockchain), while most didn't know what blockchain was or simply didn't notice. But there were a growing number of people exploring this possibility and how to make it happen—one of these early leaders was Dr. Wendy Charles.

When I was first introduced to Dr. Charles in 2019, she was well down the same path I was, having seen the obvious value of applying a transparent, tamper-resistant, decentralized audit layer to the world's most precious data system—science. Wendy had her own rich history in clinical research, with robust regulatory experience combined with a formal study of informatics. And, I soon learned, she had the most robust library of blockchain applied to healthcare and life sciences literature on the planet which she had personally curated. I immediately signed her up to be interviewed for a book I was co-authoring on the topic of blockchain for medical research. We started working together on a paper about the regulatory implications of blockchain applied to clinical research which has become a foundational piece in the industry. We have continued to write, present, develop curricula, and teach together on the topic of blockchain for scientific research as exploration and usage have skyrocketed in both public and private sectors.

Sometimes, the best way to predict the future is to create it.

I have been in awe of Wendy's nuanced understanding of the fine details of the scientific process and related information flow, especially the regulatory components of clinical research. It is this knowledge that allowed her to quickly assess the value of blockchain and related distributed data ledgers applied to the life sciences. Given her meticulous focus on the details of the alignment of these two recently intersecting worlds, she has rapidly become one of the world's foremost authorities on the topic.

Much of the early interest in the topic has been technically or scientifically deep, but not both. The academic rigor for anything we apply the scientific method to, or apply to it, must be appropriate for field. It is this very rigor that is required to understand and elicit the tremendous value we can achieve by applying blockchain to life sciences. Dr. Charles is the best person I know to deliver this rigor.

I can think of no one better than Wendy to dive deep into this topic and assemble some of the most insightful explorations of what is being done and what is possible for blockchain applied to life sciences as you will find in this book. When I co-authored Blockchain for Medical Research: Accelerating Trust in Healthcare (2020, CRC Press) with Yael Bizouati-Kennedy, its purpose was to bridge the understanding gap between researchers and technologists we were seeing in this nascent area, at an accessible introductory level. Here with Blockchain in Life Sciences: Technology that Accelerates Scientific Advancements, Dr. Charles explores the technical and scientific depths required to move the use of blockchain in life sciences to the next level.

Along with her contributing authors, Wendy covers all major categories of use cases and technical detail for blockchain in life sciences to provide scientists, administrators, providers, regulators, technologists, and other related professionals the foundation for advancing the use of this emerging technology to rapidly transform research. With a keen eye toward both scientific and technical depth, this volume goes beyond any collection before and will certainly serve as the definitive source for study and application in this area. This collection will have a major impact on life sciences operations in public, private, and academic settings for years to come.

This impact won't simply be one of scientific interest. It will deliver real-world advances to how quickly we can move new ideas to improve health into trusted and widespread use with the trust of science automated by blockchain. Whether it will be to rapidly advance the moonshot against cancer, find better ways to tackle the growing Alzheimer's epidemic, or prepare us to better respond to the next global pandemic, the acceleration blockchain can bring to the creation of actionable knowledge through enhanced life sciences practice will save lives. That's why science needs to be blockchained, and this book will help make that a reality.

Pittsburgh, PA, USA   Sean T. Manion, Ph.D.
October 2021   Chief Science Officer, Equideum

Preface

I first learned about the potential applications of blockchain for life sciences research in 2017 while attending a pharmaceutical conference directed to mobile health technologies. Each conference speaker discussed the need for blockchain in life sciences and the importance of determining how blockchain could fit into each organization's ecosystem. After spending nearly 30 years in life sciences research, this conference began my passion and pursuit of implementing blockchain to enable this technology's potential to advance research data capabilities and outcomes.

The more I became involved in the blockchain community, the more I realized that programmers were unfamiliar with the culture and regulations involved in life sciences research. Similarly, those in the life sciences industry held many misconceptions about blockchain technology and its potential to improve life sciences research. I realized the importance of assembling a book that addressed key educational concepts about blockchain in life sciences research. I was thrilled that Springer Nature Publishing agreed that there is a need for a book in this area and encouraged me to recruit my friends and colleagues as authors. This book became a labor of love as I realized the impact this knowledge will have on enhancing life sciences research.

Purpose of This Book

The purpose of this book is to educate the academic community and life sciences industries about the current uses of blockchain in life sciences research and the unique research and development opportunities enabled by blockchain. Each chapter reveals current uses of blockchain in drug discovery, drug and device tracking, real-world data collection, and increased patient engagement to unlock opportunities to advance life sciences research. The chapters also reveal possible challenges and regulatory implications for responsible implementation. In many ways, individuals and organizations involved in life sciences research must engage in a paradigm shift about blockchain opportunities to empower research participants and enable data capabilities.

Organization of This Book

This book is organized into two primary sections:

- Part: *Blockchain Uses and Real World Evidence* (Chapters "Introduction to Blockchain" through "A Blockchain-Empowered Federated Learning System and the Promising Use in Drug Discovery")
- Part: *Considerations for Ensuring Success of Blockchain in Life Sciences Research* (Chapters "Valuing Research Data: Blockchain-Based Management Methods" through "The Future of Blockchain")

The first Part, *Blockchain Uses and Real World Evidence*, introduces blockchain technologies and provides examples of uses in pillars within specific life sciences industries.

- Chapter "Introduction to Blockchain". This chapter introduces basic information about the common characteristics of blockchain technologies and the features they add to life sciences research.
- Chapter "Blockchain in Pharmaceutical Research and the Pharmaceutical Value Chain". This chapter outlines contemporary and future blockchain-integrated solutions to accelerate and optimize drug discovery and development pathways. It also explores key opportunities well-aligned with blockchain for the five main categories of the pharmaceutical value chain.
- Chapter "Blockchain-Based Scalable Network for Bioinformatics and Internet of Medical Things (IoMT)". This chapter describes blockchain-based solutions for IoMT data management as well as individual engagement and empowerment.
- Chapter "Blockchains and Genomics: Promises and Limits of Technology". This chapter explores the various existing and potential models for genomic blockchains, reviews some shortcomings and unmet needs, and explains why no technical solution alone will fulfill the promise of genomic data ownership without regulation.
- Chapter "Convergence of Blockchain and AI for IoT in Connected Life Sciences". This chapter discusses how blockchain and AI can help govern inherent and residual risks associated with IoT-enabled technologies and how these emerging technology platforms can help catalyze the transfer of scientific discovery into new biomedical products and services to improve the delivery of healthcare and patient outcomes.
- Chapter "A Blockchain-Empowered Federated Learning System and the Promising Use in Drug Discovery". This chapter discusses federated learning and proposes a blockchain-empowered coordinatorless decentralized, federated learning platform.

The second Part, *Considerations for Ensuring Success of Blockchain in Life Sciences Research*, provides organizational considerations of the components of blockchain implementation.

- Chapter "Valuing Research Data: Blockchain-Based Management Methods". This chapter encourages life sciences organizations to view their data silos differently and consider the potential value these can create for the organization. This chapter also describes common accounting principles to value and monetize health-oriented life sciences research data.
- Chapter "Blockchain Adoption in Life Sciences Organizations: Socio-organizational Barriers and Adoption Strategies". This chapter reveals socio-organizational barriers to blockchain adoption in life sciences organizations and delineates the strategies managers and executives can undertake to facilitate adoption.
- Chapter "Blockchain Governance Strategies". This chapter explores the special considerations needed to manage successful blockchain deployments for life sciences ecosystems.
- Chapter "Life Sciences Intellectual Property Through the Blockchain Lens". This chapter provides an overview of the different facets of IP protection for blockchain in life sciences and explains how life sciences organizations can utilize blockchain technologies to help procure, maintain, and enforce their IP.
- Chapter "Regulatory Compliance Considerations for Blockchain in Life Sciences Research". This chapter explores how various blockchain features could meet U.S. Food and Drug Administration (FDA) regulatory requirements for electronic records and signatures, with cautions about necessary documentation expectations.
- Chapter "The Art of Ethics in Blockchain for Life Sciences". This chapter highlights how we can design proactive digital ethics programs in life sciences that mitigate the potential negative consequences of blockchain deployments.
- Chapter "Cybersecurity Considerations in Blockchain-Based Solutions". This chapter discusses some of the most common vulnerabilities in blockchain-based solutions that can arise in the context of life sciences research. For each vulnerability, mitigating strategies are proposed to address the identified risk.
- Chapter "The Future of Blockchain". This chapter introduces the role of blockchain technologies in smart data, quantum computing, digital twins, and the emergence of the metaverse. Additional predictions and recommendations for preparing for future blockchain needs are provided.

I hope that this book provides direction for your journey into blockchain for life sciences research and guides your responsible implementation.

Denver, CO, USA

Wendy Charles, Ph.D.
Chief Science Officer, BurstIQ,
Faculty, University of Colorado

Acknowledgments

I would like to acknowledge the help of all the people involved in this project. Without their support, this book would not have become a reality.

First, I would like to thank each one of the authors for their contributions. I am sincerely grateful for your time in sharing your knowledge and experiences.

Second, I wish to acknowledge the valuable contributions of my colleagues at BurstIQ, Brooke Delgado, Leanne Johnson, and Hayley Miller to improve the quality, coherence, and content of my writing. I am also grateful for the support of the BurstIQ founders for supporting our shared efforts to educate the life sciences industries about blockchain's potential.

Last, I would like to thank my family for their patience and understanding that this book was one of my life's goals (and now accomplishments).

Denver, CO, USA   Wendy Charles

Contents

About the Editor

Dr. Wendy Charles has been involved in clinical trials from every perspective for over 30 years. She implemented institution-wide research compliance, managed two Institutional Review Boards, served as a Principal Investigator for digital health research, site Research Director and Coordinator for commercially sponsored research, worked for an SMO/CRO as a site start-up specialist, and audited research. She currently works as Chief Scientific Officer for BurstIQ, a healthcare-oriented technology company that offers blockchain and other innovative software solutions. Dr. Charles is also a faculty lecturer in the Health Administration program at the University of Colorado, Denver.

Dr. Charles augments her blockchain healthcare experience by serving on the Health Information and Management Systems Society (HIMSS) Blockchain Task Force, EU Blockchain Observatory and Forum Expert Panel, Government Blockchain Association healthcare group, and serves on IEEE Standardization Committee subgroups for Blockchain in Life Sciences Research and Governance. She also serves as Assistant Editor for Frontiers in Blockchain and a peer-reviewer for several other journals.

Dr. Charles obtained her Ph.D. in Clinical Science with a specialty in Health Information Technology from the University of Colorado, Anschutz Medical Campus. Dr. Charles is certified as an IRB Professional, Clinical Research Professional, and Blockchain Professional.

Abbreviations

AI	Artificial Intelligence
API	Application Programming Interface
BYOD	Bring Your Own Device
CAD	Computer Aided Design
CFR	U.S. Code of Federal Regulations
COMIRB	Colorado Multiple Institutional Review Board
CRO	Contract Research Organization
CYOD	Choose Your Own Device
DAO	Distributed Autonomous Organization
dbGaP	The Database of Genotypes and Phenotypes
DCT	Decentralized Clinical Trials
DDoS	Distributed Denial of Service
DLT	Distributed Ledger Technology
DSCSA	Drug Supply Chain Security Act
DTRA	Decentralized Trials and Research Alliance
DTH	Digital Twins Healthcare
EDC	Electronic Data Capture
EFPIA	European Federation of Pharmaceutical Industries and Associations
EHR	Electronic Health Record
ELN	Electronic Lab Notebooks
ESG	Environmental, Social, and Governance
ETP	European Technology Platform
EU	European Union
EUA	Emergency Use Authorization
FAIR	Findable, Accessible, Interoperable, Reusable
FASB	Financial Accounting Standards Board
FDA	U.S. Food and Drug Administration
FL	Federated Learning
FMD	Falsified Medicines Directive
GDPR	General Data Protection Regulation
GWAS	Genome-wide Association Studies

HEA	Homomorphic Encryption Algorithm
HeLa	Henrietta Lacks
HGP	Human Genome Project
HIPAA	Health Insurance Portability and Accountability Act
IMI	Innovative Medicines Initiative
IND	Investigational New Drug
INFORMED	Information Exchange and Data Transformation
IoT	Internet of Things
IP	Intellectual Property
IPFS	Interplanetary File System
IRB	Institutional Review Board
IRS	Internal Revenue Service
IT	Information Technology
KERI	Key Event Receipt Infrastructure
KPI	Key Performance Indicators
M&A	Merger and Acquisition
MELLODDY	Machine Learning Ledger Orchestration for Drug Discovery
ML	Machine Learning
MVE	Minimum Viable Ecosystem
NDA	New Drug Application
NFT	Non-Fungible Token
NHS	National Health Service
NIH	National Institutes of Health
NIST	National Institute of Standards and Technology
PhUSE	Pharmaceutical Users Software Exchange
PK	Pharmacokinetic
PKI	Public Key Infrastructure
PoA	Proof of Authority
PoC	Point of Care
PoI	Proof of Identity
PoS	Proof of Stake
PPP	Public Private Partnership
R&D	Research and Development
RBAC	Role Based Access Control
RT-LAMP	Reverse Transcription-Loop-mediated isothermal AMPlification
RWD	Real World Data
RWE	Real World Evidence
SDG	Sustainable Development Goals
SDK	Software Development Kit
SOP	Standard Operating Procedures
SQL	Structured Query Language
ST	Security Token
TMAP	Technology Modernization Action Plan
TTL	Time to Live
TTD	Therapeutic Target Database

UK	United Kingdom
US	United States
UT	Utility Token
USPTO	United States Patent Office
VC	Verifiable Credentials
VOIP	Voice Over Internet Protocol
W3C	World Wide Web Consortium

List of Figures

Introduction to Blockchain

**Blockchain in Pharmaceutical Research and the Pharmaceutical
Value Chain**

**Blockchain-Based Scalable Network for Bioinformatics and
Internet of Medical Things (IoMT)**

Convergence of Blockchain and AI for IoT in Connected Life Sciences

A Blockchain-Empowered Federated Learning System and the Promising Use in Drug Discovery

The Future of Blockchain

List of Tables

Blockchain Uses and Real World Evidence

Introduction to Blockchain

Wendy M. Charles

Abstract As life sciences research organizations explore methods to facilitate patient-centered and innovative technologies, they are increasingly exploring distributed ledger technologies ("blockchain") to address many of these needs. Blockchain is demonstrating the potential to transform life sciences research, allowing more data capabilities and innovation. Blockchain-based applications vary from audit trails for provenance to integrating remote devices to managing data for decentralized trials. As blockchain is emerging in life sciences, there are questions about the benefits and drawbacks of these technologies. This chapter introduces basic information about the common characteristics of blockchain technologies and the features they add to life sciences research. This chapter also addresses some of the uses of blockchain and lays the groundwork for the real-world applications, benefits, and drawbacks described in future chapters.

Keywords Blockchain · Distributed ledger technologies · Privacy · Trust · Audit trails · Performance

1 Introduction

Life sciences organizations use computerized systems to perform many aspects of research. Computerized systems can include laboratory processing equipment, as well as software for electronic consent, electronic signatures, electronic data capture, clinical trials management system, trial master files, statistical analysis software, image graphics, and electronic transmissions to data coordinating centers and to the regulatory agencies [1].

While life sciences research involves greater volumes of data, current electronic data management and collection methods may not be flexible enough to meet modern technological needs [2]. For example, there are increasing calls for patient-centered technologies, such as offering "dynamic consent," which involves methods to honor

W. M. Charles (✉)
Life Sciences Division, BurstIQ, Denver, CO, USA
e-mail: wendy.charles@cuanschutz.edu

© The Author(s), under exclusive license to Springer Nature Singapore Pte Ltd. 2022   3
W. Charles (ed.), *Blockchain in Life Sciences*, Blockchain Technologies,
https://doi.org/10.1007/978-981-19-2976-2_1

specific terms of individuals' consent and data access for research participants [3]. Further, few efficient or cost-effective ways exist to combine data from many sources or silos [4]. Therefore, distributed ledger technologies (collectively described as "blockchain" throughout) feature characteristics and capabilities that could address data challenges in life sciences research [5]. Most notably, blockchain offers opportunities to accelerate research innovation in ways not possible with current data technologies [6].

There is increasing interest and development of blockchain technologies in life sciences research [7]. In fact, "nearly 70% of all life sciences executives surveyed specified that they planned to implement one or more blockchain projects in 2020" ([8], p. 2). Therefore, it is necessary for stakeholders in life sciences organizations to become familiar with the nature of blockchain features that can be used to advance life sciences research.

2 Blockchain Core Characteristics

Blockchain is not technically a new technology but a set of methods that bring together standard techniques for recordkeeping. The concepts have evolved from a trusted process for time-stamping digital documents in 1991 [9] to the exchange of digital currency without intermediaries in 2008 [10]. Public interest and participation in blockchain rose with the development of Bitcoin as a "cryptocurrency," a digital currency secured by cryptography that can be exchanged by individuals ("peers") in a peer-to-peer manner without financial institutions [11].

Since 2008, the sophistication of blockchain technologies has evolved beyond the original blockchain technologies [12]. Andrianov and Kaganov [13] offer that blockchain is similar to a cloud-based service not tied to a data center, utilizing common cryptography characteristics, distributed data management, and synchronized data flows [14]. With the development of different methods and platforms, it is most accurate to consider blockchain as a set of tools and technologies rather than any single technology. As a result, there are no consistent or standard definitions of blockchain [15], including ongoing debates on whether private and/or centralized networks can constitute blockchains [16, 17].

The following are common features of most types of blockchains.

2.1 Ledgers

The first characteristic of blockchain methodology involves using "ledgers" instead of data tables or relational databases [18]. Like an audit log, an ever-growing ledger records each instance where data are created, and previous records generally cannot be modified or deleted. Modified data are instead appended to the ledger to show that the value has changed, but the original value remains for historical data purposes

[18]. With the ability to use a ledger instead of prescribed data fields, a blockchain can import and track structured or unstructured data from diverse electronic sources, depending on the data mapping and configurations [19].

When a prescribed number of entries are added to the ledger, the entries are assembled in a "block" with time stamps, validation methods, historical structure, and other selected metadata [13, 18]. When a block is formed, the entries contained in the block cannot be modified.

2.2 Cryptography

Blockchains also utilize "cryptography," a method of using codes and algorithms to secure information and communication [20]. As shown in Fig. 1, when entries of any type are added, they are represented with digital signatures comprised of unique strings of alphanumeric characters referred to as "hashes" [18]. These one-way hashes are created by complex algorithms that cannot be reversed to reveal the input [19].

Hashes are not only used to record entries onto the ledger, but also to create a digital summary of the entries in the block. As shown in Fig. 2, a block's hash is also the mechanism used to link blocks in sequential order. As a block is added to the chain, it contains the hash of the previous block.

If it were possible to modify data within a block, the modification would change the hash of that block (Fig. 3). Because blocks are linked with hashes, a change in a block's hash would change the hash in the next block, and so on in subsequent blocks—a task that is exceptionally computationally challenging [21].

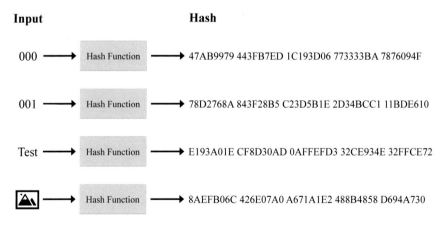

Input   **Hash**

000 ⟶ Hash Function ⟶ 47AB9979 443FB7ED 1C193D06 773333BA 7876094F

001 ⟶ Hash Function ⟶ 78D2768A 843F28B5 C23D5B1E 2D34BCC1 11BDE610

Test ⟶ Hash Function ⟶ E193A01E CF8D30AD 0AFFEFD3 32CE934E 32FFCE72

[img] ⟶ Hash Function ⟶ 8AEFB06C 426E07A0 A671A1E2 488B4858 D694A730

Fig. 1 Fictional examples of hashes. Regardless of input type, the alphanumeric hashes are unique and sophisticated so that the hashes cannot be reversed to reveal the input

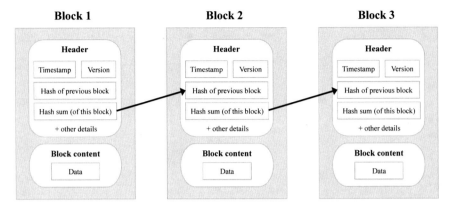

Fig. 2 Simplified depiction of block design and mechanisms of linking blocks using hashes

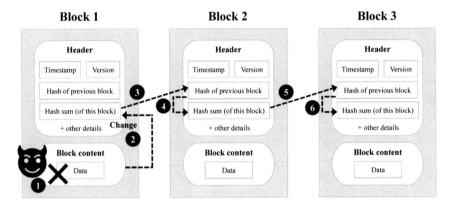

Fig. 3 Simplified depiction of how a data change in one block of an existing chain would require changing the hashes in subsequent blocks. This would be an exceptionally challenging task

2.3 Immutability (Tamper Evidence and Tamper Resistance)

While blockchain is sometimes referred to as "immutable," this book takes the position offered by the U.S. National Institutes of Standards and Technology (NIST):

> Most publications on blockchain technology describe blockchain ledgers as being immutable. However, this is not strictly true. They are tamper evident and tamper resistant, which is a reason they are trusted for financial transactions. They cannot be considered completely immutable because there are situations in which the blockchain can be modified. ([18], p. 34)

Rather than immutable, NIST encourages using the terms "tamper resistant and tamper evident" ([18], p. 34) to denote a blockchain's strong, but not absolute, security. A blockchain provides evidence that data existed at a specific time and that data were not altered [13]. This data integrity feature is particularly pertinent for life sciences research systems that do not otherwise provide a complete chain of custody or data security from data creation to data analyses.

2.4 Distribution

The last primary characteristic of blockchains involves the distribution or decentralization of storage. Instead of central servers or centralized data centers, blockchain utilizes storage distributed over many servers, referred to as "nodes" [18]. This network structure, involving peer-to-peer/organization-to-organization connections, typically creates multiple, identical copies of the ledger across the participating servers in the network. The distributed storage allows data transfer without an intermediary or risk of interference [22], preserves data integrity and availability [19], and provides data redundancy to reduce vulnerabilities to viruses, ransomware, or downtime [23]. Distributed nodes in life sciences organizations are optimally advantageous for decentralized clinical trials where data collection and management are distributed throughout the network [13]. Further data monitoring can occur from a wider variety of locations.

For the nodes to agree on which data entries are the most current and to ensure consistency across the network, blockchains use "consensus mechanisms" to reach an agreement [24]. A thorough discussion of consensus mechanisms is beyond the scope of this chapter, but the reader is encouraged to read some published overviews (e.g., [15, 25, 26]).

3 Blockchain Features

This section introduces blockchain features that may be selected for life sciences research. This section aims not to provide a comprehensive list of blockchain features but to compare and contrast standard features. This section focuses on the differences between permissioning, on- and off-chain storage designs, and smart contracts.

3.1 Permissionless Versus Permissioned

"Permissioning" involves access controls to specify which individuals, roles, or organizations are allowed to participate in a blockchain project. When blockchain permissioning was first introduced, blockchains were described as public/permissionless and private/permissioned. However, this distinction has since become more nuanced as permissioning is now available on some public blockchains [6]. For example, Enterprise Ethereum and Ethereum Private use the public Ethereum open-source code but offer private zones [27, 28]. While there are a few different ways that permissioning features could be described, this chapter characterizes approaches as "permissionless," "permissioned," and "hybrid," with a brief discussion of "consortium" blockchains.

3.2 Permissionless

The first types of blockchain platforms used for cryptocurrency were designed to offer a transparent environment for currency exchange [29]. Platforms such as Bitcoin and Ethereum permitted anyone from the public to join, review, and approve transactions [30]. These platforms use hundreds to thousands of nodes to strengthen network integrity and security [31], making it practically infeasible to corrupt a network of that size [32]. Popular permissionless blockchains are Bitcoin, Ethereum, Dash, and Monero [33].

To provide incentives for public nodes to process transactions, permissionless blockchains typically utilize consensus mechanisms where the submitter pays a transaction fee in digital currency for the nodes to process the data [34]. These transactions often use a consensus mechanism referred to as proof of work, where nodes compete against each other to complete complex computational puzzles to win the right to validate the transaction and form the block [35]. This computationally intensive process is called "mining" and is sometimes criticized for relatively slow transaction speeds, high electricity use, and pollution [36].

While large, permissionless networks are lauded for their transparency and broad decentralization, the management of patient-level information would create major privacy concerns [31]. In addition, due to the need for centralized project coordination and compliance, a completely permissionless infrastructure does not allow for the oversight required of regulated research [37]. Furthermore, permissionless networks that use slow, computationally intensive processing would not meet the requirements for high-speed processing needed for research collection and analyses [6]. Last, the costs for processing large volumes of data would likely be high and cost-prohibitive [16, 38]. Therefore, most life sciences organizations pursuing blockchain projects are starting with permissioned blockchains.

3.2.1 Permissioned

Permissioned blockchains involve a governance structure that requires individuals or organizations to receive permissions to join the network. For life sciences organizations, activities must be associated with an established, named identity for accountability [6]. Permissioned networks involve distributed and synchronized ledgers but may be restricted to nodes within a single organization or a group of organizations that invest in the governance and maintenance of the network, such as academic institutions or commercial sponsors [39]. Because these organizations provide financial support to the network, data activities do not typically require transaction fees customary of permissionless blockchains [40].

Permissioned open-source blockchains include Hyperledger Fabric, Corda, Ethereum Private, and MultiChain [33]. Permissioned blockchain companies designed for health information or life sciences research include BurstIQ's BurstChain®, Carechain, Hashed Health, and Patientory, among others [39].

Permissioned blockchains offer many advantages for privacy and flexibility but can manifest vulnerability to the limitations in protecting data integrity. Dai et al. [32] point out that there is a risk of collusion among limited nodes that may lead to excluding certain transactions or even rolling the chain back to an earlier recorded state. Along these lines, permissioned blockchains may have a controlling authority that can corrupt nodes or allow vulnerabilities that could be exploited by attackers [13].

3.2.2 Hybrid Permissioning

"Hybrid" blockchains contain features of both permissionless and permissioned blockchains. For example, a private network may manage confidential information and permissions for access and data posting but stores metadata and pursues periodic backup to a permissionless blockchain for additional data integrity [41, 42]. A hybrid blockchain design may offer a distributed network with flexibility and usability of permissioned features [41].

To ensure that permissioned blockchain research data remained trustworthy, ConsenSys adds a Hyperledger Besu module to manage a private network while connecting to the Ethereum network [43]. Similarly, Dai et al. [32] connect a private clinical trials blockchain to the Ethereum network. A snapshot of the permissioned chain is captured at periodic intervals (e.g., once per day, once per week) as a transaction on the permissionless ledger.

3.2.3 Consortium

A "consortium" blockchain involves the cooperation of separate legal entities that provide governance and support for blockchain operations [25]. A consortium is considered a semi-decentralized infrastructure with control over operations, maintenance, and regulatory compliance [13].

3.3 Off-Chain Versus On-Chain Storage

With consideration that life sciences research requires volumes of data across large networks of users, it is necessary to create data management strategies that can effectively manage data processing needs. Data can be stored in secure organizational storage with only the metadata on the ledger ("off chain"), or data could be stored together with metadata on the ledger ("on chain") [44]. These storage strategies are compared and contrasted as follows:

3.3.1 Off-Chain Storage

The first permissioned blockchains were designed to maintain traditional storage mechanisms in servers, while the blockchain was designed to record when data were added or appended. This data storage strategy is also designed to manage files, such as digital images and genomic information, too large for ledger storage [45]. This strategy could also demonstrate data integrity when the hash is unaltered [41] with time stamping by the blockchain [22].

As life sciences organizations have implemented blockchain projects, their concern for protecting intellectual property and data confidentiality has initially resulted in decisions to maintain storage off chain [38]. Such off-chain storage technology may use an InterPlanetary File System to track where each file is stored among the distributed storage [46].

However, off-chain storage may not protect the actual data or files stored off chain. Košťál et al. [47] point out that there may be a hash on the blockchain to indicate that data were deleted or altered, but the hash does not protect or restore the data in the server. As an additional consideration, the extra copies of the ledger can be expensive [22, 31].

3.3.2 On-Chain Storage

As an alternate strategy, data can be stored on the ledger with metadata and time stamping. Raw data points can be stored with tags that allow data to be mapped for grouping and aggregation. Some blockchains also allow small files to be stored on chain [6]. While there is concern that on-chain storage could reduce scalability, a measure of speed and performance, organizations using on-chain storage to create an infrastructure of separate chains and mapping [6]. For example, BurstIQ created a platform that stores data on chain with high-speed flexible mapping and access permissions [48]. This blockchain is also capable of addressing the Health Information Portability and Accountability Act (HIPAA) and the General Data Protection Regulation (GDPR) to accommodate regulated health information and individually identifiable information on the chain [49].

3.3.3 Hybrid Storage

As a hybrid storage strategy, organizations have been exploring storing non-private data on chain, such as demographic information, but storing sensitive data in off-chain servers [13]. A hybrid approach is also desirable for organizations who wish to store most data on chain, but need to manage large files off chain in data lakes [31]. This strategy offers a combination of privacy and scalability, allowing ledger length to remain more manageable [6].

Overall, blockchain-based data storage requires careful planning to ensure consistent performance for the project's duration [45].

3.4 Smart Contracts

Smart contracts are small computer programs or short code segments that execute automatically when specific conditions or rules are met [50]. Because smart contracts are designed to run automatically, smart contracts can increase efficiency and accuracy by eliminating human involvement [51]. From a computational standpoint, smart contract code can only be executed or canceled [13]. This computational strategy provides security and failover because smart contracts can be run and restarted if there is a disruption [13].

4 Blockchain Benefits for Life Sciences

Considering the unique needs of life sciences research, blockchains provide the following features-often exceeding what could be offered in a traditional data system [39]. This section recognizes that life sciences organizations already utilize many electronic systems that offer some of the features attributed to blockchain earlier in this chapter. Blockchain offers many features included in traditional commercial off-the-shelf clinical trial management systems or electronic data capture software required by U.S. Food and Drug Administration (FDA) regulations since 1997 (21 CFR Part 11). These "Part 11" systems already offer access controls, error checking, prevention of data alterations, data backups, and end-to-end audit trails of all activities.

To determine where blockchain capabilities could exceed the capabilities of conventional electronic data systems, NIST published a flowchart (initially created by the Department of Homeland Security) [18]. Common decision points pertain to (1) whether data need to be shared, (2) more than one organization is involved with creating and managing data, (3) there are high requirements for data integrity, and (4) there may not be trust among all parties. Some of these characteristics are slightly outdated and do not necessarily reflect newer blockchain platform capabilities. Life sciences organizations recognize that newer technologies, such as blockchain, are needed to enhance trust and security.

4.1 Trust

Beckstrom [52] points out that "trust is the foundational principle in clinical trials" (p. 111). The issue of trust has been a longstanding concern of patients and communities toward research institutions due to historical abuses, such as enrolling participants in research studies without their full knowledge. In the more modern era, individuals' data have been used and distributed for research purposes without their awareness or consent [53]. Unfortunately, even when individuals willingly participate in research projects, most current electronic data capture systems are designed to restrict individuals' access to the data collection or discoveries [54]. Both Benchoufi et al. [54] and Beckstrom [52] suggest that individuals would not only like to contribute to scientific advances, but would also like to gain visibility into the uses of their data to verify that the terms of their preferences are honored. Benchoufi et al. [41] argue that cases of research fraud and dubious research findings have created a growing mistrust of research institutions, stating that they can no longer be considered "trustable by default" (p. 1).

Blockchain has been introduced into the life sciences sector precisely because of the desire for more trust in data integrity, research outcomes, and collaborations among organizations that may not completely trust each other. Beckstrom [52] suggests that blockchain is a valuable addition to research collaborations where the cooperation in the blockchain network enforces honest behavior, and the transparent nature of the blockchain (within permissions) allows for better accountability [55].

Trust requires not only technology, but also coherent governance [41]. Specifically, "blockchain technologies offer a way to design governance systems: public, permissioned or private Blockchains; open- or closed-source software and smart contracts; fixed or evaluative governance rules" ([41], pp. 4–5).

4.2 Audit Trails—Provenance

The FDA defines an audit trail as "a secure, computer-generated, time-stamped electronic record that allows for reconstruction of the course of events relating to the creation, modification, or deletion of an electronic record" ([56], p. 4). Audit trails contain previous entries and metadata to associate activities with a time stamp, the person associated with the change, and the previous and new entries. This feature inherent in blockchain proves proof of existence [54].

While the FDA has required audit trails for electronic record systems since 1997, some traditional systems have been designed with insufficient ability to trace data back to original data sources—one of the top data system problems identified during

FDA inspections [37]. Benchoufi and Ravaud [57] point out that a blockchain allows additional metadata and end-to-end data provenance. This complex, enduring audit trail allows researchers, quality assurance personnel, and/or regulatory authorities to verify the authenticity of data entries [55] and is convenient for remote auditing [57].

4.3 Data Transparency Versus Privacy

With consideration that blockchains were initially designed for transparency as a method to promote trust, blockchains designed for life sciences have had to find a careful balance between transparency and privacy.

Life sciences research is structured to be a collaborative endeavor where scientists share ideas and data. Scientists who receive research funding from government agencies, such as the NIH, are required to create a Data Sharing Plan and make their data available upon request [58]. However, individuals and academic institutions are hesitant to share research datasets when there are desires to maintain a research edge in a competitive funding climate and concerns that data may be misrepresented from the context in which it was collected [59]. There are also often high costs for data administration and disputes about ownership [60].

Additionally, there is a tradeoff between the desire to provide transparency and trust among research participants [17], healthcare providers [22], and study sponsors [61]. This section describes different perspectives on the tradeoff between transparency and privacy.

4.3.1 Need for Data Privacy

Within life sciences research, the privacy of individual participation is a regulatory requirement [62]. Therefore, some life sciences organizations exposed to permissionless blockchains may be hesitant to use blockchain to process sensitive or confidential research information [63]. The emergence of private blockchains designed with permissioning capabilities has created new opportunities to utilize the features of blockchain while protecting the confidentiality of information.

As discussed earlier, some life sciences organizations utilize off-chain storage to protect private information. Benchoufi et al. [41] further advocate that data queries be designed to limit private information that recipients can receive. Angeletti et al. [63] offer the prospect that individual participants' data could even be stored in their personal computers, allowing individuals more control over how their information is used.

Blockchain also offers technological strategies to promote privacy. Differential privacy is a cryptographic method for publicly sharing a research dataset where only aggregate data are provided without allowing visibility of individual-level data. This approach is based on the premise that aggregate data do not change much if an individual is added or excluded from the data, reducing the likelihood that individual participants could be identified [63].

Organizations are also exploring opportunities to use blockchain and artificial intelligence (AI) to create synthetic data. Sometimes referred to as "decentralization intelligence" [64], data remain private within individual organizations' storage. However, the data can be modeled across organizations to create representative data sets to be analyzed and used without compromising the privacy or confidentiality of actual data.

Another privacy-preserving blockchain strategy involves "zero-knowledge proofs." Zero-knowledge proofs are cryptographic protocols that "enable one party, called prover, to prove that some statement is true to another party, called verifier, but without revealing anything but the truth of the statement" ([65], p. 204448). While promising, the technology for zero-knowledge proof capabilities is new and has not achieved wide adoption yet.

Other privacy-preserving strategies for blockchain involve edge computing for near real-time applications to manage privacy constraints associated with computing in a cloud environment [66]. Another strategy involves homomorphic encryption, where data are processed while encrypted, and the encrypted output can only be decrypted later by authorized parties [67]. It is important to note that privacy-preserving strategies continue to develop as more organizations are testing novel advances in programming.

4.3.2 Need for More Transparency

Over the past several years, there have been many investigations and media reports about research misconduct, unscientific research practices, and outright fraud within life sciences organizations [59]. In a high-profile example, a former Duke University pulmonary biology lab technician was accused of doctoring "nearly every experiment or project in which she participated" ([68], p. 978). An investigation conducted by the Office of Research Integrity found that this technician "engaged in research misconduct by knowingly and intentionally falsifying and fabricating research data included in one hundred and seventeen (117) figures and two (2) tables in thirty-nine (39) published papers, three (3) manuscripts, and two (2) research records" ([69], p. 60097). This investigation resulted in a settlement where Duke University agreed to pay the government $112.5 million for submitting falsified data to receive federal grants that may not have otherwise been awarded [70]. Sivagnanam et al. [59] noted that most questionable research findings do not involve deliberate fraud but are difficult to replicate because the data and code are not available to the research community.

Therefore, there has also been a call for more transparency of life sciences research data due to concerns about fraud and misconduct in organizations that manage data internally. While electronic data systems are designed to meet specific regulatory requirements for the submission of new drugs or devices with protections against data modifications, there is concern that other research systems could allow users or administrators to modify the primary data files, resulting in undetectable—and unrecoverable—alterations [17, 32]. Blockchain-based electronic systems provide transparency of research components critical for a study's integrity, such as data, programming code, research protocols, and statistical analysis plans [41].

Even though identifiable research participant information must be protected, blockchain technologies offer privacy-preserving strategies that allow replicating findings or sharing data without compromising research participants' identities or organizations' intellectual property. Further, projects have demonstrated that granular data sharing can be enabled that also protects intellectual property [71, 72].

4.4 Security

When blockchain technologies are used to store life sciences research data, the data must be securely maintained to protect data integrity (21 CFR Part 11), the privacy of research participants' sensitive information (21 CFR 56.111(a)(7)), and intellectual property [73]. Blockchain technologies offer several features that enhance the security of information. While these features are likely to be implemented differently, the following examples are recognized as common features. First, blockchains use cryptographic hash functions extensively that provide platform operations security and consistency [74]. Because the hash output has been created from a sophisticated algorithm, it is practically impossible to reverse engineer this information to determine the input [18]. This concept, called "collision resistance," specifies that it should be difficult for two raw text inputs to create the same output [74].

Within life sciences research, there is a frequent need to correct or update data. This capability is achieved in a blockchain with "append-only" programming where a correction/update is added as a new entry without overwriting the old entry [37]. When data are corrected or updated, many blockchain platforms design query programming that recognizes only the most current entries for the data queried, even though the ledger contains all historical changes to data [75].

In addition, the distribution of ledgers across nodes in the network creates information redundancy, preventing a single point of failure [63, 76]. Even when a node goes offline, the ledgers replicated among other nodes remain available for continued processing. Hirano et al. [77] confirmed blockchain network availability during an unplanned AWS cloud server outage in Tokyo when a node became unavailable during network testing. Because the blockchain maintained nodes in multiple locations, the redundant ledgers allowed for stable operation during the outage, and the AWS autoscaling service updated the data without errors.

Last, the nature of decentralized blockchain architecture (for some blockchain types) creates a network in which all individuals or organizations who maintain nodes must agree to follow the same protocols to prevent any entity from interfering or controlling blockchain operations [76]. This peer-to-peer environment creates a system where the nodes provide group support and oversight to ensure consistent functioning.

4.5 Performance

With consideration of the blockchain features that appear promising for adding benefit to life sciences research, the following are features that enhance capabilities for efficiencies and flexibility.

4.5.1 Automation

Smart contracts may be used to automate quality controls and safety alerts [21]. These automation can extend to enrolling patients and automating study-related visits, supplies, investigational products, and payments [13].

When clinical studies involve informed consent documents that grant or withdraw permissions, smart contracts are used to execute individuals' preferences about future uses of their data or specimens or access to private health information [13]. These smart contracts can execute granular permissions ranging from specific health values to an entire medical record and for specified periods [13, 40].

To facilitate data sharing among researchers, smart contracts are used to automate data sharing permissions among authorized parties. Specifically, smart contracts are designed to verify researchers' access to certain information and automate information transfers depending on the specified terms [6, 40].

For a sponsor or Contract Research Organization, smart contracts are not contracts, but are used to codify validation logic within legal contracts to validate transactions and rules, reducing the need for arbiters [13, 40]. Further, smart contracts can execute the terms of contracts, such as claims adjudication and billing to reduce reliance on paid staff [13]. Smart contract automation further enhances efficiencies of calculating outcomes and reports, including managing database closure [13].

4.5.2 Flexibility

Blockchain also offers electronic data system capabilities beyond the commercial off-the-shelf software available for life sciences research. Rather than purchase all-in-one commercial software, blockchain is used to create more functionality in existing software, data systems, and Internet of Things devices by using application programming interfaces to combine data streams for near real-time aggregation [6, 78].

4.5.3 Scalability

The performance of a blockchain can be measured in transactions per second, computing power, or consensus response time [79]. While cryptocurrency blockchains were designed to generate blocks slowly—an average of 10 min for Bitcoin [80]—to instill trust among the nodes, this performance is too slow for most applications [81]. Because life sciences research blockchains require high-speed read and write access, life sciences organizations utilize several features to improve speed. These may include using a consensus mechanism aligned with the governance structure of a private network, such as Proof of Authority or Proof of Stake [82], breaking files into chunks referred to as shards [83], and/or utilizing side chains [84]. Therefore, speeds for private blockchain networks have increased between 2000–20,000 transactions per second [85, 86], allowing for acceptable speed and performance for most life sciences tasks.

5 Conclusions

Blockchain is emerging to create more sophisticated and holistic data systems for life sciences research [78]. Progressing far beyond the original features of blockchain for cryptocurrency, the development of blockchain within life sciences research organizations includes many types of platforms, variations of consensus mechanisms, combinations of storage, and more capabilities for smart contracts. Electronic data systems need not be replaced by blockchain, but could be enhanced by adding these capabilities. The goal is to move the life sciences industry toward a more collaborative network with more data integrity and sharing among authorized parties while providing checks and balances among partners [87].

The following chapters of this book introduce the complexity of how blockchain is currently being used for many areas within life sciences research, with discussions of the benefits and challenges of each of these applications. While blockchain promises to create efficiencies and advancements in life sciences research, we are reminded that blockchain is software—not magic. This technology cannot solve all—or even most—problems inherent in life sciences research, but has been shown to enhance trust in life sciences data.

6 Key Terminology and Definitions

Blockchain: "A distributed digital ledger of cryptographically signed transactions that are grouped into blocks. Each block is cryptographically linked to the previous one (making it tamper evident) after validation and undergoing a consensus decision. As new blocks are added, older blocks become more difficult to modify (creating tamper resistance). New blocks are replicated across copies of the ledger within the network, and any conflicts are resolved automatically using established rules." ([18], p. 49)

Consensus mechanism: A fault-tolerant mechanism used in blockchain systems to achieve the necessary agreement on a single data value or a single state of the network among distributed nodes or multi-agent systems [24].

Dynamic consent: Dynamic consent describe personalized, online consent and communication "designed to achieve two objectives: (1) facilitate the consent process and (2) facilitate two-way, ongoing communication between researchers and research participants" ([3], p. 3).

Hash: A unique output (also called a hash digest) for an input of nearly any size (a file, text, image, etc.) by applying a cryptographic hash function to the input data ([18], p. 52).

Homomorphic encryption: A form of encryption allowing one to perform calculations on encrypted data without decrypting it first. The result of the computation is in an encrypted form. When decrypted, the output is the same as if the operations had been performed on the unencrypted data [67].

Scalability: The ability of a blockchain platform to manage increasing volumes of transactions and increase the number of nodes in the network [79].

Smart contract: A segment of code or a small computer program deployed designed to execute automatically when certain conditions are met. Nodes execute the smart contract within the blockchain network; all nodes must derive the same results for the execution, and the execution results are recorded on the blockchain [88].

Zero-knowledge proofs: "A protocol that enables one party, called prover, to prove that some statement is true to another party, called verifier, but without revealing anything but the truth of the statement" ([65], p. 204448).

Acknowledgements The author gratefully acknowledges the review and thoughtful feedback from Brooke Delgado, Leanne Johnson, and Hayley Miller.

References

1. U.S. Food and Drug Administration (2017, April 19) Program 7348.810: chapter 48—bioresearch monitoring program. Sponsors, contract research organizations and monitors. https://www.fda.gov/media/75916/download. Accessed 4 Feb 2020
2. Efanov D, Roschin P (2018) The all-pervasiveness of the blockchain technology. Elsevier Ltd., Amsterdam. http://www.sciencedirect.com/science/article/pii/S1877050918300206
3. Budin-Ljøsne I, Teare HJA, Kaye J, Beck S, Bentzen HB, Caenazzo L, Collett C, D'Abramo F, Felzmann H, Finlay T, Javaid MK, Jones E, Katić V, Simpson A, Mascalzoni D (2017) Dynamic consent: a potential solution to some of the challenges of modern biomedical research. BMC Med Ethics 18(1):4. https://doi.org/10.1186/s12910-016-0162-9
4. Angeletti F, Chatzigiannakis I, Vitaletti A (2017b) The role of blockchain and iot in recruiting participants for digital clinical trials. IEEE Communications Society, New York. https://ieeexplore.ieee.org/abstract/document/8115590
5. Hughes L, Dwivedi YK, Misra SK, Rana NP, Raghavan V, Akella V (2019) Blockchain research, practice and policy: applications, benefits, limitations, emerging research themes and research agenda. Int J Inf Manag 49:114–129. https://doi.org/10.1016/j.ijinfomgt.2019.02.005
6. Charles WM (2021a) Accelerating life sciences research with blockchain. In: Namasudra S, Deka GC (eds) Applications of blockchain in healthcare, vol 83. Springer Nature, Berlin, pp 221–252. https://doi.org/10.1007/978-981-15-9547-9_9

7. Agbo CC, Mahmoud QH, Eklund JM (2019) Blockchain technology in healthcare: a systematic review. Healthcare (Basel) 7(2):56. https://doi.org/10.3390/healthcare7020056

8. Treshock M, Fraser H, Pureswaran V (2018) Team medicine: how life sciences can win with blockchain (03013903USEN-00). https://www.ibm.com/downloads/cas/RYD0QA7G

9. Haber S, Stornetta WS (1991) How to time-stamp a digital document. J Cryptol 3:99–111. https://citeseerx.ist.psu.edu/viewdoc/download;jsessionid=1954002DCD3DC6DB6C052994 F8EF24CE?doi=10.1.1.46.8740&rep=rep1&type=pdf

10. Nakamoto S (2008, March 24) Bitcoin: a peer-to-peer electronic cash system. https://bitcoin.org/bitcoin.pdf. Accessed 11 Oct 2020

11. Fulton F (2016) In: Collins C (ed) Bitcoin for dummies. Wiley, New York. https://www.academia.edu/30046580/Bitcoin_For_Dummies_-_1st_Edition_2016_

12. Conte de Leon D, Stalick AQ, Jillepalli AA, Haney MA, Sheldon FT (2017) Blockchain: properties and misconceptions. Asia Pac J Innov Entrep 11(3):286–300. https://doi.org/10.1108/APJIE-12-2017-034

13. Andrianov A, Kaganov B (2018) Blockchain in clinical trials: the ultimate notary. Appl Clin Trials 27(7/8):16–19. http://images2.advanstar.com/pixelmags/applied-clinical-trials/pdf/2018-08.pdf#page=16

14. Lin I-C, Liao T-C (2017) A survey of blockchain security issues and challenges. Int J Netw Secur 19(5):653–659. https://doi.org/10.6633/IJNS.201709.19(5).01

15. Zheng Z, Xie S, Dai H-N, Chen X, Wang H (2017) An overview of blockchain technology: architecture, consensus, and future trends. IEEE, Piscataway. https://ieeexplore.ieee.org/document/8029379/

16. Lopez PG, Montresor A, Datta A (2019) Please, do not decentralize the internet with (permissionless) blockchains! (11) [Preprint]. https://arxiv.org/abs/1904.13093

17. Zhuang Y, Sheets LR, Shae Z, Tsai JJP, Shyu C-R (2018) Applying blockchain technology for health information exchange and persistent monitoring for clinical trials. AMIA Annu Symp Proc 1167–1175. https://www.ncbi.nlm.nih.gov/pmc/articles/PMC6371378/

18. Yaga D, Mell P, Roby N, Scarfone K (2018) Blockchain technology overview (NISTIR 8202). NIST Interagency/Internal Report, Issue. https://www.nist.gov/publications/blockchain-technology-overview

19. Charles WM (2021b) Blockchain will transform clinical research. J Clin Res Best Pract 17(2). https://www.magiworld.org/resources/journal/2_Blockchain.pdf

20. Zhao B, Huang X (2020) Encrypted monument: the birth of crypto place on the blockchain. Geoforum 116:149–152. https://doi.org/10.1016/j.geoforum.2020.08.011

21. Engelhardt MA (2017) Hitching healthcare to the chain: an introduction to blockchain technology in the healthcare sector. Technol Innov Manag Rev 7(10):22–34. https://doi.org/10.22215/timreview/1111

22. Omar IA, Jayaraman R, Salah K, Yaqoob I, Ellahham S (2021) Applications of blockchain technology in clinical trials: review and open challenges. Arab J Sci Eng 46(4):3001–3015. https://doi.org/10.1007/s13369-020-04989-3

23. Li H, Zhu L, Shen M, Gao F, Tao X, Liu S (2018) Blockchain-based data preservation system for medical data. J Med Syst 42(8):141. https://doi.org/10.1007/s10916-018-0997-3

24. Tosh DK, Shetty SS, Liang X, Kamhoua CA, Njilla LL (2017) Consensus protocols for blockchain-based data provenance: challenges and opportunities. IEEE, Piscataway. https://ieeexplore.ieee.org/abstract/document/8249088

25. Ray PP, Dash D, Salah K, Kumar N (2020) Blockchain for IoT-based healthcare: background, consensus, platforms, and use cases. IEEE Syst J 15(1):85–94. https://doi.org/10.1109/JSYST.2020.2963840

26. Shahaab A, Lidgey B, Hewage C, Khan I (2019) Applicability and appropriateness of distributed ledgers consensus protocols in public and private sectors: a systematic review. IEEE Access 7:43622–43636. https://doi.org/10.1109/ACCESS.2019.2904181

27. About Enterprise Ethereum Alliance (2020) Enterprise ethereum alliance. https://entethalliance.org/about/. Accessed 31 July 2020

28. Private Ethereum Networks (2019) Go Ethereum. https://geth.ethereum.org/docs/interface/pri vate-network. Accessed 31 July 2020
29. Calvaresi D, Calbimonte J-P, Dubovitskaya A, Mattioli V, Piguet J-G, Schumacher M (2019) The good, the bad, and the ethical implications of bridging blockchain and multi-agent systems. Information (Basel) 10(12):363. https://doi.org/10.3390/info10120363
30. Labazova O (2019) Towards a framework for evaluation of blockchain implementations. Bepress/Elsevier, Inc., Amsterdam. https://aisel.aisnet.org/icis2019/blockchain_fintech/blo ckchain_fintech/18/
31. Jung HH, Pfister FMJ (2020) Blockchain-enabled clinical study consent management. Technol Innov Manag Rev 10(2):14–24. https://doi.org/10.22215/timreview/1325
32. Dai H, Young HP, Durant TJS, Gong G, Kang M, Krumholz HM, Schulz WL, Jiang L (2018) TrialChain: a blockchain-based platform to validate data integrity in large, biomedical research studies (1807.03662) [Preprint]. National Center for Cardiovascular Disease. https://arxiv.org/abs/1807.03662
33. Kumar Sharma T (2019) Permissioned and permissionless blockchains: a comprehensive guide. Blockchain Council. https://www.blockchain-council.org/blockchain/permissio ned-and-permissionless-blockchains-a-comprehensive-guide/. Accessed 25 July 2021
34. Wang Y, Wang H (2020) Using networks and partial differential equations to forecast bitcoin price movement. Chaos 30(7):073127. https://doi.org/10.1063/5.0002759
35. McGinn D, McIlwraith D, Guo Y (2018) Towards open data blockchain analytics: a Bitcoin perspective. R Soc Open Sci 5(8):180298. https://doi.org/10.1098/rsos.180298
36. Köhler S, Pizzol M (2019) Life cycle assessment of bitcoin mining. Environ Sci Technol 53(23):13598–13606. https://doi.org/10.1021/acs.est.9b05687
37. Wong DR, Bhattacharya S, Butte AJ (2019) Prototype of running clinical trials in an untrustworthy environment using blockchain. Nat Commun 10(1):917. https://doi.org/10.1038/s41 467-019-08874-y
38. Steinwandter V, Herwig C (2019) Provable data integrity in the pharmaceutical industry based on version control systems and the blockchain. PDA J Pharm Sci Technol 73(4):373–390. https://doi.org/10.5731/pdajpst.2018.009407
39. Essén A, Ekholm A (2020) Centralization vs. decentralization on the blockchain in a health information exchange context. In: Larsson A, Teigland R (eds) Digital transformation and public services: societal impacts in sweden and beyond. Routledge, London, pp 58–82. https://doi.org/10.4324/9780429319297
40. Choudhury O, Sylla I, Fairoza N, Das AK (2019) A blockchain framework for ensuring data quality in multi-organizational clinical trials. IEEE, Piscataway. https://ieeexplore.ieee.org/doc ument/8904634
41. Benchoufi M, Altman DG, Ravaud P (2019) From clinical trials to highly trustable clinical trials: blockchain in clinical trials, a game changer for improving transparency? Front Blockchain 2(23). https://doi.org/10.3389/fbloc.2019.00023
42. Sato T, Himura Y (2018) Smart-contract based system operations for permissioned blockchain. Curran Associates, Inc. https://ieeexplore.ieee.org/document/8328745
43. ConsenSys (2020) Enterprise Ethereum: 5 reasons why Enterprise Ethereum is so much more than a distributed ledger technology. ConsenSys. https://consensys.net/enterprise-ethereum/best-blockchain-for-business/5-reasons-why-enterprise-ethereum-is-so-much-more-than-a-distributed-ledger-technology/. Accessed 31 July 2020
44. Joshi P, Gokhale P (2021) Electronic health record using blockchain and off chain storage: a systematic review. IT Ind 9(1):247–253. http://www.it-in-industry.org/index.php/itii/article/view/125
45. Zhang P, Schmidt DC, White J, Lenz G (2018) Blockchain technology use cases in healthcare. In: Raj P, Deka GC (eds) Advances in computers Blockchain technology: platforms, tools and use cases, vol 111. Academic Press, Cambridge, pp 1–41. https://doi.org/10.1016/bs.adcom.2018.03.006
46. Sun J, Yao X, Wang S, Wu Y (2020) Blockchain-based secure storage and access scheme for electronic medical records in IPFS. IEEE Access 8:59389–59401. https://doi.org/10.1109/acc ess.2020.2982964

47. Košťál K, Helebrandt P, Belluš M, Ries M, Kotuliak I (2019) Management and monitoring of IoT devices using blockchain (dagger). Sensors (Basel) 19(4):856. https://doi.org/10.3390/s19 040856
48. Pennec F (2018, February 23) Healthcare blockchain startup BurstIQ secures $5M investment. HIT Consultant Media. https://hitconsultant.net/2018/02/23/healthcare-blockchain-startup-bur stiq-secures-5m/. Accessed 26 July 2020
49. Srivastava G, Parizi RM, Dehghantanha A, Choo K-KR (2019) Data sharing and privacy for patient IoT devices using blockchain. Springer, Berlin. https://doi.org/10.1007/978-981-15-1301-5_27
50. Chamber of digital commerce (2018) "Smart contracts" legal primer. https://digitalchamber. org/wp-content/uploads/2018/02/Smart-Contracts-Legal-Primer-02.01.2018.pdf
51. McKinney SA, Landy R, Wilka R (2018) Smart contracts, blockchain, and the next frontier of transactional law. Wash J Law Technol Arts 13(3):313–347. http://hdl.handle.net/1773.1/1818
52. Beckstrom K (2019) Utilizing blockchain to improve clinical trials. In: Metcalf D, Bass J, Hooper M, Cahana A, Dhillon V (eds) Blockchain in healthcare: innovations that empower patients, connect professionals and improve care. CRC Press, Taylor & Francis Group, pp 109–121. https://www.routledge.com/Blockchain-in-Healthcare-Innovations-that-Empower-Patients-Connect-Professionals/Dhillon-Bass-Hooper-Metcalf-Cahana/p/book/978036703 1084
53. The National Commission for the Protection of Human Subjects of Biomedical and Behavioral Research (1979) The Belmont report: ethical principles and guidelines for the protection of human subjects of research. https://www.hhs.gov/ohrp/regulations-and-policy/belmont-report/
54. Benchoufi M, Porcher R, Ravaud P (2018) Blockchain protocols in clinical trials: transparency and traceability of consent. F1000Res 6. https://doi.org/10.12688/f1000research.10531.5
55. Albanese G, Calbimonte J-P, Schumacher M, Calvaresi D (2020) Dynamic consent management for clinical trials via private blockchain technology. J Ambient Intell HumanIz Comput. https://doi.org/10.1007/s12652-020-01761-1
56. U.S. Food and Drug Administration (2018, December 7) Data integrity and compliance with drug CGMP: questions and answers guidance for industry. https://www.fda.gov/regulatory-information/search-fda-guidance-documents/data-integrity-and-compliance-drug-cgmp-que stions-and-answers-guidance-industry. Accessed 19 Jun 2021
57. Benchoufi M, Ravaud P (2017) Blockchain technology for improving clinical research quality. Trials 18:335. https://doi.org/10.1186/s13063-017-2035-z
58. National Institutes of Health (2003) Final NIH statement on sharing research data. https://gra nts.nih.gov/grants/guide/notice-files/NOT-OD-03-032.html
59. Sivagnanam S, Nandigam V, Lin K (2019) Introducing the open science chain. Association for Computing Machinery, New York. https://doi.org/10.1145/3332186.3332203
60. Glicksberg BS, Burns S, Currie R, Griffin A, Wang ZJ, Haussler D, Goldstein T, Collisson E (2020) Blockchain-authenticated sharing of genomic and clinical outcomes data of patients with cancer: a prospective cohort study. J Med Internet Res 22(3):e16810. https://doi.org/10. 2196/16810
61. Kumari M, Gupta M, Ved C (2021) Blockchain in Pharmaceutical Sector. In: Namasudra S, Deka GC (eds) Applications of blockchain in healthcare, vol 83. Springer Nature, Berlin, pp 199–220. https://doi.org/10.1007/978-981-15-9547-9_8
62. Charles WM, Marler N, Long L, Manion ST (2019) Blockchain compliance by design: regulatory considerations for blockchain in clinical research. Front Blockchain 2(18). https://doi. org/10.3389/fbloc.2019.00018
63. Angeletti F, Chatzigiannakis I, Vitaletti A (2017a) Privacy preserving data management in recruiting participants for digital clinical trials. ACM, New York. https://dl.acm.org/citation. cfm?id=3144733
64. Singh SK, Rathore S, Park JH (2020) Block IoT intelligence: a blockchain-enabled intelligent IoT architecture with artificial intelligence. Futur Gener Comput Syst 110:721–743. https:// doi.org/10.1016/j.future.2019.09.002

65. Tomaz AEB, Nascimento JCD, Hafid AS, De Souza JN (2020) Preserving privacy in mobile health systems using non-interactive zero-knowledge proof and blockchain. IEEE Access 8:204441–204458. https://doi.org/10.1109/ACCESS.2020.3036811

66. Jayasinghe U, Lee GM, MacDermott Á, Rhee WS (2019) TrustChain: a privacy preserving blockchain with edge computing. Wirel Commun Mob Comput 2019:2014697. https://doi.org/10.1155/2019/2014697

67. Zhou L, Wang L, Ai T, Sun Y (2018) BeeKeeper 2.0: confidential blockchain-enabled IoT system with fully homomorphic computation. Sensors (Basel) 18(11):3785. https://doi.org/10.3390/s18113785

68. McCook A (2016) Duke fraud case highlights financial risks for universities. Science 353(6303):977–978. https://doi.org/10.1126/science.353.6303.977

69. U.S. Department of Health and Human Services (2019) Findings of research misconduct. Fed Regist 84(219):60097–60098. https://ori.hhs.gov/sites/default/files/2019-11/2019-24291.pdf

70. Office of Public Affairs (2019) Duke University agrees to pay U.S. $112.5 million to settle false claims act allegations related to scientific research misconduct. U.S. Department of Justice. https://www.justice.gov/opa/pr/duke-university-agrees-pay-us-1125-million-settle-false-claims-act-allegations-related. Accessed 23 Aug 2021

71. Burki TK (2019) Pharma blockchains AI for drug development. Lancet 393(10189):2382. https://doi.org/10.1016/S0140-6736(19)31401-1

72. Warr WA (2021) National Institutes of Health (NIH) workshop on reaction informatics. https://chemrxiv.org/engage/api-gateway/chemrxiv/assets/orp/resource/item/611cf1a6ac8b499b36458d19/original/national-institutes-of-health-nih-workshop-on-reaction-informatics.pdf

73. Wang J, Wang S, Guo J, Du Y, Cheng S, Li X (2019) A summary of research on blockchain in the field of intellectual property. Elsevier, Amsterdam. http://www.sciencedirect.com/science/article/pii/S187705091930239X

74. Dasgupta D, Shrein JM, Gupta KD (2019) A survey of blockchain from security perspective. J Bank Financ Technol 3:1–17. https://doi.org/10.1007/s42786-018-00002-6

75. Banga R, Juneja M (2018) Clinical trials on blockchain. PhUSE, Broadstairs. https://www.lexjansen.com/phuse/2018/tt/TT11.pdf

76. Wang Y, Li J, Yan Y, Chen X, Yu F, Zhao S, Yu T, Feng K (2021) A semi-centralized blockchain system with multi-chain for auditing communications of wide area protection system. PLoS ONE 16(1):e0245560. https://doi.org/10.1371/journal.pone.0245560

77. Hirano T, Motohashi T, Okumura K, Takajo K, Kuroki T, Ichikawa D, Matsuoka Y, Ochi E, Ueno T (2020) Data validation and verification using blockchain in a clinical trial for breast cancer. J Med Internet Res 22(6):e18938. https://doi.org/10.2196/18938

78. Learney R (2019) Blockchain in clinical trials. In: Metcalf D, Bass J, Hooper M, Cahana A, Dhillon V (eds) Blockchain in healthcare: innovations that empower patients, connect professionals and improve care. CRC Press, Taylor & Francis Group, pp 87–108. https://www.routledge.com/Blockchain-in-Healthcare-Innovations-that-Empower-Patients-Connect-Professionals/Dhillon-Bass-Hooper-Metcalf-Cahana/p/book/9780367031084

79. Eklund PW, Beck R (2019) Factors that impact blockchain scalability. Association for Computing Machinery, New York. https://doi.org/10.1145/3297662.3365818

80. Rathore H, Mohamed A, Guizani M (2020) A survey of blockchain enabled cyber-physical systems. Sensors (Basel) 20(1):282. https://doi.org/10.3390/s20010282

81. Burchert C, Decker C, Wattenhofer R (2018) Scalable funding of bitcoin micropayment channel networks. R Soc Open Sci 5(8):180089. https://doi.org/10.1098/rsos.180089

82. Lee H-A, Kung H-H, Udayasankaran JG, Kijsanayotin B, Marcelo AB, Chao LR, Hsu C-Y (2020) An architecture and management platform for blockchain-based personal health record exchange: development and usability study. J Med Internet Res 22(6):e16748. https://doi.org/10.2196/16748

83. Ricotta F, Jackson B, Henry T (2019) Secure adaptive data storage platform. United States Patent No. US 2019/0012466 A1. https://patentimages.storage.googleapis.com/74/97/75/8d9604b1b85a5d/US20190012466A1.pdf

84. Merena S, Thangadurai E, Shankar M (2021) Electronic health care record using blockchain technology. Int J Eng Res Appl 11(1):10–13. https://doi.org/10.9790/9622-1101031013
85. Gorenflo C, Lee S, Golab L, Keshav S (2020) FastFabric: scaling hyperledger fabric to 20 000 transactions per second. Int J Network Manage 30(5):e2099. https://doi.org/10.1002/nem.2099
86. Nakaike T, Zhang Q, Ueda Y, Inagaki T, Ohara M (2020) Hyperledger fabric performance characterization and optimization using GoLevelDB benchmark. IEEE, Piscataway. https://iee explore.ieee.org/document/9169454
87. Meyyan P (2018, January 16) Decrypting the utility of blockchain in clinical data management. VertMarkets. https://www.clinicalleader.com/doc/decrypting-the-utility-of-blockchain-in-clinical-data-management-0001. Accessed 23 Oct 2018
88. Alharby M, Aldweesh A, van Moorsel A (2018) Blockchain-based smart contracts: a systematic mapping study of academic research. IEEE, Piscataway. https://ieeexplore.ieee.org/document/8756390

Dr. Wendy Charles has been involved in clinical trials from every perspective for 30 years, with a strong background in operations and regulatory compliance. She currently serves as Chief Scientific Officer for BurstIQ, a healthcare information technology company specializing in blockchain and AI. She is also a lecturer faculty member in the Health Administration program at the University of Colorado, Denver. Dr. Charles augments her blockchain healthcare experience by serving on the EU Blockchain Observatory and Forum Expert Panel, HIMSS Blockchain Task Force, Government Blockchain Association healthcare group, and IEEE Blockchain working groups. She is also involved as an assistant editor and reviewer for academic journals. Dr. Charles obtained her PhD in Clinical Science with a specialty in Health Information Technology from the University of Colorado, Anschutz Medical Campus. She is certified as an IRB Professional, Clinical Research Professional, and Blockchain Professional.

Blockchain in Pharmaceutical Research and the Pharmaceutical Value Chain

Kevin A. Clauson, Rachel D. Crouch, Elizabeth A. Breeden, and Nicole Salata

Abstract Pharmaceutical research can yield life-changing agents for treating and curing disease, improving quality of life, extending life, and enhancing innovation in the broader healthcare ecosystem. However, the historical processes and approaches for drug discovery and development are fraught with high costs, low success rates, and enduring challenges—from preclinical research to Phase IV surveillance. Overall, the pharmaceutical value chain, consisting of (1) research and discovery, (2) clinical development, (3) manufacturing and supply chain, (4) launch and commercial considerations, and (5) monitoring and health records, suffers from pain points at a variety of stages across multiple vector types. The strengths and characteristics of distributed ledger technology (DLT) (e.g., blockchain), in conjunction with other established and emerging technologies, map extraordinarily well to many of the most substantial challenges in pharmaceutical research and the pharmaceutical value chain. This chapter outlines contemporary and future blockchain-integrated solutions to accelerate and optimize drug discovery and development pathways. It also explores key opportunities well-aligned with blockchain for the five main categories of the pharmaceutical value chain. Finally, this chapter debunks the misconception that technical challenges are the chief obstacle for the conception and implementation of blockchain-based solutions in the pharmaceutical industry while alerting the reader to other challenges and approaches to navigate them.

Keywords Blockchain · Drug discovery · Distributed ledger technology · Pharmaceutical research · Pharmaceutical value chain · Supply chain

K. A. Clauson (✉) · R. D. Crouch · E. A. Breeden
College of Pharmacy and Health Sciences, Lipscomb University, Nashville, TN, USA
e-mail: kevin.clauson@lipscomb.edu

R. D. Crouch
e-mail: rachel.crouch@lipscomb.edu

E. A. Breeden
e-mail: beth.breeden@lipscomb.edu

N. Salata
PharmD Live, Washington, DC, USA
e-mail: nsalata@pharmdlive.com

© The Author(s), under exclusive license to Springer Nature Singapore Pte Ltd. 2022
W. Charles (ed.), *Blockchain in Life Sciences*, Blockchain Technologies,
https://doi.org/10.1007/978-981-19-2976-2_2

1 Brief Overview of Pharmaceutical Research

1.1 Drug Delivery and Discovery

Taking a therapeutic from a concept to a marketed drug molecule is an extensive process that typically requires a decade or more of research and costs more than $2 billion to complete [1]. The process begins with target identification, where basic research is conducted to identify a biological entity (e.g., gene, signaling molecule, etc.) associated with a particular disease that can be modulated by a small molecule or biologic (i.e., a "druggable" target) [2]. Once a druggable target has been identified, the target must undergo a series of tests to confirm that regulation of the target is associated with modification of the disease state, which typically includes the use of in vitro studies and animal models of the disease. In the next stage, compounds are identified that can modulate the target's activity, which in many cases can include hundreds or even thousands of molecules to be screened using assays designed to detect target engagement. Compounds exhibiting target engagement are then subjected to additional screening to identify a single "lead" compound or a few lead compounds possessing drug-like characteristics such as high potency and selectivity, aqueous solubility, and metabolic stability. The final stage in what is considered to be the discovery phase of the drug discovery and development process is lead optimization, where the structure of the lead compound(s) is/are modified to increase the safety and efficacy of the drug by improving properties such as off-target binding or oral absorption [3]. The discovery stage alone takes 1–3 years and $200 million on average to complete [2].

Following the discovery phase is a period of preclinical development, which involves extensive animal testing to further evaluate the safety and efficacy of the drug prior to advancement into clinical trials. These studies are required by the United States Food & Drug Administration (FDA) and provide critical information regarding the potential for the drug to successfully progress through clinical trials [1]. An additional 1–2 years and $100 million or more are typically required to complete preclinical development studies [2].

While the discovery and preclinical stages of drug development are costly, the clinical trial stage is by far the most costly phase of the entire process, averaging $1–2 billion to complete phase I, II, and III trials [2]. Phase I clinical trials are primarily intended to evaluate the drug's safety and determine an appropriate dose in humans. These studies are typically conducted on healthy individuals and comprise less than 100 participants. Phase II trials further evaluate drug safety, but in this stage, drug efficacy is also evaluated through studies in 100–300 individuals who have the disease state. Phase III trials dive deeper into the safety and efficacy of the drug by evaluating different patient populations, doses, and drug combinations in several hundred to several thousand individuals with the disease. While a phase I trial may only take several months to complete, phase II and III trials typically last several years. Only 12% of drugs successfully progress through phase III trials to receive FDA approval and reach the market [1]. After FDA approval, the safety and efficacy of the drug continue to be monitored in what is referred to as phase IV. This post-approval monitoring phase provides additional information that may not have become apparent in smaller cohorts of clinical trial subjects, such as adverse events, drug–drug interactions, and necessary dose adjustments in certain patient populations. This final phase generally lasts for several months and can cost an additional several hundred million dollars.

1.2 Challenges Associated with Drug Delivery and Discovery

Numerous challenges exist across all stages of the drug discovery and development process, from the discovery and preclinical phases extending into the clinical trial and post-marketing phases. The nature of these challenges is varied, pertaining to aspects that may be scientific, logistical, financial, ethical, and/or legal. To a large degree, financial challenges associated with drug discovery and development stem from difficulties related to the other stated problem areas (e.g., scientific or logistical issues). Accordingly, the cost of delivering a new drug to market has gradually increased over time, despite major advances in science and technology to enable the potential development of previously unattainable therapies. Consequently, high costs combined with the uncertainty that a drug will successfully reach the market represent a significant barrier to drug development.

1.3 Challenges Associated with Preclinical (i.e., In Vitro, In Vivo) and Phase 0/I–IV Studies

Challenges in drug discovery and development begin with gaps in the science that inform drug discovery efforts [4]. While deficiencies in understanding the pathophysiology of diseases make it difficult to identify drug targets, these deficiencies may not be realized until a drug reaches phase III clinical trials when a drug fails to demonstrate clinical efficacy. Further complicating this dilemma are animal models that insufficiently represent human disease, leading to drugs that demonstrate efficacy in preclinical studies but not in clinical trials. Likewise, animal studies sometimes fail to identify toxicities that arise in humans during clinical studies.

Partly contributing to these gaps in knowledge of disease mechanisms and failures in preclinical to clinical translation are deficiencies in published data [4]. This problem spans all stages of drug discovery and development—from understanding disease pathophysiology to identifying disease biomarkers and drug targets to translating preclinical models into human disease. There are complexities at every step with each of these aspects of drug discovery and development dependent on the reliability and reproducibility of published data.

Both the production and the dissemination of published data have their limitations. For example, deficiencies among both investigators and reviewers in conducting and interpreting statistical analyses can result in the publication of statistically insignificant data. Furthermore, due to the "publish or perish" culture of academia, investigators may be motivated to cut corners or outright falsify data for the sake of publication. Alternatively, the volume of data that can now be generated due to advancements in instrumentation and technology creates complexity in the storage, maintenance, and retrieval of data, potentially leading to innocent mistakes in data conversion, processing, and/or reporting. From a dissemination standpoint, the lack of reporting of raw data and detailed experimental methods can make reproducibility from one lab to another challenging. In addition, because negative results are typically not published, time and resources are likely wasted on studies destined to fail.

Finally, disregarding potential problems with the published data in and of themselves, the sheer volume of published data available makes searching the literature for relevant and comprehensive information an arduous and time-consuming task. Because all publications are not open access, accessibility to published data for some investigators may be limited by cost.

Increased collaboration among academia and the pharmaceutical industry can help remediate some of these challenges by facilitating data sharing, sharing costs, and expanding the pool of expertise contributing to a given drug discovery and development effort [4]. Collaboration between academia and industry is particularly beneficial in bridging the basic biomedical research required for the early stages of drug discovery (i.e., academia) with the costly later stages of drug development (i.e., industry). While academic drug discovery programs can identify drug targets and drive "hit-to-lead" campaigns during the discovery stages, these programs generally have to rely on partnerships with industry to fund and facilitate late-stage preclinical development and, especially, clinical trials. Drug development is also facilitated by those in the academic sector via the provision of consultancy services and in roles like key opinion leaders [5]. However, collaboration creates its own challenges. With expanded collaboration comes increased complexity in the storage, maintenance, retrieval, and, particularly, data sharing due to the introduction of multiple sources of information in physically distanced locations. In addition, as intellectual property (IP) is critical to developing a revenue-generating drug product, the involvement of multiple entities in the discovery and development of a drug introduces an additional layer of complexity to the ownership and protection of IP and the distribution of royalty payments.

Similar challenges arise in the later stages of clinical development (e.g., phase III and IV clinical trials) when more patients and multiple clinical trial centers are typically involved in collecting data. Not only is the storage, maintenance, retrieval, and sharing of data a logistical concern, but with clinical studies, it is also an ethical and legal concern, as the personal information of trial participants must be protected. Furthermore, specific information at times must be blinded to patients and/or investigators to avoid introducing bias into the study. Likewise, prior to initiation of a clinical trial at any stage (phase I–IV), informed consent must be collected from trial participants (generally at multiple locations). This information might also be shared with auditors and regulatory review boards while maintaining patient privacy. Similarly, "big data" (e.g., medical records, genomic databanks, clinical trial results, etc.) can be a useful source of information across all aspects of the drug discovery and development process, but also come with logistical and ethical/legal challenges regarding collecting, maintaining and distributing these data, as well as protecting the privacy of individuals involved.

1.3.1 Adaptive Trial Design

One strategy developed to improve the efficiency of clinical trials is the implementation of an adaptive trial design. The FDA defines an adaptive trial design as "a clinical trial design that allows for prospectively planned modifications to one or more aspects of the design based on accumulating data from subjects in the trial" [6]. By permitting adjustments to the trial based upon data that were not yet available at the start of the trial, adaptive trial designs can potentially improve statistical efficiency, ethical conduct, data interpretation, and general risk reduction for both trial sponsors and trial participants. However, while an adaptive design has several potential advantages, maintaining trial integrity becomes more challenging when evaluating interim data. Care must be taken to preserve the blinding of investigators and patients intended to remain blinded throughout the study. Regardless, adaptive trial design offers particular promise for personalized/precision medicine and allows for a more rapid response to epidemics due to viruses (e.g., COVID-19, Ebola hemorrhagic fever, Middle Eastern Respiratory Syndrome) [7].

2 Introduction of the End-To-End Pharmaceutical Value Chain

2.1 Five Main Categories: (1) Research and Discovery, (2) Clinical Development, (3) Manufacturing and Supply Chain, (4) Launch and Commercial Considerations, and (5) Monitoring and Health Records

For the purposes of this chapter, the following five phases comprise the pharmaceutical value chain: research and discovery; clinical development; manufacturing and supply chain; launch and commercial considerations; and monitoring and health records [8]. Before delving deeper into each phase, it is critical to understand that each phase serves as a funnel for future phases and that there may be temporal overlap of phases throughout progression down the chain.

As the first phase in the pharmaceutical value chain, **research and discovery** represents a significant challenge. Responsible for the discovery and preliminary understanding of eligible pharmaceutical compounds, success in this phase is not only required for further progression in the chain but also cyclically relies on and contributes to past and future successes, respectively. The Therapeutic Target Database (TTD), a collection of documented protein and nucleic acid targets, reports that 427,262 potentially active target drug structures have been identified. Additionally, 33,598 have been profiled for potential drug properties, and only 2,797 have successfully become approved drugs [9]. For scope, these numbers are winnowed from the million or more compounds that undergo initial screening [3].

The pain points in the research and discovery phase are fairly easy to identify. Siloed information guarded by a small number of corporations prevents collaborative learning and improved discovery processes. The sheer volume of available data is prohibitive to thorough exploration and cataloging, potentially hiding value in plain sight simply due to inadequate exploratory resources.

As the second phase of the value chain, **clinical development** comprises everything from preclinical evaluation to phase III clinical trials and includes the submission of an Investigational New Drug application (IND) to the FDA. Supplied with successful targets identified during research and discovery, this phase consists entirely of rigorous, reproducible, and regulated testing. According to Hughes and colleagues, only approximately 1 in 10 compounds that make it to this phase continue to approval and the pharmaceutical market [3]. This is corroborated by the data supplied by TTD, which suggests that approximately 50% of identified compounds make it to clinical development, and only 15% of those are successfully approved [9].

An interesting variation of the typical approval process has been identified due to the COVID-19 pandemic: the Emergency Use Authorization (EUA). Used to help hasten the clinical trial process while also making potentially lifesaving medications available to the public before full approval is granted, EUAs were granted to three COVID-19 vaccinations and one COVID-19 treatment [10]. Currently, EUAs represent a rare mechanism to accelerate the time associated with this phase.

Pain points and problems in the clinical development phase echo several of those from research and discovery, but on a different scale. Whereas research dollars were spread out to maximize the number of discoverable compounds, clinical development dollars are focused on a comparatively small number of projects. This concentration of funding is further exacerbated by the time spent on each project. With a proclaimed need to "fail faster" so that resources can be reallocated to other projects, time is a critical factor in this phase. Clinical trial issues also abound here, including those with data sharing, data integrity, informed consent, recruitment, and retention.

The next phase in the pharmaceutical value chain is **manufacturing and supply chain**. As this phase is the first that is truly visible and has the most immediate impact on the general population, it is also the most noticeably impacted by supply chain disruptions, like those caused by the COVID-19 global pandemic. It is also the most publicly scrutinized, particularly with respect to drugs already on the market. Other recent issues tied to this phase include counterfeit drugs (despite extensive regulation); the presence of carcinogenic contaminants in products manufactured in international facilities; and an abundance of drug shortages attributable to a variety of causes, including global climate events, shortages of raw materials, political instability, and others.

On a less visible front, this is also the phase that includes the lobbying, discussion, monitoring, and the filing mechanisms of getting a drug approved, including submitting a New Drug Application (NDA). Inspection of manufacturing facilities and processes, official materials associated with the drug, any treatment/benefit claims, and the accuracy and validity of the clinical trials are evaluated here to ultimately determine if the drug will be brought to market [11].

Manufacturing and supply chain pain points are different from the previous two phases and generally fall into two categories: logistics and bureaucratic regulatory processes. Logistics encompasses standard supply chain woes and the necessary communication to properly execute the final stages of the drug approval process. Regulatory pain points include outdated, cumbersome, and even analog processes necessary to pursue approval, as well as the lack of organization and access to data from previously approved medications to make more educated decisions on newer treatments [12].

The penultimate phase of the pharmaceutical value chain is **launch and commercial considerations**. It begins with considerations regarding the release of the newly approved pharmaceutical agent to the public, including the prevalence of the target disease or condition, unique storage and preparation requirements, prescribing restrictions, etc. These factors are then extrapolated to marketing, packaging, commercial coverage, and consumer uptake, ultimately dispatching tightly coordinated efforts to begin recouping much of the funding spent to arrive at this point. At the very least, there are three layers of effort to this phase: one for consumers, one for providers, and one for payors, each with its own intricacies and regulatory framework.

The launch and commercial considerations phase brings an interesting turn to previous problems and pain points. While logistical issues still play a role in distributing the new drug product, revenue becomes the primary focus. Maximizing the efficient use of communication and distribution channels is vital to success. Trust, drug properties, and perceived utility are just a few examples of barriers that must be considered and thoroughly addressed in the efforts tied to this phase.

Lastly, once a new drug product has made it to market, the **monitoring and health records phase** commences. Commonly referred to as phase IV or post-marketing studies, this is the phase where continued surveillance of the drug, its performance, and any associated information is collected, scrutinized, and published, as needed [11]. However, just because a drug product has made it this far does not ensure success. Like others before it, this phase also carries the potential for failure. Several notable examples of drugs that have made it to this phase only to ultimately be withdrawn from the market due to safety concerns include Vioxx (rofecoxib), an anti-inflammatory medication, and Meridia (sibutramine) and Belviq (lorcaserin), two weight-loss medications.

Pain points in this final phase are largely related to accessing and recording reliable data. In a perfect world, these issues would have surfaced during the clinical development phase; however, due to the numerical and characteristic limitations on the populations studied during that phase, it is common for issues to be dismissed due to statistical insignificance or being overlooked in the targeted population. Adverse effects, special populations, misuse and abuse, quality issues, and even safety issues must then come through specific channels to be properly recorded and explored. The lack of access to and acceptance of many sources of real-world data (RWD) and real-world evidence (RWE), including social media, wearables, and electronic health records (EHRs), limit manufacturers' ability to quickly identify and understand the scope and severity of some issues [12]. Conversely, the same deficiencies can also hinder the recognition of unintended or unstudied benefits of a drug, delaying the steps necessary to offer expanded access and indications.

Increasingly, both technology giants (e.g., Amazon, Apple, Google, Microsoft) and digital health startups (e.g., BurstIQ, EncrypGen, Equideum Health, Patientory) are attempting to capitalize on their resources and expertise in areas including data analytics, predictive modeling, decentralized artificial intelligence (AI) and blockchain to innovate, improve efficiencies, and address pain points in life sciences and pharmaceutical value chain (Fig. 1) [8].

Blockchain and distributed ledger technologies (DLT) are being explored and employed across all five components of the pharmaceutical value chain. Specifically, technology-accelerated approaches for research and discovery include harnessing the potential of data mining, predictive modeling, and AI to identify and prioritize "druggable" targets and candidate medications [13], as well as for drug repositioning—also referred to as repurposing or reprofiling [14]. Clinical development-related targets include digital twins [15] and enhancing clinical trial management and remote participant monitoring [16]. Innovations such as 3D printing medications in drug manufacturing [17] and drug supply chain optimization [18] range from early to late-stage efforts. Combatting challenges with clinical decision support (e.g., alert fatigue) [19], ePrescribing [20], virtual tools [21], and enabling personalized medicine [22] are notable areas within launch and commercial. For monitoring and health records, decentralized medication management systems [23], identity and remote care [24], prescription delivery [25], and adherence [26] are among the efforts to date.

To reiterate, the pharmaceutical value chain has an underlying cyclic nature—one or more steps may be revisited at any time during a drug's life cycle to maintain viability. Manufacturing processes may need to be changed or relocated; off-label

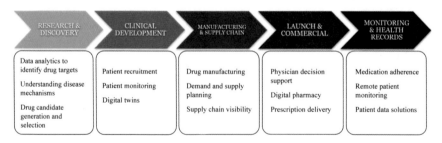

Fig. 1 Technology-focused solutions for the pharmaceutical value chain (Adapted from [8])

use may identify a new indication worth studying, or the brand name and marketing materials may need to be changed due to consumer confusion or common misuse. However, regardless of the circuitous path that may define a specific use case, the pharmaceutical value chain carries high costs, inefficiencies, gaps, and opportunities for technology to improve the process.

2.2 Differentiating Pharmaceutical Value Chain from Pharmaceutical Supply Chain

Before proceeding, the distinction between the pharmaceutical value chain, pharmaceutical supply chain, and medicine value chain is worth noting. The pharmaceutical supply chain, often incorrectly referred to as the pharmaceutical value chain, consists of all elements of the manufacture and distribution of pharmaceutical products. It is most succinctly represented in the third phase of the pharmaceutical value chain, aptly named manufacturing and supply chain, and is considered by most to include the regulatory requirements that accompany the physical manufacture and movement of pharmaceuticals.

Conversely, the medicine value chain deals with the specific monetary pricing and associated 'real' value of medicines. Comparatively, it represents a truncated or alternately aligned version of the full chain. Outlined by Aitken in "Understanding the pharmaceutical value chain," this oft-cited but narrow interpretation of the full value chain combines the research and development phases with the manufacturing and supply chain [27]. Because Aitken's article focuses on easily traceable costs, the less directly attributed costs of research, discovery, and development are obscured through this reorganization. While simplifying the cost structure is sufficient for his needs, it diminishes the ability to effectively see and address each phase and its pain points. To fulfill the purpose of demonstrating the potential value of blockchain across the entirety of the life cycle of a pharmaceutical product, this narrowed view is therefore discounted in favor of the more comprehensive alternative.

3 Blockchain Efforts Within Pharmaceutical Industry

Despite the relative novelty of blockchain and DLT, numerous efforts are already underway to identify and explore the benefits they may offer to the pharmaceutical industry. Consider the following use cases. The pain points that they specifically address will be further discussed in a later section.

3.1 Pharmaceutical Users Software Exchange (PhUSE) Blockchain Project

The Pharmaceutical Users Software Exchange (PhUSE) was founded in 2004 in the United Kingdom as a community where pharmaceutical programmers could discuss ideas and shepherd the industry's future direction [28]. Now a global, independent, and volunteer-run non-profit organization, PhUSE has taken on issues like data transparency, open-source technology, data standards optimization, and frequently assessed emerging trends and technologies, not the least of which is blockchain [28].

In 2017, their efforts began with the question of how blockchain can offer solutions across the entirety of the pharmaceutical value chain; in 2018, PhUSE published its first report on the transformative promise offered by blockchain to both the pharmaceutical and healthcare industries. After identifying blockchain models that could be useful in healthcare, the report gave several examples of how blockchain could improve existing pain points. Two use-case projects were outlined that had the potential to quickly illustrate and capture the benefits of blockchain while simultaneously serving to lay the necessary groundwork for the multitude of changes that must be undertaken to fully embrace the technology. The two projects recommended were: the use of smart contracts to maintain efficiency and quality in the supply chain (modeled from other industry uses and adapted) and increased access to and transparency for patient data through a blockchain-facilitated patient portal-type function [29].

In 2020, PhUSE published Phase 2 of the project, focusing on blockchain applications in the pharmaceutical industry, explicitly emphasizing the improvements it could bring to the clinical trial process. They set about to deliver a proof-of-concept solution that addressed the following four needs: patient identification, data infrastructure, eConsent tools, and architecture specialists [30]. Using Ethereum, they were successfully able to build proof-of-concept, however, they highlighted the inefficiency of the platform for real-time data sharing needs, stopped short of calculating the return on investment, and noted complications with integration and the various user interfaces. Still, the successful deployment of both patient identification and consent/enrollment tools, as well as the validated model for protected, shareable data, are vital steps in developing industry interest and trust in the application of blockchain.

3.2 Innovative Medicines Initiative (IMI) Blockchain-Enabled Healthcare

The Innovative Medicines Initiative (IMI) is a byproduct of the European Technology Platform on Innovative Medicines (aka INNOMED) [31]. Established in 2007, the first IMI Initiative (IMI1), executed as a public–private partnership (PPP) between the European community and the European Federation of Pharmaceutical

Industries and Associations (EFPIA), was created to improve the drug development process and ultimately create safer and more effective medicines [32]. After 7 years of prolific, breakthrough research and documented progress on the project initiatives, IMI2 was created in 2014 to continue the undisputed success of IMI1 and build on the advancement and momentum it generated [31].

One of the more ambitious projects of IMI2 was "Blockchain-Enabled Healthcare." Designed to include and represent stakeholders from the entirety of the healthcare system, the blockchain-enabled healthcare program built upon the associated momentum of PhUSE and endeavored to establish an incentive-based blockchain ecosystem that could be used unilaterally by the pharmaceutical industry for development, manufacturing, and distribution needs [33]. One of the notable outputs of this industry group was their conceptual approach to a blockchain-enabled healthcare system, which they delineated across a three-layer proposal (Fig. 2).

Understanding the difficulty in maintaining oversight of a distributed service, IMI2 also established a Healthcare Foundation feature as both a governance and integration structure [33]. Their primary development was PharmaLedger, a blockchain-enabled consortium currently comprised of 29 members, including ten European Union Member States, Switzerland, Israel, and the U.S., along with representatives from the EFPIA, subject matter experts, research centers, hospitals, patient organizations, etc. [34].

PharmaLedger is focused on leveraging blockchain for the supply chain (e.g., anti-counterfeiting, clinical product traceability, e-leaflet, finished goods traceability), clinical trials (e.g., recruitment, e-consent), and health data (e.g., connected health devices, networked Internet of Things (IoT) medical devices, remote patient monitoring). Complimentary efforts include the combination of blockchain, machine

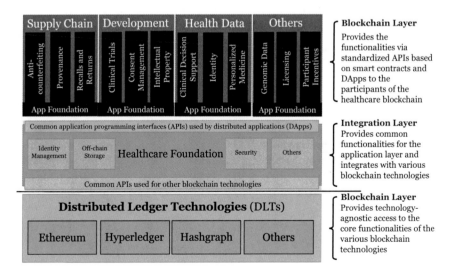

Fig. 2 Layered proposal for a blockchain-enabled healthcare system (Adapted from [33])

learning (ML), and AI to help realize value-based healthcare delivery via personalized medicine. The project is ongoing as its roadmap outlines a three-stage approach of design and foundations, followed by development and deployment, culminating in validation and sustainability. PharmaLedger has also committed to privacy and transparency in health data, a secure and trustworthy supply chain, improved patient ownership of health data, and accelerated clinical development through innovation in the clinical trial process.

3.3 The MELLODDY Project and Millions of Molecules Blockchain + Smart Contracts for Human Participant Regulations and Consent Management

The MELLODDY project (Machine Learning Ledger Orchestration for Drug Discovery) is another endeavor from IMI2. A blockchain-based, federated ML platform, MELLODDY is designed to leverage proprietary data without compromising it; the aim is that the program will be able to glean insights from the independent drug discovery efforts of multiple private entities without disclosing the source [35]. A unique component of this model is that the analysis includes not only data points from millions of biochemically-active small molecules but also several hundred terabytes of image data—the insights from which companies could use both retroactively and with future discoveries to help accelerate and improve drug development [36].

3.4 Information Exchange and Data Transformation (INFORMED) Initiative

Whereas PhUSE is a predominantly private-backed initiative, and IMI2 is a PPP, the Information Exchange and Data Transformation (INFORMED) Initiative is a modified PPP endeavor operated by the FDA and comprised of government, academic, non-profit, and industry members [12]. INFORMED was started in 2016 by the Office of Hematology and Oncology Products to improve data aggregation, organization, and mining for oncology products. While oncology was the initial driver of the program, the draw of similar open access data-sharing technology provided through blockchain expanded the project to include ways to leverage all existing data, including images, test results, RWD, RWE, and other digital or digitized sources [37].

3.5 Moneyball Medicine

'Moneyball medicine' is a phrase that entered the lexicon in 2012 [38]. Drawing on the lessons in Michael Lewis' 2003 book, Moneyball: The Art of Winning an Unfair Game, the authors posit that the moneyball concepts of evidence-based decision-making and associated value determination apply to baseball and medicine. Based on this association, they offer the accountable care model of healthcare as the true test of moneyball medicine, with the added caution that all cost-effectiveness models are only as solid as the data and the assumptions on which they are built.

Fast-forward five years to the publication of the book MoneyBall Medicine, a commentary expanding previous points but with a more specific perspective: data and analytics are the true drivers of evidence and transformative value in healthcare [39]. Unfortunately, the ubiquity and volume of data now involved in daily healthcare operations and accumulated over the last decade demand better data-oriented tools and improved models. Glorikian specifically notes the importance of AI and ML tools for drug discovery and repositioning, emphasizing the importance of deploying advanced data-driven technology to provide increased value at a reduced cost [40]. While he did not specifically mention the use of blockchain, his forward thinking identifies with one of the many identified use cases for applying blockchain in data analysis and management.

In summary, both the realized and unrealized value of blockchain in healthcare are becoming evident. With pioneering efforts coming from both the public and private sectors, sometimes in unique collaborative partnerships, healthcare's costly and inefficient pain points are being reimagined and resolved through the innovative avenues offered by DLT.

4 Mapping Blockchain Characteristics to Pain Points in the Pharmaceutical Value Chain

4.1 Adapted Fit-For-Purpose Framework and Design Elements

As many of the projects in the previous section can attest, blockchain and the characteristics of DLT are well-suited or easily adapted to specifically address known pain points in the pharmaceutical value chain. This can, at first, be confusing; the known problems and pain points from each of the five phases of the chain are varied and intrinsically different. Clearly, one solution cannot solve them all. However, the flexibility and adaptability of the DLT fit-for-purpose framework [16] allows for the manipulation of individual design elements to arrive at a perfect-fit solution.

Looking back to the PhUSE example, the well-known and established frameworks, Ethereum and Hyperledger Fabric, were compared for appropriateness-of-fit for prospective projects. While neither framework was ideal for all the intended use cases, different design elements of each were selected to create a best-fit-for-purpose framework. Borrowing the smart contracts, tokenization/incentivization, and benchmarked automation of Ethereum, and the private permissioned setup of Hyperledger Fabric, a health system could set up an incentivized blockchain security network protecting access to internal data, ensuring only approved parties can access the private data within and keeping out malicious entities like ransomware [29].

This example also highlights the importance of feature tradeoffs. Different features and design elements impact the end functionality of the framework. Therefore, it is important to analyze which blockchain elements are essential for a project, allowing others to be modified or displaced in favor of functionality. This will be addressed further in the following subsection.

4.2 Matching Characteristics (e.g., Decentralized, Distributed, Conditionally Immutable, Scalable, Cryptographically Secured) to Identified Pain Points in Each of the 5 Categories

Looking back to the pain points and problems identified in the pharmaceutical value chain, there are many opportunities for technological intervention and improvement, specifically through features of DLT.

Starting with the research and discovery phase, pain points have been identified in dealing with the sheer volume of available data, limited financial and exploratory resources, data siloing to protect proprietary interests, and inhibited research progress due to the resulting data separation. In short, a faster, less expensive method is needed to examine proprietary data distributed among many sources without compromising ownership. A quick glance back at projects already underway indicates that a solution is already being tested for this: MELLODDY. Applying the decentralized, distributed, security, traceability, authentication, practical immutability, and scalability attributes of blockchain to a federated ML model, MELLODDY accomplishes the following [35]:

- Improved speed of analysis at a reduced cost
- Access to protected IP without compromising it
- Collaborative learning without data or learning centralization
- Instant applicability of insights to proprietary data
- Decreased lead generation time.

The next chain phase, clinical development, is probably one of the more heavily researched areas of applicability for blockchain elements. It is also one where trade-offs will need to be considered based on project priorities. Common complaints during clinical trials include lack of data sharing capabilities, both real-time and delayed, possible data duplication due to inter-site blinding, recruitment, retention, duration, and the inability to efficiently analyze data for meaningful observations. Summarily, all of these issues also result in high costs.

Several examples have already been presented both in previous sections and the literature to address these issues. Starting with consent and enrollment, PhUSE recommended using a tokenized voiceprint recorded on blockchain to establish unique identities. The decentralized nature of blockchain, consensus authority, incentivization, and near-immutability of the data not only made it hack-resistant but also allowed patients to provide access to their information across multiple different organizations with minimal effort [29]. They then used this system with the addition of smart contracts and automated benchmarking to manage informed consent documentation, ensuring the correct form was signed and accessible to the right site at the right time and that necessary procedural cascades were consistently performed per protocol. Similarly, models leveraging blockchain for improved research participant recruitment, retention, and research data sharing have been proposed [41].

Approaching the problem from a different perspective, PharmaLedger embraces the view that token incentives can be used to anchor large volumes of information to the blockchain without actually being incorporated into the blocks. This key difference maintains the security and tamper resistance of the anchored information through standard blockchain principles but affords the user control over their information by requiring the token key for access. In this example, a universal health identifier and record would be stored for each patient, decentralized across the network. Any time the patient's record needed to be accessed, the owner's key can be provided via an encrypted interface, allowing access only for as long as needed. The previous data cannot be compromised. Any new data entered in the record are anchored in new blockchains linked to the same patient identifier and EHR. Because real-time access is desired across this system, block size is minimized, and transaction speed is maximized by anchoring information to the chain instead of directly incorporating it. This essentially allows the network to act as a high-level health information exchange with no interoperability issues because blockchain contract rules set the standards for data.

In these cases, as well as many others, recruitment, retention, monitoring, protocol compliance, data management, data analysis, and data transparency are aided by the decentralized and secure nature of blockchain while consensus mechanisms (e.g., proof-of-stake (PoS), proof-of-authority (PoA)), tamper resistance, and the distributed nature of blockchain protect data from corruption, integrity, and privacy concerns [42]. Additionally, the use of smart contracts can appropriately display identifiable or de-identified data depending on the user access permissions, maintaining blinding, aiding in faster recruitment of eligible patients, and ensuring proper follow-up with primary care teams regarding participation and monitoring.

The pharmaceutical value chain's manufacturing and supply chain phase is already benefitting from advancements in other industries that deal with logistics in both manufacturing and distribution. While regulatory processes have been implemented to improve track and trace functions, blockchain's near-immutability, consensus mechanisms, transparency, and distributed nature make it unparalleled for use in this sector. Counterfeit medications, unapproved source products, chain-of-custody traceability, audit trails, and even false claims are all addressed with blockchain elements [43]. Because counterfeit and unapproved products and source materials lack the transparent and traceable audit trails that establish provenance and chain-of-custody, it is easy to ferret them out or avoid them entirely. False claims are also easily spotted, as corresponding invoice audit trails easily corroborate or refute possession of the allegedly dispensed product. Plus, blockchain can easily be traced and verified through a marker as simple as a QR code or similar encoded tracker for in-network entities [44].

The predominant benefits of blockchain on the launch and commercial considerations phase of the chain are improved ease of dissemination and transparency of data and reduced cost recovery. As previously discussed, commercial efforts related to drug launches revolve heavily around positioning, pricing, and maximizing revenue recoupment from the previous phases of the chain. The reduced costs from all previous phases contribute directly to reduced revenue needs here, resulting in potentially industry-altering methods and rates for pricing newly released drugs. Payment and reimbursement models governed by smart contracts would speed up payments and approvals by decentralizing an unnecessarily slow, centralized process [45]. Additionally, the security, scalability, and transparency of blockchain allow for rapid and more cost-effective dissemination of drug-related information, improving regulatory-based efforts as well as marketing and medical knowledge.

Finally, blockchain features can be immensely helpful in post-launch monitoring and health records. Aligned with the aforementioned improvements in clinical trial enrollment and tracking, blockchain-based programs could have faster and more comprehensive access to patient-reported issues and events; all kept secure, private, and immutable through the core principles of blockchain. In the same line, blockchain-based record and data management would make it easier to trace and notify affected parties of problems or concerns that arise with a drug. In the event of a drug recall, blockchain-based tracking systems would facilitate rapid and complete notification and collection or disposal of recalled lots.

5 Blockchain—But Not in a Vacuum

5.1 Blockchain-Complementary Established and Emerging (e.g., Machine Learning, Artificial Intelligence) Technologies for the Pharmaceutical Value Chain

Blockchain offers a great deal of transformative promise to healthcare and the pharmaceutical industry. However, as the projects highlighted in previous sections have shown, it is not a standalone, one-size-fits-all solution. Rather, it is a tool among many that can be collectively leveraged to improve known problems and issues facing healthcare. The MELLODDY project combines blockchain with federated ML and AI to efficiently and securely deliver on its promise. Without applying the

ML model and the AI capable of using it, even blockchain could not effectively overcome the issues that accompany overwhelming volumes of siloed proprietary data. Similarly, PhUSE acknowledged that a major component of any blockchain-based technology needed to include a team of architecture experts, ensuring that the data shared, captured, and analyzed on the network was available and usable to the parties who need it [30]. Having information on the blockchain is useless without the applications and interfaces necessary to read and interact with it. Members of the Decentralized Trials and Research Alliance [46] have also launched efforts leveraging blockchain-complementary technologies and approaches, including the use of privacy-preserving, federated ML with the Veterans Incentivized Coordination and Integration initiative as a Data Integrity and Learning Network focused on veteran well-being [47]. Arguably, the most positively radical approach in the quest for "faster medical miracles" is Distributed Autonomous Science, which would theoretically require either a complex distributed autonomous organization (DAO) or a "complex series of more simplistic DAOs" to achieve those goals [48].

Consider a project with multiple stakeholders who all need access to the same set of information but with different access permissions, like a clinical trial. Middleware, interfaces, application programming interfaces (APIs), and associated infrastructure must be developed to recognize different stakeholder roles, request appropriate tokens/keys, and display only relevant or approved information. Patients should not have access to global study data, just as researchers should not have access to blinded study information, and data analysts should not have the ability to add new data. Blockchain frameworks can secure, transmit, approve, store, track, and audit data. However, even smart contracts are limited in how they can be applied, as any rules built into them apply unilaterally to every node and data point on the network.

Blockchain is also being used in telehealth and virtual reality spaces (Equideum) [47]. Outfitted with AI programming, it is helping rehabilitate patients dealing with addiction, psychosocial, and other mental health issues, as well as pain management, communication, and even incarceration [49]. Other applications include natural language analysis and processing, neural networks, evidence-based non-pharmacological, clinical intervention, and behavior modification models.

Ultimately, the important element is this: solutionism must be avoided. Blockchain alone is just a tool, not a solution. As in many other circumstances, it is important to match the best solution to fit the need. Combining blockchain with other established and emerging technologies is essential to successful implementation and solutions.

6 Debunking Myths Around Challenges with Blockchain

6.1 The Myth of the Technical Challenge

There are undoubtedly barriers to fully realizing the potential benefits of blockchain utilization in the pharmaceutical industry. However, the most resounding and easily debunked is that the technology is the primary roadblock and is too challenging to implement.

This myth is perpetuated on three fronts: user interaction, infrastructure, and integration.

As demonstrated by both PhUSE and PharmaLedger, it is possible to leverage other technologies, including middleware, APIs, and custom user interfaces, as well as interprofessional communication, to effectively create solutions that deliver a user-friendly experience. In fact, PharmaLedger is built on the premise that the technology can be seamlessly and artfully applied to increase consumer confidence in technology generally beyond the scope of comprehension of the general public [34].

This myth also likely originates in the generally accepted knowledge that customization is associated with confusion and cost [50]. As one of the most useful attributes of blockchain is its flexibility and adaptability to any set of circumstances or problem, customizability is essentially another core feature of the technology. Nevertheless, because the most well-known and discussed use cases for blockchain happen to be cryptocurrencies, their frameworks can be perceived as immutable as the data they contain, further stressing reconciliation with the idea of adaptable architecture. Similarly, across almost any industry, healthcare notwithstanding, customization leads to increased costs. Based on the assumption that there is a 'base model,' customization or personalization is equated with 'upgrading.' Specifically in the world of technology, improving transaction speed, scalability, traceability, data management, and data analysis all generally equate to higher cost and greater resource utilization in previously established models.

So, it comes as little surprise then that the guise of technological challenges is superimposed over other outdated and inapplicable standards from the very historical programs and procedures that blockchain helps overcome.

Fortunately, this myth is fairly easily dispelled, not only through the diligent preliminary work done by multiple pioneering entities but also through the understanding that the very basis of blockchain is founded on the concepts of trust, traceability, and security modified to meet the needs of its users.

6.2 The Reality of Challenges Tied to Change Management, Resource Allocation, Paradigm Shift, and Reaching Consensus

Instead of reflecting true technology challenges in the use of blockchain, the technology myth is more likely founded on the fearful understanding that underlying and disparately managed systems will need to agree upon and conform to specific standards to maximize benefits.

Technological and procedural entrenchment is often a hallmark of large organizations and institutions, so much so that an entire field of study has been dedicated to change management or the successful transition from one system to another.

User buy-in, training, education, planning, commitment, and follow-through are all necessary for successful change management. The culture within the organization is also a necessary element to identify and navigate. Properly implemented, change management can go very smoothly; but poorly attempted or implemented change management can not only result in failure and further entrench the organization in outdated practices and ideals. Perhaps one of the most recent areas where this has played out is the implementation of EHRs. Multi-million dollar projects that took years and countless resources to execute revolutionized the practice of evidence-based medicine across the world. The improved efficiency, availability of and access to data, and clarity and organization of information were a stark contrast to the slow, disorganized, and cumbersome practice of paper charting.

So, it is with blockchain; planning, resource allocation and management, consensus understanding of underlying needs related to both input and output, and a desire for change will all be necessary to face the fear and defeat the myth.

Fortunately, recognizing the inherent challenge in realizing this paradigm shift, forward-thinking organizations and individuals have built partnerships and collaborative efforts to identify and address these needs. Through their continued efforts, the world is gaining an understanding of how adaptable and controllable blockchain implementations are, the realized benefits of blockchain utilization, and the next challenges to tackle to improve its application.

7 Blockchain and The Idea Pipeline

New use cases for blockchain in healthcare and pharmaceuticals are constantly being discovered and explored. Some are building off the success of established uses, while others are forging new pathways that take advantage of blockchain's unique features. The following are examples of areas where blockchain is being explored or could prove beneficial.

7.1 Pharmacogenomics

Blockchain applications are already being successfully combined with other technologies and applied to massive data stores to efficiently and thoroughly analyze existing data, particularly in new drug discovery. However, much remains to be discovered about the underlying mechanisms of existing medications and known compounds, including comprehensive biological action, interactions, and alterations based on genetic factors. With improved computing power, distributed networks, privacy assurances, and the appropriate application of ML algorithms and AI, blockchain could usher in a new era in personalized medicine through an immense expansion of the understanding of the interplay between pharmaceutically active compounds and variations in the human genome and genetic expression. Early efforts for this include the use of Ethereum for storage and querying of pharmacogenomic data via smart contract capabilities [22], as well as commercial interests seeking to address core pain points of sequencing costs, regulatory costs, and privacy [51].

7.2   Collaborative Pharmaceutical Development

One of the stated goals of several initiatives, including MELLODDY, is to improve the analysis of existing compound knowledge, facilitate identification of existing molecules for pharmaceutical development, and enhance discovery of new molecules. However, in doing so, MELLODDY aims to preserve the origin of information to protect IP rights. While this is not inherently problematic, it does eliminate the possibility of uniting researchers on parallel but unequal research paths.

In the present market, many new molecules and products are being discovered or created by small companies lacking the resources for large-scale study and development. To remedy this, many seek to partner with large-scale pharmaceutical manufacturers to hasten the journey down the pharmaceutical value chain and hopefully create a mutually beneficial partnership. However, this model still relies on one company conducting the preliminary research and a second company leveraging their experience and resources to improve the process.

Meanwhile, it is not uncommon for different pharmaceutical companies to expend resources in research and development only to abandon a project similar to, but behind that, of a promising competitor. Realistically, the resources of all parties involved could have been optimized if the concurrent research was complementary instead of parallel. The combined efforts of both teams could have arrived at any number of conclusions faster and through reduced resource expenditure had they been working together.

To that end, the question must be asked if there is a way for a blockchain-based program to encourage collaboration between entities doing similar but unequal research? This approach could develop similarly to the 'coopetition' model [52, 53] that blockchain enabled with the [54] via the use of Quorum (i.e., enterprise, permissioned version of Ethereum). Coopetition allowed companies with traditional competing interests in the healthcare provider data management space (i.e., Aetna, Humana, MultiPlan, Quest Diagnostics, UnitedHealthcare) to form a consortium and work together for the mutual benefit of all participants [55]. The use of smart contracts to negotiate agreements between parties could also be used to create a standardized path for collaborative research, further accelerating clinical development. Regardless of the outcome, it would improve and decrease resource utilization while reducing the time needed for a fully developed discovery.

7.3 Patient Access, Medication Reclamation, and Prescription Waste Reduction

The cost of individual cancer drugs and biological agents commonly have prices surpassing $10,000 each per month [56], which functionally limits patient access to these lifesaving agents. Perhaps counterintuitively, prescription waste has also been observed to occur in up to 41% of patients receiving oral cancer drugs—chiefly due to cancer progression, death, and toxicity [57]. This disconnect highlights that financially driven health disparities in vulnerable patient populations are concurrent with an avoidable waste of sealed, single-dose packaged oral oncolytics.

One novel application of blockchain technology in the pipeline aims to address challenges around patient access and financial toxicity associated with high-cost cancer medications. Their approach to medication reclamation and redistribution also may provide ancillary benefits, including reducing prescription waste and environmental pollutants, as well as yielding valuable supply chain data and related indicators for this category of surplus medication. On the front end, RemediChain [58] encourages and incentivizes citizens to text #FlipYourScrip to donate unopened, unexpired medications along with a picture of the medication packaging. Upon receipt of the picture and information, RemediChain will either: (1) direct individuals to a safe drug disposal facility if the medication is not suitable for redistribution—including providing a gift card for those who opt to text a picture from the drug disposal unit, or (2) provide free shipping via a partnership with FedEx if the medication can be donated to a patient in need. After launching this campaign, in 2021, RemediChain was able to match high-cost cancer medications with nearly 100 patients who otherwise would have gone without these lifesaving treatments (Fig. 3).

On the back end, the RemediChain platform is leveraging blockchain technology to create a surplus medication database with shared governance via forming an international consortium of research universities, cancer centers, and other stakeholders.

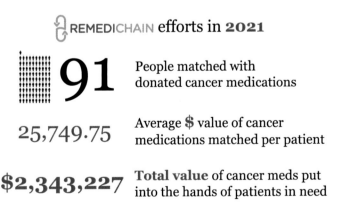

REMEDICHAIN efforts in 2021

91 People matched with donated cancer medications

25,749.75 Average $ value of cancer medications matched per patient

$2,343,227 Total value of cancer meds put into the hands of patients in need

Fig. 3 Number of patients matched with high-cost cancer medications and value in US dollars of matched medications by RemediChain, 2021 [58]

This platform also helps fill a related need, as no single organization is responsible for oversight of prescription waste, nor is tasked with tracking the precursor of prescription waste—surplus medications; consequently, the net impact of prescription waste (e.g., financial, health, and environmental) is unknown. Surplus medication research enabled by this platform could have implications in pharmaceutical sciences, population health, and environmental sciences. While RemediChain primarily focuses on the US pharmaceutical industry, its research could also be localized for low- and middle-income countries. Importantly, while concurrently putting medications into the hands of patients who need them, this research into surplus medication can lead to innovative processes, methods, and systems that prevent the conversion of surplus medication to prescription waste and increase the conversion of surplus medication into reclaimed medication for distribution across low-resource settings at scale.

7.4   The Evolution of the Traditional Retail Pharmacy

The traditional pharmacy dispensing model is under immense strain to maintain the line between safety and profitability; specifically, retail pharmacies find it increasingly difficult to do both. While unpopular, the idea of repositioning pharmacists in a dispensing model is essential to restoring profitability and improving patient safety and service.

Blockchain has already been shown to streamline the pharmaceutical supply chain and payment systems through smart contracts, traceability, and the use of micropayments [45]. While pharmacists are necessary to assess the safety and appropriateness of highly variable prescription orders, their participation in product preparation is a misuse of resources in a system where blockchain can be implemented. Freeing pharmacists from these tasks enables them to engage in cognitive- and service-based efforts (e.g., patient counseling, education, immunization, therapy reviews). This, in turn, fundamentally alters the systemic perception of pharmacist value, allowing value-based models to reimburse them for targeted health outcomes and care provided, as opposed to product dispensed.

While regulatory challenges make this opportunity challenging to capitalize on, it should be noted that an opportunity does, in fact, exist. Reimagining the pharmacist's role to exclude supply-chain activities that can easily be automated, tracked, and audited safely, an end-to-end value-based model could be put into play, improving patient safety, patient satisfaction, and pharmacists' capacity to apply their skills and knowledge where it is most needed.

These are only a few areas where blockchain could profoundly impact the pharmaceutical industry in the near future, based on existing utilization and anticipated opportunities. However, only the surface has likely been scratched concerning the positive and transformative impact blockchain can have on the future of pharmaceuticals.

8 Future Directions

The World Economic Forum highlighted that COVID-19 has emphasized the need for a cross-sector approach to collaboration in healthcare, necessitating new models (e.g., coopetition) and approaches [59]. The timing may also be right for the innovation-minded, as 76% of business executives recently indicated the need for new ways of collaborating (i.e., data sharing) with ecosystem partners and other stakeholders. Those sentiments align with findings from Gartner that collaborative data use via decentralized approaches is a major strategic trend across industries going forward [60]. This functional groundswell of support for both the philosophical underpinnings of blockchain and what it enables from a pragmatic perspective suggests a heretofore unseen openness to change from the most disruption-resistant sector on the planet—healthcare.

This same theme of change has been realized in pharmaceutical research and across each component of the pharmaceutical value chain, potentially buoyed by applying emerging concepts like the Internet of Behaviors [60]. Just as the IoB has been suggested as a means to harness wearable technology and "digital dust" to influence decision-making in other sectors, its use of data, incentives, and disin-centives could be applied to various stages along the pharmaceutical value chain. The approach itself is not particularly novel, as decades ago, texts like "Persuasive Technology: Using Computers to Change What We Think and Do" [61] illustrated this type of potential. However, the ubiquity of data, development of DLT-supported smart contracts, opportunities for privacy-preservation (e.g., zero-knowledge proofs, homomorphic encryption, federated learning) [62], and impetus for change have never been more pronounced than now. What remains to be seen is how tools for inno-vative change like this might be optimally employed while recognizing the critical need to address the accompanying ethical and societal challenges.

9 Conclusions

The pharmaceutical value chain encompasses much more than the subset of the pharmaceutical supply chain, which is limited to the elements of manufacturing and distribution. The five phases comprising the full pharmaceutical value chain: research and discovery; clinical development; manufacturing and supply chain; launch and commercial considerations; and monitoring and health records each suffers from pain points that have collectively resulted in a ponderous 17 years from "new idea to treatment" [48]. Blockchain is a relatively new addition to the arsenal for advancing pharmaceutical research, but when combined with other established and emerging technologies (e.g., AI, ML), has already begun to address these impediments via industry and PPP efforts (e.g., MELLODDY, PharmaLedger). While technology-related obstacles have slowed the adoption of some blockchain-facilitated solutions, challenges with change management and awareness deficits of future-facing models like coopetition can prove even more problematic to leveraging DLT to address the desired pain points. Those who can successfully shepherd their laboratories, organizations, universities, systems, and consortia to maximize the benefits of DLT while navigating hurdles will be well-positioned to make a positive impact on the lives of patients, industry participants, and global heath.

References

1. PhRMA (2015) Biopharmaceutical research & development: the process behind new medicines. http://phrma-docs.phrma.org/sites/default/files/pdf/rd_brochure_022307.pdf. Accessed 13 Jan 2022
2. Horizney C (2019) The drug discovery process. https://www.taconic.com/taconic-insights/quality/drug-development-process.html. Accessed 9 Jan 2022
3. Hughes JP, Rees S, Kalindjian SB, Philpott KL (2011) Principles of early drug discovery. Br J Pharmacol 162(6):1239–1249. https://doi.org/10.1111/j.1476-5381.2010.01127.x
4. Forum on Neuroscience and Nervous System Disorders, Board on Health Sciences Policy, Institute of Medicine (2014) The National Academies Collection: reports funded by National Institutes of Health. In: Improving and accelerating therapeutic development for nervous system disorders: workshop summary. National Academies Press (US), Washington (DC). https://doi.org/10.17226/18494
5. Flier JS (2019) Academia and industry: allocating credit for discovery and development of new therapies. J Clin Invest 129(6):2172–2174. https://doi.org/10.1172/JCI129122
6. U.S. Food and Drug Administration (2019) Guidance for industry: adaptive design clinical trials for drugs and biologics. https://www.fda.gov/media/78495/download. Accessed 9 Jan 2022
7. Singer DR, Zaïr ZM (2016) Clinical perspectives on targeting therapies for personalized medicine. Adv Protein Chem Struct Biol 102:79–114. https://doi.org/10.1016/bs.apcsb.2015.11.003
8. CB Insights (2021) The big tech in pharma report: from digital pharmacies, AI for drug discovery, & apps for medical records. How Amazon, Microsoft, Apple, and Google will reimagine the industry value chain. https://www.cbinsights.com/research/report/big-tech-pharma-amazon-microsoft-apple-google/. Accessed 14 Aug 2021
9. Zhou Y, Zhang YT, Lian XC, Li FC, Wang CX, Zhu F, Qiu YQ, Chen YZ (2022) Therapeutic target database update 2022: facilitating drug discovery with enriched comparative data of targeted agents. Nucleic Acids Res 50(D1):1398–1407. http://db.idrblab.net/ttd/. Accessed 9 Jan 2022
10. U.S. Food and Drug Administration (2022) Emergency use authorization. https://www.fda.gov/emergency-preparedness-and-response/mcm-legal-regulatory-and-policy-framework/emergency-use-authorization. Accessed 9 Jan 2022
11. U.S. Food and Drug Administration (2015) FDA's drug review process: continued. https://www.fda.gov/drugs/information-consumers-and-patients-drugs/fdas-drug-review-process-continued. Accessed 8 Jan 2022
12. U.S. Department of Health and Human Services (n.d.) Information exchange and data transformation (INFORMED) initiative. https://www.hhs.gov/cto/projects/information-exchange-and-data-transformation-initiative/index.html. Accessed 9 Jan 2022
13. Burki T (2019) Pharma blockchains AI for drug development. Lancet 393(10189):2382. https://doi.org/10.1016/S0140-6736(19)31401-1
14. Boniolo F, Dorigatti E, Ohnmacht AJ, Saur D, Schubert B, Menden MP (2021) Artificial intelligence in early drug discovery enabling precision medicine. Expert Opin Drug Discov 16(9):991–1007. https://doi.org/10.1080/17460441.2021.1918096

15. Trenfield SJ, Awad A, McCoubrey LE, Elbadawi M, Goyanes A, Gaisford S, et al (2022) Advancing pharmacy and healthcare with virtual digital technologies. Adv Drug Deliv Rev 114098. https://doi.org/10.1016/j.addr.2021.114098

16. Mackey TK, Kuo TT, Gummadi B, Clauson KA, Church G, Grishin D et al (2019) 'Fit-for-purpose?'—challenges and opportunities for applications of blockchain technology in the future of healthcare. BMC Med 17(1):68. https://doi.org/10.1186/s12916-019-1296-7

17. Elbadawi M, McCoubrey LE, Gavins FKH, Ong JJ, Goyanes A, Gaisford S et al (2021) Harnessing artificial intelligence for the next generation of 3D printed medicines. Adv Drug Deliv Rev 175:113805. https://doi.org/10.1016/j.addr.2021.05.015

18. Clauson KA, Breeden EA, Davidson C, Mackey TK (2018) Leveraging blockchain technology to enhance supply chain management in healthcare: an exploration of challenges and opportunities in the health supply chain. Blockchain Healthc Today 1. https://doi.org/10.30953/bhty.v1.20

19. Gangula R, Thalla SV, Ikedum I, Okpala C, Sneha S (2021) Leveraging the hyperledger fabric for enhancing the efficacy of clinical decision support systems. Blockchain Healthc Today 4. https://doi.org/10.30953/bhty.v4.154

20. Aldughayfiq B, Sampalli S (2022) Patients', pharmacists', and prescribers' attitude toward using blockchain and machine learning in a proposed ePrescription system: Online survey. JAMIA Open 5(1):ooab115. https://doi.org/10.1093/jamiaopen/ooab115

21. Krittanawong C, Aydar M, Hassan Virk HU, Kumar A, Kaplin S, Guimaraes L, et al (2021) Artificial intelligence-powered blockchains for cardiovascular medicine. Can J Cardiol S0828-282X(21)00912-0. https://doi.org/10.1016/j.cjca.2021.11.011

22. Gürsoy G, Brannon CM, Gerstein M (2020) Using ethereum blockchain to store and query pharmacogenomics data via smart contracts. BMC Med Genomics 13(1):74. https://doi.org/10.1186/s12920-020-00732-x

23. Li P, Nelson SD, Malin BA, Chen Y (2019) DMMS: a decentralized blockchain ledger for the management of medication histories. Blockchain Healthc Today 2. https://doi.org/10.30953/bhty.v2.38

24. Zhang P, White J, Schmidt DC, Lenz G, Rosenbloom ST (2018) FHIRChain: applying blockchain to securely and scalably share clinical data. Comput Struct Biotechnol J 16:267–278. https://doi.org/10.1016/j.csbj.2018.07.004

25. Jackson A, Srinivas S (2021) A simulation-based evaluation of drone integrated delivery strategies for improving pharmaceutical service. In: Srinivas S, Rajendran S, Ziegler H (eds) Supply chain management in manufacturing and service systems. International Series in Operations Research & Management Science, vol 304. Springer, Cham. https://doi.org/10.1007/978-3-030-69265-0_7

26. Gonzales A, Smith SR, Dullabh P, Hovey L, Heaney-Huls K, Robichaud M et al (2021) Potential uses of blockchain technology for outcomes research on opioids. JMIR Med Inform 9(8):e16293. https://doi.org/10.2196/16293

27. Aitken M (2016) Understanding the pharmaceutical value chain. Pharm Policy Law 18(1–4):55–66. https://doi.org/10.3233/ppl-160432

28. PhUSE (n.d.) The global healthcare data science community. In: About PhUSE. https://phuse.global/About_PHUSE. Accessed 7 Jan 2022

29. PhUSE (2018) How blockchain can transform the pharmaceutical and healthcare industries. In: Emerging trends archived projects. https://phuse.s3.eu-central-1.amazonaws.com/Deliverables/Emerging+Trends+%26+Technologies/How+Blockchain+Can+Transform+the+Pharmaceutical+Health+Industry.pdf. Accessed 7 Jan 2022

30. PhUSE (2020) Blockchain technology: phase 2 report. In: Emerging trends archived projects. https://phuse.s3.eu-central-1.amazonaws.com/Deliverables/Emerging+Trends+%26+Technologies/Blockchain+Technology+Phase+2+Report.pdf. Accessed 7 Jan 2022

31. Innovative Medicines Initiative (n.d.) History—the IMI story so far. In: About IMI. https://www.imi.europa.eu/about-imi/history-imi-story-so-far. Accessed 7 Jan 2022

32. Innovative Medicines Initiative (2021) PharmaLedger. In: Projects and results: project fact-sheets. https://www.imi.europa.eu/projects-results/project-factsheets/pharmaledger. Accessed 7 Jan 2022

33. Innovative Medicines Initiative (2018) Topic: blockchain enabled healthcare. https://www.imi.europa.eu/sites/default/files/uploads/documents/apply-for-funding/future-topics/IndicativeText_BlockchainHealthcare.pdf. Accessed 7 Jan 2022
34. PharmaLedger (2021) Blockchain enabled healthcare. PharmaLedger introduction. https://pharmaledger.eu/wp-content/uploads/PharmaLedger-Official-Presentation.pdf. Accessed 13 Jan 2022
35. MELLODDY (n.d.) Machine learning ledger orchestration for drug discovery. https://www.melloddy.eu/. Accessed 7 Jan 2022
36. Innovative Medicines Initiative (2020) MELLODDY. In: Projects and results: project factsheets. https://www.imi.europa.eu/projects-results/project-factsheets/melloddy. Accessed 7 Jan 2022
37. Khozin S, Pazdur R, Shah A (2018) INFORMED: an incubator at the US FDA for driving innovations in data science and agile technology. Nat Rev Drug Discov 17(8):529–530. https://doi.org/10.1038/nrd.2018.34
38. Greene JA, Podolsky SH (2012) Moneyball and medicine. N Engl J Med 367(17):1581–1583. https://doi.org/10.1056/NEJMp1211131
39. Glorikian H, Branca MA (2017) MoneyBall medicine: thriving in the new data-driven healthcare market. Routledge. https://www.routledge.com/MoneyBall-Medicine-Thriving-in-the-New-Data-Driven-Healthcare-Market/Glorikian-Branca/p/book/9781138198043
40. Bean R (2020) Moneyball medicine: data-driven healthcare transformation. Forbes, 26 April 2020. https://www.forbes.com/sites/ciocentral/2020/04/26/moneyball-medicine-data-driven-healthcare-transformation/?sh=529eae8c58ab. Accessed 9 Jan 2022
41. Zhang P, Downs C, Le NTU, Martin C, Shoemaker P, Wittwer C et al (2020) Toward patient-centered stewardship of research data and research participant recruitment with blockchain technology. Front Blockchain 3:32. https://doi.org/10.3389/fbloc.2020.00032
42. Omar IA, Jayaraman R, Salah K, Simsekler MCE, Yaqoob I, Ellahham S (2020) Ensuring protocol compliance and data transparency in clinical trials using blockchain smart contracts. BMC Med Res Methodol 20(224). https://doi.org/10.1186/s12874-020-01109-5
43. Rayan RA, Tsagkaris C (2021) Blockchain-based IoT for personalized pharmaceuticals. In: Cardona M, Solanki VK, Garcia Cena CE (eds) Internet of medical things. CRC Press, pp 51–62. https://www.routledge.com/Internet-of-Medical-Things-Paradigm-of-Wearable-Devices/Cardona-Solanki-Cena/p/book/9780367272630
44. Norfeldt L, Botker J, Edinger M, Genina N, Rantanen J (2019) Cryptopharmaceuticals: increasing the safety of medication by a blockchain of pharmaceutical products. J Pharm Sci 108(9):2838–2841. https://doi.org/10.1016/j.xphs.2019.04.025
45. Attaran M (2020) Blockchain technology in healthcare: challenges and opportunities. Int J Healthc. https://doi.org/10.1080/20479700.2020.1843887
46. DTRA (2022) Decentralized Trials and Research Alliance. https://www.dtra.org/. Accessed 19 Jan 2022
47. Equideum Health (2022) Equideum Health appoints Colonel Joseph Wood, MD, PhD (US Army, Retired) as Executive Director of Veterans-Facing VICI initiative, also to serve as Vice President, Decentralized Trials and Virtual Health. https://equideum.health/press-releases/equideum-health-appoints-colonel-joseph-wood-md-phd-us-army-retired-as-executive-director-of-veterans-facing-vici-initiative-also-to-serve-as-vice-president-decentralized-trials-and-virtual-hea/. Accessed 19 Jan 2022
48. Manion ST, Bizouati-Kennedy Y (2020) DAO of science. In: Manion ST, Bizouati-Kennedy Y (eds) Blockchain for medical research: accelerating trust in healthcare. Productivity Press, New York. https://doi.org/10.4324/9780429327735
49. Hort J, Valis M, Zhang B, Kuca K, Angelucci F (2021) An overview of existing publications and most relevant projects/platforms on the use of blockchain in medicine and neurology. Front Blockchain 4:14. https://doi.org/10.3389/fbloc.2021.580227
50. Siiskonen M, Watz M, Malmqvist J, Folestad S (2019) Decision support for re-designed medicinal products—assessing consequences of a customizable product design on the value chain from a sustainability perspective. In: Siiskonen M, Watz M, Malmqvist J, Folestad S (eds) Proceedings of the design society: international conference on engineering design, vol 1(1), pp 867–876. https://doi.org/10.1017/dsi.2019.91

51. Grishin D, Obbad K, Estep P, Quinn K, Wait Zaranek S, Wait Zaranek A, et al (2018) Accelerating genomic data generation and facilitating genomic data access using decentralization, privacy-preserving technologies and equitable compensation. Blockchain Healthc Today 1. https://doi.org/10.30953/bhty.v1.34
52. Brandenburger A, Nalebuff B (2021) The rules of co-opetition. Rivals are working together more than ever before. Here's how to think through the risks and rewards. Harvard Business Review. https://hbr.org/2021/01/the-rules-of-co-opetition. Accessed 19 Jan 2022
53. Pawczuk L, Wiedmann P, Simpson L (2019) Deloitte development LLC. So you've decided to join a blockchain consortium: defining the benefits of 'coopetition'. https://www2.deloitte.com/content/dam/Deloitte/us/Documents/process-and-operations/us-defining-the-benefits-of-coopetition.pdf. Accessed 11 Jan 2022
54. Synaptic Health Alliance (2022) https://www.synaptichealthalliance.com/resources. Accessed 11 Jan 2022
55. Hashed Health (2019) The seven major consortia. https://hashedhealth.com/consortia-july-2019-2/. Accessed 11 Jan 2022
56. National Cancer Institute (2018) Financial toxicity and cancer treatment (PDQ®)–Health professional version. https://www.cancer.gov/about-cancer/managing-care/track-care-costs/financial-toxicity-hp-pdq. Accessed 13 Jan 2022
57. Monga V, Meyer C, Vakiner B, Clamon G (2019) Financial impact of oral chemotherapy wastage on society and the patient. J Oncol Pharm Pract 25(4):824–830. https://doi.org/10.1177/1078155218762596
58. RemediChain (2021) RemediChain. #FlipYourScrip. Donate unused meds. https://www.remedichain.org/. Accessed 13 Jan 2022
59. Ceulemans H, Galtier M, Boeckx T, Oberhuber M, Dillard V (2021) From competition to collaboration: how secure data sharing can enable innovation. https://www.weforum.org/agenda/2021/06/collaboration-data-sharing-enable-innovation. Accessed 19 Jan 2022
60. Panetta K (2020) Gartner top strategic technology trends for 2021. https://www.gartner.com/smarterwithgartner/gartner-top-strategic-technology-trends-for-2021. Accessed 13 Jan 2022
61. Fogg BJ (2002) Persuasive technology: using computers to change what we think and do. Morgan Kaufmann, San Francisco. https://doi.org/10.1145/764008.763957
62. Miller R (2019) Emerging privacy preserving technologies are game changing. https://bertcmiller.com/2019/05/25/privacy-preserving-technologies.html. Accessed 19 Jan 2022

Kevin A. Clauson, PharmD, is a Professor at the Lipscomb University College of Pharmacy & Health Sciences in Nashville, TN. He began investigating the potential for blockchain technology as follow-up to his previous work as founding director of a World Health Organization (WHO) Collaborating Center. Dr. Clauson has explored blockchain for clinical trial data sharing at the White House, CryptoKitties and CryptoZombies as hands-on assignments with students, and as a core implementation for a charity pharmacy. He currently serves on the editorial board of Blockchain in Healthcare Today and Frontiers in Blockchain, the Advisory Board for Equideum Health (formerly ConsenSys Health) and RemediChain, and as a distributed ledger technology subject matter expert for HIMSS and IEEE. Additionally, he serves as a member of the Roster of Experts, Department of Digital Health, WHO and the Network of Digital Health Experts, Digital Health Center of Excellence, FDA. Dr. Clauson's work has generated coverage by the New York Times, Forbes.com, Wall Street Journal, and BBC Radio. He received his Doctor of Pharmacy from the University of Tennessee—Memphis and completed a Research Fellowship at the University of Missouri—Kansas City.

Rachel D. Crouch, PharmD, PhD is an Assistant Professor at the Lipscomb University College of Pharmacy & Health Sciences and Adjunct Assistant Professor in the Department of Pharmacology at Vanderbilt University; she previously served as a Postdoctoral Research Fellow at the Vanderbilt Center for Neuroscience Drug Discovery. Dr. Crouch combines her PhRMA Foundation sponsored fellowship training, PhD in Pharmacology, and clinically-focused PharmD degrees to inform her evidence-based, pragmatic research in drug metabolism and pharmacokinetics (DMPK). Her research with aldehyde oxidase was supported by the National Institutes of Health (NIH), and her scholarly work has been published in numerous peer-reviewed biomedical journals, most recently on an effort to develop inhibitors targeting the Severe Acute Respiratory Syndrome Coronavirus 3CL Protease (SARS-CoV-2 CLpro), led by collaborators at the Cleveland Clinic. She has also presented the results of her research nationally and internationally. Dr. Crouch earned her Doctor of Philosophy from Vanderbilt University and her Doctor of Pharmacy from Lipscomb University.

Elizabeth A. Breeden, DPh, MS, is an Associate Professor at Lipscomb University College of Pharmacy & Health Sciences. Breeden serves as founding director of the Master of Science in Health Care Informatics (MHCI) and dual PharmD/MHCI degree programs, and served as the PGY2 Pharmacy Informatics Residency Program Director. She led development for the Lipscomb University/IBM Watson Analytics Collaboration, which resulted in the first college of pharmacy to incorporate Watson Analytics within research and curricular offerings. Her experience includes positions in academia, health system pharmacy, and the health system software development industry. Breeden served as lead in defining product strategy, managing the software development life cycle, and meeting regulatory requirements for clinical products. Breeden completed the Bachelor of Arts from the University of Tennessee, Bachelor of Science in Pharmacy from Samford University and Master of Science in Biology from Austin Peay State University.

Nicole Salata, PharmD, MS, BCGP is Lead Clinical Informatics Pharmacist and freelance medical writer who is passionate about patient health and safety. Dr. Salata currently works as a clinical informaticist at PharmD Live, where she applies her clinical acumen to frontline health care technology to improve patient health and outcomes in the digital space, and as a freelance medical writer for The Med Writers. Dr. Salata is also a board-certified geriatric pharmacist with extensive community pharmacy, automation, and management experience across 15 years in patient-facing roles. Additionally, she contributes to the profession and sector in a variety of ways, most recently serving as a grant reviewer for the American Society of Health-System Pharmacists (ASHP) Foundation. Dr. Salata earned her Doctor of Pharmacy and Master of Science in Health Informatics from the University of Illinois at Chicago.

Blockchain-Based Scalable Network for Bioinformatics and Internet of Medical Things (IoMT)

Ned Saleh

Abstract The major problem in today's data creation and monetization is that the data creators (individual people trading, traveling, and interacting on social media) are not the data aggregators (the Googles, Facebooks, and Amazons of the world). As such, the full potential of the personal data value in the age of informatics has yet to fully materialize. This leads to constant conflict within the data ecosystem regarding who has the right to own and monetize data; the creators or aggregators. It has also led to a protracted debate on data sovereignty and expanded legislation for data privacy that we deal with every day when we navigate any website. The holy-grail solution for such a problem is vertical integration, i.e., integrating the data value chain by combining and ensuring that data creators and aggregators are the same in the data value stack. Until recently, this was deemed technologically impossible because individuals in society cannot be their own bank, e-commerce platform, their own search engine, and their own social media. However, the advent of miniaturized sensors driven by advancements in device engineering and miniaturization ushered in a new age of multifunctional sensors, often called the Internet of Things (IoT). In particular, the distributed miniaturized devices that measure the biological attributes of individuals are called the Internet of Medical Things (IoMT). This chapter describes an end-to-end ecosystem that offers a solution to this problem and the commercial pilot model it has implemented utilizing the nascent but promising blockchain technology.

Keywords Internet of Medical Things (IoMT) · Quantified wellness · Proof of identity · Homomorphic algorithms · Data monetization · Bioinformatics

N. Saleh (✉)
Synsal Inc, San Jose, USA
e-mail: ned.saleh@synsal.com

© The Author(s), under exclusive license to Springer Nature Singapore Pte Ltd. 2022
W. Charles (ed.), *Blockchain in Life Sciences*, Blockchain Technologies,
https://doi.org/10.1007/978-981-19-2976-2_3

1   Introduction

The major problem in today's data ecosystem harvested from individuals is creation, control, and monetization. A core component of the problem is that data creators (individual people trading, traveling, and interacting on social media) are not the data aggregators (e.g., Google, Facebook, Amazon, etc.). Because of this split, the full potential of the personal data value in the age of informatics has yet to fully materialize. This resulted in constant conflict within the data ecosystem of who has the right to own, control, and monetize data, the creators or aggregators, and led to a protracted debate on data sovereignty along with expanded, and sometimes conflicting, multi-jurisdiction legislations for data privacy that we deal with every day. Our approach to addressing this conundrum is the holy-grail solution represented in vertical integration, i.e., integrating the data value chain by combining and making the data creators and aggregators the same entities in the data value stack. Until recently, this was deemed technologically impossible since individuals in society cannot be their own bank, e-commerce platform, search engine, and social media. However, the advent of miniaturized sensors (through which individuals can create their own data) driven by advancements in device engineering and miniaturization ushered in a new age of multifunctional sensors, often called the Internet of Things (IoT). In particular, the distributed miniaturized devices that measure the biological attributes of individuals are called the Internet of Medical Things (IoMT). This manuscript describes an end-to-end ecosystem that offers the solution to this problem through a scalable network and the commercial pilot model in which it has been implemented.

1.1   Data Ownership

Data ownership is a complicated topic that has been recently addressed globally in academic, commercial, and legal circles. The European Union General Data Protection Regulation [1] and the California Consumer Privacy Act [2] are examples of legal frameworks intended to protect individuals' data. However, these regulations are complicated, challenging to enforce, and could be circumvented through loopholes that individuals do not realize [3]. This situation may be experienced several times a day as we navigate the internet and are asked to accept cookies, often with limited choices [4]. (As a tip, website cookies can be manually removed at any time from the site information icon in the URL field). Self-sovereignty is a concept that is almost entirely addressed in connection with identity and credentialing leading to a wide debate on Self-Sovereign Identity (e.g., [5]). However, within these elaborate debates, the essential question of who owns the data has not received much attention, even though there is an overarching assumption that individuals in society own their data (e.g., [6]). Within the context of life sciences research, three arguments are made. First, do researchers "create" data from individuals' devices? Second, do individuals "control" their data? Finally, do individuals monetize their data? These concepts are analyzed below.

1.1.1 Data Creation

Data creation from IoMT is often defined as the digital rendering of a reality or actions taken by an individual. This capture or rendering usually involves proprietary technology, and the resulting data are stored or represented in a digital format [4]. For example, when epidemiology researchers plot disease outbreaks using Google maps mobile application, a digital route is generated using Google's proprietary technology (satellite-based GPS, software, algorithms, other sensors, etc.) and proprietary digital format [7]. As such, this digital route becomes the property of Google.

Similar examples can be drawn from transacting with various commercial IoMT devices, so long as this digital rendering uses proprietary means. For example, Google purchased Fitbit, a consumer-grade fitness monitor, in 2021, resulting in Google's ownership of all Fitbit services and user data [8]. However, in other situations like capturing direct digital photography of a person, the images are owned by the photographer [9], medical images are often owned by healthcare facilities or providers [10]. The complicated ownership context suggests focusing more on privacy and control of IoMT and device data.

It should be noted that the absence of direct or active permission does not change the fundamental analysis that these technologies have knobs and controls that users can utilize to prevent data capture. Further, users benefit from interactive products and willingly agree to trade off data privacy for such free benefit [4].

1.1.2 Data Control from Devices

For individuals to exclusively control their data, there must be a protocol that ensures a secure chain of custody of the data from the point of collection or rendering to the endpoint where the data may be stored, utilized, or monetized in a marketplace. The author argues that the data collection technology used must also be part of the continuous chain of custody, owned and controlled by the individual, referred to as vertical integration of the total value stack [11]. Thus, in most cases and especially for bioinformatics, a hardware device or a "dongle" is most suitable for the inception of data at the point of interaction. The collected data must be tamper-proof and traceable, this condition is best served by a trusted network based on blockchain. Encryption can be integrated into the process to guard against data copying, especially in open (permissionless) blockchains. This security adds another layer of guaranteed control by the individual. Finally, for such a blockchain-based network to fully function as a truly decentralized network, a consensus protocol is needed, such as Proof-of-Identity (PoI). PoI requires the initial registration of biometrics to ensure that the user identity is confirmed before the hardware device collects bioinformatic data [12]. With PoI securing the authenticity of the data, the entire blockchain-based network becomes trusted and self-sovereign, and the provenance of data is ensured.

1.1.3 Data Ownership from Devices

One exception to this analysis is medical/clinical data, which are regulated under different laws. The analysis above is not to be confused with DNA data and personal genome. These data are based on a body part (albeit nanoscale), and accordingly, the sequence is more about measuring the ingredients of human tissue. Using the aggregator's proprietary means, any digital rendering of reality becomes its sole property. No other party (including the person subject of data collection) has privacy rights to someone else's property.

Furthermore, different aggregators might simultaneously render reality in different yet proprietary formats. For example, for a person making purchases on an e-commerce platform using a credit card, both the credit card company and the platform will have different digital profiles using their own proprietary software. As a remedy to this significant problem, this chapter offers a solution based on a breakthrough device owned by the individual. Individuals collect data with a model that makes the owner both the creator and aggregator of their data. Further details are offered in the next section.

The final element in building a genuinely user-owned network beyond data control is the users' exclusive right to monetize the data. This element stems from the fact that if data are indeed considered private property, conditions in a free market imply that individuals have the right to transfer this property (or rent it) to another owner for compensation by fair market value. Irrespective of the scenarios that allow for data monetization, which will be addressed later, the fundamental right of ownership transfer follows immediately from true data ownership in a free society. As proposed in this manuscript, the concept of true data ownership makes the individual's data property and capital.

1.2 Data in Blockchain-Based Network

Like any individually owned property, data can only be traded if processes and standards for data valuation may fluctuate by supply and demand and retain a fundamental baseline. Data valuation remains one of the most intriguing business practices today. The initial approach to data valuation was in bulk data sales for advertising and marketing purposes (as Google and Facebook do). It then evolved as data inputs to artificial intelligence and machine learning algorithms. Data valuation recently became a rapidly expanding field in economics; there are several valuable treatises on this topic (e.g., [13, 14]), and there are attempts to create semi-automated personal data value calculators (e.g., calc.datum.org/, ig.ft.com/how-much-is-your-personal-data-worth/).

In this chapter, a blockchain-based data network—based on the vertical integration approach proposed—also benefits from the same principles of data valuation. Furthermore, if structured as a Decentralized Autonomous Organization based on these governance principles, such a network offers its participants the maximum data value return on their data collected vertically integrated within the network. The final point in this section involves handling encrypted data in the network described above. Homomorphic Encryption Algorithms (HEAs) are readily available and able to handle encrypted data, creating monetizable analytics that contribute to data valuation to the extent that the valuation models are still valid [15]. For example, in the Synsal network, an innovative miniaturized device [16, 17, 18] is accessed through a combination of retinal and fingerprint identity verification, satisfying the PoI requirement. The device directly collects bioinformatics, converts raw data to encrypted data, and uploads to a blockchain-based network. HEA-run analytics are sent back to the user or are monetized. More details about the Synsal network are described in the following section.

2   Case Implementation of Internet of Medical Things (IoMT) with Real Ownership

This section demonstrates how the network described above is implemented in a realistic pilot commercial ecosystem. As shown in Fig. 1, the Synsal ecosystem [16, 17, 18]) starts with users accessing the device using biometrics necessary for the PoI protocol requirement. Data authenticity is guaranteed, as the Synsal device can only be accessed if the pre-registered user is interacting with the device. The device was engineered to make it challenging to maneuver around this step. For example, if the device is operated in a blood drop collection mode, a miniaturized proprietary cartridge is inserted into the device [16, 17, 18]. A micro-needle is mounted on the device upward so that a finger prick and fingerprint actions are performed simultaneously. Another example involves a sputum sample deposited in a cartridge when the user approaches the inserted cartridge. As shown in Fig. 2, a retina scanner on the device simultaneously verifies the user's identity. This level of engineering control

Fig. 1 Dr. Ned Saleh demonstrates how a user would interact with a prototype Synsal device

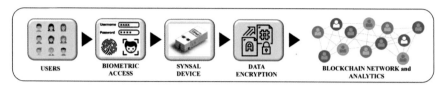

Fig. 2 The data process flow from users through the Synsal ecosystem

guarantees the fidelity of the data in the network and secures data ownership in the entire network based on the vertical integration approach discussed earlier.

The Synsal device is an engineering innovation that can handle several bodily fluids down to a few droplets (100 μl) thanks to breakthroughs in microfluidics engineering. The Synsal device is designed to be a non-invasive, general wellness Class-I medical device. It falls under the U.S. Food and Drug Administration (FDA) regulations related to Over The Counter and Direct-To-Consumer devices [19, 20]. Various assays (e.g., metabolic, cardiac, infectious, etc.) were validated on this device. The Synsal device creates a bioinformatics measurement from the electro-optical sensors in the device that analyze the pre-calibrated raw signal then encrypt it before it is transmitted to the network. The user-specific bioinformatics data enter the network encrypted. The network operators cannot access the encrypted data, nor can they decrypt it; only the original user can decrypt the data from a private key programmed at the initial unpacking and device set up. The user—and the user alone—is the data owner.

The data at this stage reside within the blockchain-based network, whether on-chain or off-chain, as in an InterPlanetary File System. The HEA can now access the data, perform analytics, and integrate these analytics with the rest of the network. It should be noted that the near-immutability attribute built-in in the blockchain architecture alone does not guarantee data ownership in a blockchain network. While the data are cryptographically tamper-free, it may still be transparent and visible to other users in the network—especially if the network is public and/or permissionless. Hence, encryption is necessary for exclusive control (and thus ownership) of the data. The ability of the user to decrypt their data and the analytics-based derivatives of the data generated by the network algorithms offers higher valuation of the users' data and unlock the maximum potential of monetization in the data marketplace connected to the network. It is also possible to program the encryption of the data to be portable and interoperable with other blockchain networks; this further maximizes the data value and offers more flexibility—especially in cases where dynamic consent is implemented. Fortunately, HEAs are themselves interoperable and can be implemented universally. Today we are witnessing a plethora of programmable blockchains and smart contract languages beyond Ethereum and Solidity that can sustain this level of flexibility (e.g., Casper, Polygon, etc.)

2.1 The Synsal Network

As stated in earlier sections, vertical integration of users' data stack relies on owning the initial aggregation layer, i.e., the hardware level that collects raw signals or a form of digital rendering of a specific attribute of the user-related reality. In that sense, one may question the availability of the technology, devices, or product practically accessible to the users to own, rent, or utilize in any other way that does not impact their exclusive ownership (including control and monetization) of their data. For example, suppose a person would like to perform an MRI and exclusively own and monetize his/her data. In that case, he/she would have to buy a next-generation portable MRI scanner such as Hyperfine that costs approximately $50k [21]. Incidentally, this is a remarkably inexpensive price point for an MRI device, yet it may not be accessible to most ordinary middle-class individuals.

The market for miniaturized IoMT now includes home ultrasound scanners at price points accessible to retail consumers and readily available on e-commerce platforms like Amazon. Even whole genome sequencers are now available in portable home devices priced around $1k thanks to advances in genomics technology like the Nanopore [22]. Furthermore, the market for home-based Point of Care (PoC) diagnostics and wearables is rapidly improving and covering vast areas of vitals measurement. In the age of the COVID-19 pandemic, PoC home testing now covers infectious diseases.

We are witnessing a blurring between wearables, home PoC diagnostics, and home (semi-)clinical devices, including home hemodialysis [23]. Despite all the home-based IoMT and PoC devices, no data ownership model is designed structurally to guarantee absolute control and security of user data, this is especially true if these devices are connected to databases or networks for data analysis straight to medical records. The realm of patient clinical data ownership or medical records remains outside the scope of this manuscript. However, it suffices to mention that a complicated web of laws regulates the ownership of patient data and medical records, which vary from one state to another within the USA and from one country to another. The Health Insurance Portability and Accountability Act (HIPAA) regulates health care providers' patient records privacy and permissions. However, there are no legal provisions on medical data monetization, especially if the data are deidentified [24].

Furthermore, disease reporting laws may override the HIPAA provisions per state or federal law [25] under the U.S. Centers for Disease Control and Prevention. To the best of the author's knowledge, there are no known networks to date that involve IoMT or home devices collecting clinical data or bioinformatics that comprise *simultaneously* engineering (as opposed to administrative) guarantee of legal ownership (dynamic control and monetization) and maximum potential for collective and individual analytics. Additionally, without explicit data encryption, blockchain-based networks alone do not guarantee absolute ownership of user data—even in private and permission-based networks. There remain risks of identity disclosure through differential privacy [26] or zero-knowledge proof loopholes [27, 28]. As shown in Fig. 3, a vertically integrated bioinformatics network is architected to achieve real

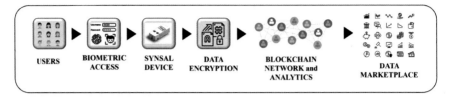

Fig. 3 The creation and protection of data value through the Synsal Network

ownership of data where users exclusively control their data from the point of inception by guaranteed engineering encryption. The data reside on a blockchain-based network with the maximum potential of data monetization through HEA that unlocks the maximum potential of data value. If this network is connected to a marketplace, it constitutes an end-to-end ecosystem for data ownership and monetization, a Data As A Capital model.

2.2 Sensors, Device Engineering, and Scaling in the Synsal Network

This section offers some technical glimpses into the Synsal device engineering [16, 17, 18]. The device addresses an essential aspect of the true miniaturization of a home-based or decentralized analytical test lab. The approach taken in engineering the device is not a "brute-force" miniaturization [29], where all device elements are mechanically scaled-down and cramped in compact space. Such an approach only scales down the physical sub-modules but does not scale down the raw sample handling—especially if the samples are only a few drops of bodily fluid (blood, tears, etc.). In addition, such an approach does not proportionately scale down mechanical tolerances and thermal loading and becomes not feasible from the system engineering design analysis. Proper scaling of an analytical home-based device must also include scaling-down of the sample handling chemistry. Without such technological breakthrough, the device will eventually have to analyze a diluted sample of the input fluid. The biomarker concentration will drop significantly and will likely be below the detection limits of any known sensor. The Synsal device is designed to implement the miniaturization and scale down on the microfluidics level, thanks to the breakthrough in the proprietary surface activation technology [30, 31] that leads to micro-splitting of fluids samples while preserving the biomarker concentration [16, 17, 18]. Specifically, the technology allows for creating programmable selective fluid flow due to the ability to create different polarities of surface energy. When interacting with lyophilized assays, this makes for natural chemistry scaling down. The Synsal device is equipped with electrochemical and optical sensors designed and validated for 1picoAmp-1pmol with high frequency ultrabright LED illumination to perform particle counting/cytometry and time/frequency-domain optical signal processing, with usable signal-to-noise ratio.

Besides the ability to be programmed to measure various assays, the device can also perform Nucleic Acid Amplification Tests, which are critical for viral detection assays. However, unlike Polymerase Chain Reaction protocol, the device uses a well-known isothermal protocol called Reverse Transcription-Loop-mediated isothermal AMPlification (RT-LAMP). The device also has a biometric activation input user interface and built-in encryption chips [32].

3   Tokenization and Value Scaling in the Blockchain-Based Network of Hardware Devices

This section illustrates how a blockchain network can be tokenized to maximize user-owned data valuation. The Synsal network is designed to implement the true ownership model discussed above based on vertical integration of the data value stack. Furthermore, data are processed to create maximum potential value from analytics and monetizable reports. This model uses an innovative and proprietary network effect based on Reed's modified law in scaling [16, 17, 18]. Figure 4 displays the scaling properties when following Reed's modified law in scaling [33].

In a classical database where network effects are absent, each additional user contributes to the growth of the network depending on the size of the network (Sarnoff

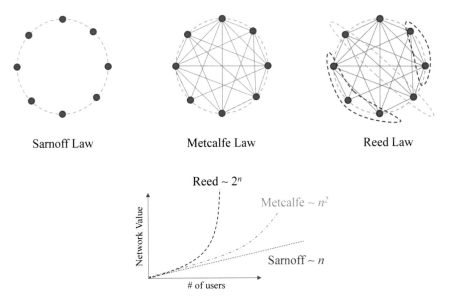

Fig. 4 Differences in scaling properties when using Reed's Law versus other laws used to predict value in scaling

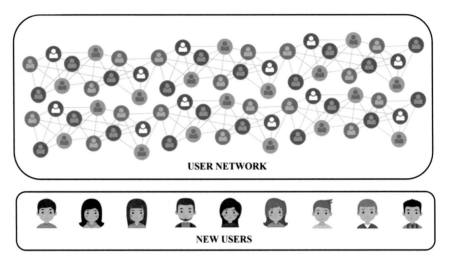

Fig. 5 A depiction of the relationship between users and the total Synsal Network

Law). An additional small group of users increases the value of the network by a factor proportional to $\Delta n/N$, where n is 'number of users added' and N is the total number of users in the network (Fig. 5). The network's scaling follows a logarithmic pattern with the integration of such increments according to the formula:

$$\int_{N}^{n+N} \frac{dn}{N} = \log\frac{N+n}{N}$$

Accordingly, the network value increases logarithmically, but its growth stagnates as it becomes larger. For example, the value of the data in a network of siloed users (as in a network of protected medical records) follows such a logarithmic trend. However, suppose the user's data can be processed collectively. In that case, the network effect will kick in and drive the value of the network higher than the logarithmic scaling, depending on the way the users are clustered in inter-related groups. There are various ways to model network effects discussed in the literature (e.g., [33]), the details of which are outside the scope of this manuscript. It is of *paramount* importance to notice that if the identity of the users is protected, then the clustering actually cannot happen. The maximum value of the network remains logarithmic. However, for the network to scale under the network effects depicted in Fig. 4, the identity of the users must be known to the network operators. In most practical cases, the network operators have access to the user's identity and become the beneficiary of the data valuation. This is very common in social media platforms where the users' identity is known to the data aggregators (owners or operators of the network) who are the ultimate beneficiaries of the network value, as in advertising revenue.

To recap, the dilemma is that the network cannot simultaneously scale the value of the data utilizing the network effects and keep the identity of the users protected. One way to resolve this dilemma—perhaps the only way possible—is using encryption like HEA to cluster users in interrelated groups. In this case, the network operators (data aggregators) could not monetize the data. However, the network can be set up as is the case in the Synsal Network so that the individual users, and only the individual users, can monetize the data and yet benefit from network effect driven scaling since they can decrypt their profiles.

3.1 Tokenization and Value Scaling

This section offers additional means that the data value in the network can be further scaled if the value is captured in the form of a token, typically and more suitably a Utility Token (UT). In a blockchain-based network, the UT can be created using smart contracts that are built-in automated code in the network that can offer users more functionality and control and, perhaps most importantly, carry parameters of dynamic consent. The tokenized data can be tightly or loosely tethered to the UT, which can be traded on a platform or interoperably on a secondary market. The UT must be carefully programmed in the smart contract regarding its utility rights. If the UTs are finite while the number of new users is not, a condition of scarcity can be created that drives the valuation of the UT in addition to the rules of network effect that continue to apply to the UT. An additional feature unique to the Synsal Network (and any other network based on acquiring a piece of hardware to participate in the network) can be added. The device can be purchased using a value-based token considered a Security Token (ST) similar to a stable coin. The advantage of incorporating an ST is that network can operate with a dual token. This is particularly important to stabilize the UT—especially if the latter is finite, making it otherwise prone to speculations often called "pump-and-dump" actions. The coupling of the ST to the UT adds stability to the network. This chapter does not, however, address the technology and regulatory requirement to operate such a dual token network.

3.2 Basic Stabilization Tokenomics

Finally, it is important to provide some guidance on the token economics (tokenomics) of the dual token system described above, making for a stable and scalable blockchain hardware-based network of bioinformatics data. It is fair to state that many or most of the Initial Coin Offerings in the past several years irrecoverably lost a significant portion of their initial value due to a lack of certain tokenomic guardrails and bootstrapping mechanisms. It is exceptionally challenging to recover value once the native UTs slump on a trading platform. To make a UT representing a network like the Synsal Network, there should be some or all of the following factors:

(1) Intrinsic value creation and growth mechanism from data created by users. UT value is tethered to data
(2) Value creation from a single user in the network without scaling from the network effect
(3) Security Token coupled to the UT to offer robust UT stabilization mechanism, and liquidity exit
(4) UT participation is mandatory by users and data buyers, leading to high UT velocity
(5) UT derivatives on the platform (options, futures, swaps) to be used when applicable to offer another layer of valuation.

4 Future Directions

Engineering miniaturized IoMT devices improved on the heels of the COVID-19 pandemic. Home monitoring of vitals and other wellness needs provides crucial indicators of underlying conditions behind elevated COVID-19 illness; telemedicine also makes strides within the same context. Also, clinical, wellness, and medical data are expected to accumulate significantly [34]. With increases in data volumes, we will likely witness broader debates on data privacy, ownership, and monetization. There are also questions regarding the rights of governments and medical health record keepers to share data in semi-real-time to get insight on possible outbreaks of a new pandemic—especially if new Coronavirus variants or other viruses may be more contagious than the variants we witnessed in the course of the pandemic.

We also are witnessing expansion in the blockchain domain and utility-based cryptocurrency offerings. Examples include Algorand (algorand.com/; [35], Avalanche (https://www.avax.network/), Solana (solana.com/), Harmony (harmony.one/), and in particular, BurstIQ (burstiq.com), a HIPAA-compliant blockchain network. This massive momentum in Blockchain is also taking place as virtual reality applications are moving to a new plateau with the emergence of the Metaverse [36]. Alongside this momentum, we are also witnessing world governments, including the U.S. federal and state governments [37], trying to stay abreast with the legislative demands to keep the U.S. as a jurisdiction friendly to innovative blockchain companies and venture capital to operate and grow. In total, it is hoped that all this momentum will help materialize and widen the adoption of vertically integrated blockchain hardware-based networks similar to the one described in this manuscript by Synsal. Life sciences, clinical research, and market analysts will benefit as more bioinformatics data, quantified wellness, and analytics reports are generated thereof from such networks.

5   Conclusions

This chapter addressed a rapidly increasing topic of importance in the age of bioin-formatics and IoMT, namely, data ownership and monetization. In order to guarantee true ownership of data (privacy, control, and monetization), vertical integration of the full data stack is needed. As such, the initial point of data aggregation usually involves hardware. This chapter presented a case study of the Synsal Network, where an innovative device collects data from a network of participating users. The device acts as a "dongle" to participate in the network. The network uses encryption and algorithms to maintain ownership of the data and maximize the scaling of data value to the users' interest. As an added benefit, data tokenization can accelerate data valu-ation and liquidity—especially when a data marketplace is part of this end-to-end ecosystem.

References

1. European Parliament and Council of the European Union (2016) General Data Protection Regulation 2016/679. https://gdpr.eu/
2. California Consumer Privacy Act (2018) Title 1.81., Sections 1798.100—1798.199.100. https://leginfo.legislature.ca.gov/faces/codes_displayText.xhtml?division=3.&part=4.&law Code=CIV&title=1.81.5
3. Check Hayden E (2013) Privacy loophole found in genetic databases. Nature. https://doi.org/10.1038/nature.2013.12237
4. Vezyridis P, Timmons S (2021) E-Infrastructures and the divergent assetization of public health data: expectations, uncertainties, and asymmetries. Soc Stud Sci 51(4):606–627. https://doi.org/10.1177/0306312721989818
5. StClair J, Ingraham A, King D, Marchant MB, McCraw FC, Metcalf D, Squeo J (2020) Blockchain, interoperability, and self-sovereign identity: Trust me, it's my data. Blockchain Healthc Today https://doi.org/10.30953/bhty.v3.122
6. Banterle F (2018) Data ownership in the data economy: a European dilemma (No. 3277330). SSRN. https://doi.org/10.2139/ssrn.3277330
7. Monir N, Abdul Rasam AR, Ghazali R, Suhandri HF, Cahyono A (2021) Address geocoding services in geospatial-based epidemiological analysis: a comparative reliability for domestic disease mapping. Int J Geoinformatics https://doi.org/10.52939/ijg.v17i5.2029
8. Frew J (2021) Should you worry about your health data now that Google owns Fitbit? MUO. https://www.makeuseof.com/google-owns-fitbit-health-data/
9. Lewis A (2021) Who owns a photograph in the social media age? JD Supra; Fenwick & West LLP. https://www.jdsupra.com/legalnews/who-owns-a-photograph-in-the-social-9457360/
10. Who owns medical records: 50 state comparison (2015) Health Information and the Law Project. http://www.healthinfolaw.org/comparative-analysis/who-owns-medical-records-50-state-comparison http://www.healthinfolaw.org/comparative-analysis/who-owns-medical-records-50-state-comparison
11. Orszag P, Rekhi R (2020) The economic case for vertical integration in health care. NEJM catalyst 1(3). https://doi.org/10.1056/cat.20.0119
12. Cerezo Sánchez D (2019) Zero-knowledge proof-of-identity: Sybil-resistant, anonymous authentication on permissionless blockchains and incentive compatible, strictly dominant cryptocurrencies. SSRN Electron J. https://doi.org/10.2139/ssrn.3392331

13. Bendechache M, Limaye N, Brennan R (2020) Towards an automatic data value analysis method for relational databases. In: Filipe J, Smialek M, Brodsky A, Hammoudi S (eds) Proceedings of the 22nd international conference on enterprise information systems. SciTePress, Science and Technology Publications, Lda, pp 833–840. https://doi.org/10.5220/0009575508330840
14. Robinson SC (2017) What's your anonymity worth? Establishing a marketplace for the valuation and control of individuals' anonymity and personal data. Digit Policy Regul Gov 39:88. https://doi.org/10.1108/DPRG-05-2017-0018
15. Catak FO, Aydin I, Elezaj O, Yildirim-Yayilgan S (2020) Practical implementation of privacy preserving clustering methods using a partially homomorphic encryption algorithm. Electronics 9(2):229. https://doi.org/10.3390/electronics9020229
16. Khalid W, Saleh N, Saleh F (2017) Standalone microfluidic analytical chip device (United States Patent Application). https://pimg-faiw.uspto.gov/fdd/64/2017/38/033/0.pdf
17. Saleh N, Khalid W, Saleh F (2017a) Apparatus and method for programmable spatially selective nanoscale surface functionalization (United States Patent Application). https://pimg-faiw.uspto.gov/fdd/80/2017/80/033/0.pdf
18. Saleh N, Khalid W, Saleh F (2017b) Self-flowing microfluidic analytical chip (United States Patent Application). https://pimg-faiw.uspto.gov/fdd/98/2017/38/033/0.pdf
19. U.S. Food and Drug Administration (2019) General wellness: policy for low risk devices—Guidance. Center for Devices and Radiological Health Guidance. https://www.fda.gov/regulatory-information/search-fda-guidance-documents/general-wellness-policy-low-risk-devices
20. U.S. Food and Drug Administration (2020) Classify your medical device. Overview of Device Regulation. https://www.fda.gov/medical-devices/overview-device-regulation/classify-your-medical-device
21. O'Connor M (2020) FDA clears "world's first" portable, low-cost MRI following positive clinical research. Health Imaging; Innovate Healthcare. https://www.healthimaging.com/topics/healthcare-economics/fda-clear-worlds-first-portable-mri
22. de Rojas C (2020) Portable sequencing is reshaping genetics research. Labiotech.eu; Labiotech. https://www.labiotech.eu/in-depth/portable-sequencing-genetics-research/
23. Home hemodialysis (2015) National Kidney Foundation. https://www.kidney.org/atoz/content/homehemo
24. Sharma R (2018) Who really owns your health data? Forbes. https://www.forbes.com/sites/forbestechcouncil/2018/04/23/who-really-owns-your-health-data/
25. Reportable diseases. (2022) Medline Plus; National Library of Medicine. https://medlineplus.gov/ency/article/001929.htm
26. Nissim K, Steinke T, Wood A, Altman M, Bembenek A, Bun M, Gaboardi M, O'Brien DR, Vadhan S (2017) Differential privacy: a primer for a non-technical audience (Grant No. 1237235). Center for Research on Computation and Society, Harvard University. https://privacytools.seas.harvard.edu/files/privacytools/files/nissim_et_al_-_differential_privacy_primer_for_non-technical_audiences_1.pdf
27. Al-Aswad H, El-Medany WM, Balakrishna C, Ababneh N, Curran K (2021) BZKP: blockchain-based zero-knowledge proof model for enhancing healthcare security in Bahrain IoT smart cities and COVID-19 risk mitigation. Arab J Basic Appl Sci 28(1):154–171. https://doi.org/10.1080/25765299.2020.1870812
28. Tomaz AEB, Nascimento JCD, Hafid AS, De Souza JN (2020) Preserving privacy in mobile health systems using non-interactive zero-knowledge proof and blockchain. IEEE Access 8:204441–204458. https://doi.org/10.1109/access.2020.3036811
29. Nourse MB, Engel K, Anekal SG, Bailey JA, Bhatta P, Bhave DP, Chandrasekaran S, Chen Y, Chow S, Das U, Galil E, Gong X, Gessert SF, Ha KD, Hu R, Hyland L, Jammalamadaka A, Jayasurya K, Kemp TM, Holmes EA (2018) Engineering of a miniaturized, robotic clinical laboratory. Bioeng Transl Med 3(1):58–70. https://doi.org/10.1002/btm2.10084
30. Saleh N, Sahagian K, Ehrlich PS, Sterling E, Toet D, Levine LM, Larne M (2014) Maskless plasma patterning of fluidic channels for multiplexing fluid flow on microfluidic devices. Lab-on-a-Chip, Microfluidics & Microarray World Congress Agenda, San

Diego, CA. https://www.academia.edu/42946895/Maskless_Plasma_Patterning_of_Fluidic_Channels_for_Multiplexing_Fluid_Flow_on_Microfluidic_Devices

31. Saleh N, Sterling E, Toet D (2015) Non-contact current measurements for AMOLED back-planes using electron-beam-induced plasma probes. Dig Tech Pap 46(1):118–121. https://doi.org/10.1002/sdtp.10306

32. Silicon Valley startup produces device for detecting Coronavirus (2020) Business; Bloomberg. https://www.bloomberg.com/press-releases/2020-03-21/silicon-valley-startup-produces-device-for-detecting-coronavirus

33. Currier J, NFX team (2018) The network effects bible. Guides Publishing. https://guides.co/g/the-network-effects-bible/121715

34. Pastorino R, De Vito C, Migliara G, Glocker K, Binenbaum I, Ricciardi W, Boccia S (2019) Benefits and challenges of big data in healthcare: an overview of the European initiatives. Eur J Public Health, 29(Supplement_3):23–27. https://doi.org/10.1093/eurpub/ckz168

35. Chen, J., & Micali, S. (2016). *Algorand* (arXiv:1607.01341). http://arxiv.org/abs/1607.01341

36. Jeon H-J, Youn H-C, Ko S-M, Kim T-H (2021) Blockchain and AI meet in the metaverse. In: Fernández-Caramés TM, Fraga-Lamas P (eds), Blockchain Potential in AI [Working Title]. IntechOpen. https://doi.org/10.5772/intechopen.99114

37. Botella, E. (2021, June 28). Wyoming wants to be the crypto capital of the U.S. Slate. https://slate.com/technology/2021/06/wyoming-cryptocurrency-laws.html

38. Highest intensity focused laser (2008) Guinness World Records. https://www.guinnessworldrecords.com/world-records/highest-intensity-focused-laser

Dr. Ned Saleh is the Founder of Synsal, Inc. He received his engineering doctoral degree from the University of Michigan in 2004 under the supervision of Prof. Gérard Mourou, who won the Physics Nobel prize in 2018. Dr. Saleh's academic work played an important role in bringing this award to the world front stage when the laser system he developed with a handful of scientists was cited in the Guinness Book of World Records for the "Highest Intensity Focused Laser" in 2008 [38]. Dr. Saleh's career afterwards took him to a series of senior academic and industry positions, including Berkeley Lab, Intel, IBM, Applied Materials and Apple, and others in the Silicon Valley.

Blockchains and Genomics: Promises and Limits of Technology

David Koepsell and Mirelle Vanessa Gonzalez Covarrubias

Abstract One of the early, non-financial uses of blockchain technologies around which several startups have developed was to help manage, monetize, and make the sharing of genomic data more private. Because deidentified genomic data are excluded from HIPAA and many other regulatory contexts worldwide—and is already a widely traded commodity for science valued in the hundreds of millions over the past decade—genomic blockchains proved a promising entry point for using the benefits of blockchains for dissemination and remuneration of data. Several models have been tried, and most have touted their abilities to make users the "owners" of the genomic data in ways in which current models fail. This paper will explore the various existing and potential models for genomic blockchains, review some shortcomings and unmet needs, and explain why no technical solution alone will fulfill the promise of genomic data ownership without regulation.

Keywords Genomics · Blockchain · Non-fungible tokens · Genomic privacy · Regulation

1 Introduction: A Brief History of Capitalization on Genes

During the course of the multi-billion dollar, international effort known as the Human Genome Project (HGP), a for-profit entrant sought to change the game for sequencing—forever revolutionizing the science and business of genotyping and sequencing of genomes. Celera, founded by Craig Venter, developed the revolutionary process of shotgun sequencing, which made large-scale, rapid sequencing of genomes possible [1]. The effort was costly, and Celera and Venter entered the race to finalize a map of the human genome, attempting to beat the international

D. Koepsell (✉)
Department of Philosophy, Texas A&M, EncrypGen, Inc, Mexico City, Mexico
e-mail: drkoepsell@tamu.edu

M. V. G. Covarrubias
Instituto Nacional de Medicina Genómica, Mexico City, Mexico
e-mail: vane.mx@gmail.com

© The Author(s), under exclusive license to Springer Nature Singapore Pte Ltd. 2022
W. Charles (ed.), *Blockchain in Life Sciences*, Blockchain Technologies,
https://doi.org/10.1007/978-981-19-2976-2_4

HGP to the finish line. To finance this effort, Celera consulted with its attorneys and decided they could make a value statement for their capital needs via the eventual patenting of genes found during the race. Patents of genes found could prove valuable enough to make the tremendous expenditures on the research and development of new sequencing technologies and equipment worth it [2].

The standard scientific description of a human gene as the unit of heredity is a sequence of DNA that "codes" for a functional macromolecule such as a protein or RNA. At the time of the HGP, many believed that humans had as many as 100,000 genes. Now, 20 years after the first broad "map" of the genome, we know there are likely between 22,000 and 30,000 genes in the human genome [3]. Until recently, there were thousands of patents on human genes due to the efforts begun under Celera, which fueled genetic testing companies like Myriad's profits for nearly twenty years. Myriad patented specific mutations on BRCA1 and BRCA2 genes to evaluate the risk, diagnosis, and prognosis of several types of cancer including breast and ovarian cancers that earned them billions of dollars before the American Civil Liberties Union challenged the legality of patents on naturally occurring genes [4]. In 2013, the U.S. Supreme Court invalidated all existing and future patents on naturally occurring genes [5].

Until 2013, patents on naturally occurring genes were seen as a potentially lucrative way to lay claim to parts of genomes, but since then, there has been no way to legally claim any right to genetic code. This is because—except for copyright—there is no way to "own" data. Copyright only extends to original expressions, and part of the reasoning behind the Supreme Court's rejection of gene patents was that genes are naturally occurring. They are found in nature, not invented by humans [4]. For the same reasons, they are not expressions capable of copyright protection. The data that express a gene—or any part of a naturally occurring genome—record things found in nature, are not a protectable original expression. As such, like any other natural law or other naturally occurring thing, the data about what is found cannot be owned in any legal sense. It could be hidden for some time, but if anyone else discovers that data, they can do with it what they will.

In our post-Myriad world, naturally occurring genomic data cannot be owned. While privacy laws seek to make biometric data, including genetic data, protected against unauthorized dissemination, they do nothing to alter this fact: no one can own a naturally occurring genome or parts of it. Those data are inherently shareable and legally, universally so when deidentified [6]. However, that does not mean the data are not extremely valuable, nor that billions of dollars of capital and commerce are flowing into its discovery and use in science. One can make a lot of money transacting genomic data and do a lot of good science with them—even absent legal regimes allowing monopolization of the data. The entire business plans of the two largest consumer genetic testing companies are built upon this [7].

2   The Scientific and Market Value of Genomic Data

Genomic data are highly valued because either alone or as part of a cohort, data provide information on health and disease risk, biological sex and can even confirm ethnicity. Genetic data can be uploaded to published algorithms to impute additional genetic information and infer non-apparent phenotypes including, IQ, ancestry, biological age, and paternity relationships. When clinical information accompanies genetic data, the value increases as part of the analyses mentioned above but also can be the primary source of revenue for large genetic testing companies and the pharmaceutical industry seeking drug targets for global or specific populations [8]. Scientists seek to identify and use genetic data driven by the quest for markers that could lead to precision medicine. The advent of genome-wide association studies (GWAS) and next-generation sequencing has procured the scientific community with a vast collection of genetic information for human diseases in thousands of scientific reports [9]. These data have been mined, deposited, and curated in public genetic databases. For example, the 1000 Genomes Project has made genetic information available of different populations representing most of our world [10]. Specifically, the Exome Aggregation Consortium has released exome sequence data from at least 60,706 individuals, and GnomAD has aggregated 125,748 exome sequences and 15,708 whole-genome sequences from unrelated individuals with disease and population research goals. This genetic information has been released for the "benefit of the wider biomedical community" without restriction on use (https://gnomad.broadinst itute.org/). The ALFA database from the National Center for Biotechnology Information goes a step further. ALFA is an allele frequency aggregator integrated with dbSNP collecting data from the database of Genotypes and Phenotypes (referred to as "dbGaP") from millions of individuals and trillions of variants [11]. ALFA aims to annotate, interpret, and compare frequencies among populations, identify rare variants, and report its clinical significance. This database is publicly available at https://www.ncbi.nlm.nih.gov/snp/docs/gsr/alfa/.

In parallel, direct-to-consumer DNA testing has doubled for five consecutive years, reaching about 90 million tests sold according to GeneticsDigest.com [12]. Companies can be ranked based on their reputation, testing techniques, service, reviews, and price. Among these highly valued companies are CRI Genetics, AncestryDNA, and 23andMe. Unlike public databases, these companies are privately owned. For example, 23andMe has faced allegations of misleading marketing, limited health reports, and return policies with fees.

As evidence for the value of aggregated genetic data, the two large testing companies that have monopolized the gathering and sale of those datasets have recently seen an investment/valuation of over $4 billion each in the past year [13]. Studies have also looked into how these private companies profit from, and the possible value of, the data business they have cornered, finding that each of their customers' datasets has been sold over 200 times [14]. The average amounts earned on those datasets are above $200 each [13]. To put that into perspective, 30 million people have bought a DNA test from the biggest two genomic testing companies [15]. Based on those figures so far, the value of their data is about $10 billion, which is consistent with the valuation and investments seen over the past year in those companies by Wall Street [16].

Data purchasers include pharmaceutical companies, research institutes, private companies developing new genetics-based therapeutics, governments, and other life-science businesses and organizations. Ordinarily, to gather genetic data for basic research from scratch, one must have a funded human subjects study, gather samples, sequence or genotype them, and hope the expense and time generate the needed data. However, due to exceptions regarding the use of deidentified data, having access to a large amount of searchable, pre-sequenced or genotyped, deidentified data are an invaluable starting point for GWAS that can prove essential for building unique new studies upon without having first to do the costly onboarding of human subjects.

There is clearly a market for genetic data. Even without legal monopolies, data are generating revenue and fueling basic science. Increasingly, there appear to be calls for greater involvement of human subjects whose data are being sold in decisions about its use and the value chain for the data, including some compensation, transparency, and involvement in decision-making about its use. Thus the emergence of numerous companies creating platforms and mechanisms aimed at providing "ownership" of genetic data, greater privacy and control over the sales of the data, and remuneration for those sales.

2.1   On the Nature of Data, and the Data of Nature

Data represent observations. They differ from accounts of observations, which may be prosaic, elaborative, interpretive, or illustrative. To fulfill its usefulness in science, the underlying data behind accounts must be as "unadulterated" as possible. When data are collected, they should objectively describe reality at its most granular level and be verified by other observations. Temperature readings, x-rays, EEG readings, heart rate readings, wind speed measurements, and air pressure readings are all examples of measurements and recordings of data. Fingerprint scans, retinal scans, and genotyping or genome sequencings are measurements and recordings of biometric data.

Data cannot be monopolized. Unlike expressions or inventions—a distinction that is suspect as treated in the laws of copyright and patent, respectively—data cannot be held to the exclusion of its uses by others, by law. Privacy laws are relatively new, historically speaking, and have recently begun to be expanded to protect personal data, including personal medical data. New state, national, and international privacy laws for personal data mainly work to provide greater transparency and individual control over the use of personal data [17]. These laws do not create a new form of property but prevent commercial entities and states, to a degree, from trading or otherwise using such data in ways that can prejudice individuals. There is still no manner in which data can be owned and, consistent with a rational ontology of property with the aims of science, that should remain the case [17].

Information wants to be free. Until now, our laws have recognized the necessity for the free and abundant flow of data. While individuals may be uncomfortable with certain uses of their data and feel that they should be more involved in the manner of dissemination of data about themselves, we should first recognize what genetic data are, why they cannot be property, and why they do not fall into other categories that have been granted monopolies, such as copyright and patent. Simply put: data are about natural facts, and those facts are not exclusive nor rivalrous, nor are they original expressions or inventive. Bottling nature up, excluding others from access to it, attempting to monopolize and even profit from such nonexclusive, intangible, and ephemeral information is anathema to science. Science progresses best through the free dissemination of information, and data are the foundation of all scientific progress. Freedom of inquiry, a bedrock value of modern liberal democracy, demands free access to data as well [18].

2.2 Fair and Sustainable Data Use

A seminal case in bioethics that underscores the injustice that can result when a subject is used without fair remuneration for their contribution to science is that of Henrietta Lacks. Her cancerous cells were extracted with her consent as part of her treatment. However, after some laboratory procedures and their immortalization, these cells were widely spread as a tool mostly in cancer research and hence proved to be extraordinarily valuable [19]. The cervical cancer cell line HeLa (short for Henrietta Lacks) is currently sold to hundreds of research labs worldwide as a standard assay for thousands of studies. Their use has facilitated the development of numerous technologies, papers, inventions, and many tons of HeLa cells still exist, driving science and innovation to this day. However, Henrietta Lacks died in poverty from the cancer whose cells proved so valuable to so many. For decades, she went both unrecognized and unremunerated for the value she gave to the world [19].

Only recently has the National Institutes of Health settled with the family of Henrietta Lacks for an undisclosed sum after a book, movie, and subsequent public outcry caused social sentiment to favor some remuneration for the HeLa cells' contributions to public health and industry [20]. Although the letter of the law regarding human tissues and subjects—and the consent received by Henrietta Lacks—were sufficient, in retrospect, there was a failure of justice and imbalance of goods that prompted many to agree that she should have been compensated. Compensating her children and their families may not have completely achieved the justice that was lacking. However, it helps and signals that perhaps we need to reevaluate the nature of compensation for those whose tissue or data benefit science and industry like this [20].

Data are essential for scientific progress and never before have we been so awash in data of so many types, delivered instantly through networks, from "nodes" around the world. Having ready access to data is necessary for nearly any type of scientific study. Data should remain plentiful and easy to obtain so that science can progress, but as we found in the case of the HeLa cells, some data are extremely valuable. Compensating people for helping to provide scientists with valuable data could help alleviate growing wariness by a public that feels like the Henrietta Lacks case is both unjust and common. Individuals wish to help science but also share in the wealth.

The harvesting and use of personal data for commerce have become both common and frowned upon in the past decade, although these data fund the numerous "free" internet platforms. Personal, biometric, and medical data are highly valued not just for their great scientific uses but also because they help drive billions of dollars worth of commerce and because people willingly give up access to those data. It is likely that most people do not realize the extent to which the data they give away is valuable nor that there are alternative models for its gathering and use that might include them in the revenue stream.

Currently, most genomic data in the stream of science and commerce is provided by users quite willingly, though not necessarily with full knowledge of its value, uses, or extent. The biggest accumulators of genomic data are the largest consumer genetic testing companies, like 23andMe and Ancestry.com, which have now tested about 50 million people. Those customers have purchased tests typically with interest in learning more about their heritage and with some possibility of finding out some medically useful or at least interesting facts about their genes. For most customers, these direct-to-consumer tests—also considered "recreational genetics"—provide a one-time and limited interpretation of genetic information in a customed report, concluding the relationship between the testing company and the customer. However, these companies make most of their profits selling not kits and a report, but data. Since about 80% of their customers agree to have their data "used for science," the testing companies repeatedly sell those customers' data. While they have been privately held, there is little transparency about data sales.

The days of large-scale, freely acquired, and highly profitable data harvesting could well be nearing an end. Increasingly, legislators are being called upon to regulate the use of personal data. Laws are being enacted that proscribe how personal data can be gathered, used, or sold. Without more specific consent, transparency, or oversight, doing business with genetic data will likely grow even more difficult. This poses a risk for science, and an opportunity for those who want to transform the ways that data are gathered and used, including by blockchain-based platforms, so that everyone can benefit and so that an era of data sustainability can be developed, increasing justice and promoting the fair generation of science and wealth.

3 Democratize, Decentralize, and Disintermediate Data (The Three Ds)

Blockchains are well known for their promise to decentralize and democratize capital and commerce. The first digital blockchain, Bitcoin, was developed specifically to provide access to wealth and its transfer without needing specific "fiat" currency or banks. The radical decentralization, democratization, and disintermediation of money through Bitcoin has been demonstrated to provide anyone a means to hold and spend money without state-sponsored currencies or banks to provide them accounts [21]. Billions of people worldwide are unbanked because they lack access through social and legal institutions to the trust-providing mechanisms needed in most cases. However, if they have a smartphone, computer, or even just a "paper" bitcoin wallet, they can hold and transfer money [22].

Numerous companies, individuals, teams, and decentralized organizations are now using the radical power of blockchains to democratize, decentralize, and disintermediate transactions beyond money. Because blockchains provide mechanisms for recording and ensuring the trust of transactions and settlements, they are considered promising for storing and securing medical data to be used both in personalized medicine and science. Dozens of medical record blockchains are currently under development—all attempting various facets of the three Ds (democracy, decentralization, and disintermediation)—ordinarily promised fulfilled by blockchain technologies and infrastructures [23]. For medical data, this is complicated somewhat by the ways that political institutions and nations treat that data (as private, protected, and limited in its use) and by blockchains in general, which are not primarily used for enhanced privacy given that they typically maintain public and permanent records [24].

Genomic data are excluded from much of the regulatory burden that prevents their free dissemination. Deidentified genomic data can generally be transferred, bought, and sold without violating privacy regulations aimed at personal data. This is because once data are deidentified, it becomes very difficult to reidentify their donors. However, it is not become impossible to re-identify individuals, as a number of studies have shown [25, 26]. In combination with medical data or other user-supplied data, re-identifying genomic data—associating it with a particular person—becomes possible, though not necessarily easy.

If ever a market was ripe for the three Ds of blockchain, the genomic data market is it. That market is worth well within the billions of dollars, judging from the $10 billion spent by Wall Street firms this past year to buy up the companies testing and selling data [27]. Disintermediating, democratizing, and decentralizing that market would ideally provide greater transparency, greater justice, and better access to deidentified data at a lower cost. It would change the game in genomics as a science and as an industry. It would also pave the way for similar user-centric data provenance and payments for other types of user data, medical and commercial.

3.1 Blockchain Genomics: The Current Slate

Most of those companies who have launched or are launching "blockchain genomics" products have promised to allow users to "own" their genomes. The author has observed a half-dozen serious contenders that have entered the space since 2017, including in order of appearance: EncrypGen, LunaDNA, Zenome, Nebula

Genomics, Shivom, Genomes.io, and Genobank.io. Others have arisen and disappeared, but most of these remain operational in some sense. At various times, they have offered tokens (EncrypGen, Zenome, and Shivom) for use in trading genetic data and getting paid for it. At various times, each has also suggested that their products will provide greater security for genetic data, and more frequently, that their platforms will allow users to "own" their genetic data [28].

A number of these companies state that they use blockchains, though it is not always clear how they are being used. Only a couple have public blockchain explorers or tokens that can be explored via Etherscan (http://etherscan.io). Claims that blockchains provide greater security for the genomic data are generally orthogonal to the technology itself, which does not secure anything. They do not prevent hacking a database; they are merely less susceptible to being hacked than other sorts of databases, which is not helpful since no blockchain currently can support large data files on-chain (written into the blockchain itself) like genomes [29]. Few blockchain genomics companies are putting genomic data on blockchains themselves, though blockchains do not provide pointers to such data in other databases. This is often considered necessary since most privacy rules demand the ability to delete data even if blockchains could accommodate three billion entries per user without becoming hopelessly bogged down [30].

Blockchains, as described above, are ledgers and provide transparency through publicly scannable audit trails for transactions (like sales, transfers, queries, etc.). They do not typically protect against hacking nor offer greater anonymity. In fact, because blockchain explorers or browsers often accompany blockchains, they provide a full accounting of all the transactions they host and may be public rather than private. Blockchains like Bitcoin have been used very successfully by law enforcement, for instance, to track illicit currency exchanges (like Bitcoin used for ransomware attacks), and forensic analysis of blockchain transactions has been used to uncover money laundering and other crimes [31]. The nature of blockchain platforms requires careful planning and technology design to provide anonymity, greater privacy, and security.

Companies in the blockchain genomics space continue to advertise their products to provide security, anonymity, or other features typically not provided by blockchains, or for which blockchains are not necessarily the ideal mechanisms of delivering these goals. Our concern here is not with those claims but rather with the bold claims that every one of these companies has made at some point to provide users with "ownership" of their genomic data.

4 Why Genomes Cannot Be Owned

No one "owns" genes unless they have been deliberately edited, in which case one might be able to patent that invention. However, patents are not property per se; they are state-supported monopolies to practice some art for a limited time. Typical indicia of ownership for property are missing in the relationship between individuals and their unique genetic code. Setting aside intellectual property rights for a moment, the three billion base pairs that make up one's genetic code are 99.9% similar to everyone else's. One's uniqueness genetically is limited to a tiny portion of the genome in one's cells, as well as the unique history one has experienced as a living being. While the DNA that holds that code is physically embedded in cells in one's body, it is not always confined there. When hairs get washed down the sink, the follicles contain cells that have that code in them. Individuals are poor proprietors of the physical property that makes them up, not to mention the data in the cells they constantly cast away. It is not reasonable to be a strict curator of one's physical self, much less the data involved.

Historically, property consists of things over which we can exert both exclusion of others and public indicia of ownership. Property rights obtain over real estate and moveables: land and things we can hold to the exclusion of others. However, property laws do not apply to bodies or their parts. Nor do property or intellectual property laws apply to the data that expresses our genomes or other parts of us. Data are typically not covered by intellectual property (IP) laws, including laws like copyright and patent. IP law protects original expressions or inventions, and nothing about the data gleaned from our bodies is an original expression.

We may be "owners" of our physical bodies at a particular moment, but we are not owners in any historically justified sense of the bits of ourselves that we leave around in our wakes. That detritus contains our DNA, and anyone sufficiently motivated could gather and purify it and sequence our genomes from it. We cannot reasonably be called the owners of things we leave lying around in public, nor would the law recognize that we are as centuries of property law make clear that ownership requires exclusionary and responsible custody. More so than things, when we generate data and it becomes public in some way, anyone may use it, with very few exceptions arising from privacy law [32]. Therefore, we cannot be owned, and we cannot own ourselves beyond our current, corporeal existence, to the degree that we can exclude others from it.

5 How Shall We Treat Genes?

There is clearly sentiment that we should be the owners of our "own" genomes, but this is not legally or practically possible given longstanding constraints on owning data—the practical impossibility given that those cells contain copies of genomes. Genomic selves cannot be contained nor kept secret without some new, very strict laws that would be anathema to individual freedoms. Copyright and patent do not apply and cannot provide monopolistic rights to the data from genomes either because those data are not original expressions but rather natural phenomena.

If goals are based upon the justice we perceive relating to fairly participating in research, having a say over how data are used, and engaging in the wealth that our data help produce, we can achieve this without the pretense of ownership. Part of the solution is technological, involving perhaps blockchains or other similar platforms that allow tagging, tracking, indicia of origin, ability to reject buyers or users, and participating in settlements and profits. However, while they may be necessary to solve problems of genetic justice, these technical requirements are not sufficient to solve them fully. Networks involving these requirements will still be wholly voluntary, and there is nothing to constrain or prevent the use of genomic data outside such voluntary networks.

The European General Data Protection Regulation and similar laws cover genetic data but carve out large and necessary research exceptions for deidentified data [33]. Depending on the scope of informed consent, there is still broad leeway for researchers to use genetic data without requiring subjects from whom data have been gleaned to consent to specific future uses or profit from its commerce. As discussed above, access to deidentified genomic data is necessary and fruitful for medical science. Restricting data uses must balance concerns of privacy with the public good. While it has been shown that deidentified genetic data are technically possible to re-identify in the presence of other data, it has not been shown that this is being done, that it will be done to any significant degree, nor that such activities would be maliciously done nor likely to lead to personal harms [34]. Public data breaches of millions of personal, identifiable, and prejudicial data have already occurred, exposing people's private lives a lot more surely than genetic data do or likely could [35].

Individual genetic variations are, nonetheless, biologically and necessarily associated with individuals, so it seems reasonable to treat that data as special, much as individuals are guardians of their public personae. Some rights have arisen in various jurisdictions governing the use of public data, like laws that prohibit commercializing on another's likeness without their consent. Nevertheless, there are numerous exceptions in light of free speech and research concerns. Can laws guide something similar for genomic data—establishing a state-sanctioned right to be arbiters of the uses of genetic data and realize the fruits of wealth even while not insisting that some new, difficult-to-conceive property right should make individuals "owners" of things over which property claims are simply not appropriate, nor legally permissible, and maybe even not technically feasible? [36].

The values that regulations should capture, and that cannot be solved by technology alone, include a respect for individual interests over what happens with both matter and data that originate, respect for the value and necessity for sharing deidentified data that are hard to re-identify as broadly as possible for basic science, respect for the just distribution of wealth generated from contributing data for science, and respect for individual privacy.

6 What About Non-Fungible Tokens, NFTs?

In 2021, Non-Fungible Token (NFT) technology became the bastion of hopes and dreams about laying claim to all sorts of new types of property. NFTs place a digitally unique stamp on a public blockchain, and only the keyholder can lay claim to the item referenced by that stamp [37]. However, NFTs do not create property, they can only serve as a reference or potentially official title to types of property already recognized by law. Their only power lies in providing a public, verifiable, encrypted indicia of a claim to something, but the claims are only as good as the property law that backs them up [38]. If a life sciences organization holds an NFT with a claim to ownership of that DNA, it will not be valid as a claim of ownership under any existing legal jurisdiction [39]. The organization will only own a token that serves as the title to nothing.

Laying claim to one's genome via an NFT suffers a complete absence of legal strength; it provides no ownership over the code nor rights to disseminate with restrictions, nor property or copyright claims to the genetic code because, as discussed above, no such rights exist [39]. Those claims cannot exist without laws, and currently, no laws exist that will provide ownership over genomic data. NFTs are simply a new manner of establishing title over property, they cannot create new property laws and will not solve the problems noted above: the need to disseminate genomic data for science and to simultaneously preserve these values mentioned above: justice, remuneration, consent, and privacy.

6.1 The Need for Regulation

Technology alone cannot solve issues relating to control, remuneration, and greater scientific demand and use for deidentified genomic data. Science needs more data; that data should be deidentified and combined with useful metadata, where possible. Further, people should be remunerated for its use as well as provided transparency and some degree of control over the uses of the data. No technical solution can accommodate all those needs perfectly, nor can any provide ownership of genomic data. Moreover, it is not clear that there is a tremendous social demand for people to own their data. Numerous startups have promised it, but there is little evidence of wide-scale buy-in by the genetically tested public.

Genomic blockchains can be very useful for providing users with payments and greater control of where their data may go, as well as decentralizing storage, sharing, and hosting of data. We can best apply blockchain technology to these uses now while simultaneously being honest about their limitations. Ownership is a legal category that has been strictly defined and well understood for centuries while limited historically to things over which people can exert exclusive and rivalrous control. In the case of intellectual property, where the objects of patent and copyright provide state-sanctioned limited monopolies over original expressions and inventions, technologies have provided only limited means of preventing "misuse" and no practical means of preventing the dissemination of unauthorized copies. No one "owns" even their original expressions in any real sense. They can simply call upon the legal system to try to punish illicit uses and reproductions post hoc.

Because genomic data are not part of any category for intellectual property protection, if we wish to provide even the barest analogies to ownership—like the exclusivity of access and the ability to limit its spread—we will need not only new technologies but new laws. Only regulations enacted by states can start to recognize anything like an ownership interest in genetic data. Such regulations should not interfere with the clear scientific value of access to lots of deidentified genetic data by scientists worldwide.

Current regulations provide some protections stemming from privacy concerns to prevent misuse of genetic data for discriminatory purposes. However, laws could help to encourage and even mandate that contributors to the genetic databanks of the future, likely incorporating blockchain technologies, share in the wealth and help direct the uses of that data. Just as some privacy laws currently prevent unwanted commercial uses of one's image, so too could uses of genetic data be guided by tort laws that enable people to sue for misuse and require any transfer of data to be logged in a publicly accessible ledger.

Meanwhile, we should be honest about the limits of blockchain and other technologies, none of which can currently create new legal categories or rights. None of which can guarantee against misuses of genetic data.

7 Future Directions

As demand for genomic data increases—and as concerns about allowing people to better control, audit, and profit from the use of their deidentified data grow—blockchains can only achieve their promise of alleviating these concerns if they are accompanied by greater legal certainty. Specifically, blockchain genomics companies should be forthright about the limits of technologies and work together to advocate for policies that could help clarify the relations between individuals and their genomic data. In the past, states have adopted new laws recognizing new types of property, beginning with the creation of intellectual property rights a couple hundred years ago. If we wish culturally to recognize something like "ownership" over any data, including genomic data, we will need to agree on the legal frameworks that could create such new laws and not rely merely on technical attempts to improve control and benefits.

Improved mechanisms and communities dedicated to sharing deidentified data widely, allowing for greater access while enhancing both privacy and transparency, are some of the benefits that could be realized through selective, careful uses of genomic blockchains. These goals would be furthered by better interoperability among existing databases, improved tracking of usage, and associating donors/subjects with the data used so that they can (a) know how their data are used, (b) perhaps choose to either opt-in or out of studies, and (c) get compensation for the use of data in some fair, equitable manner. It may well also be time for the state to consider passing new laws that grant new rights to data so that one could have some legal right over one's unique genetic data. Without such laws, technology can only go so far to provide some sort of custodial relationship, and breaches will have limited means of recompense if discovered.

8 Conclusions

Genomic blockchains can help solve some issues regarding conflicting needs of science and justice, providing mechanisms for better tracking, management of transfers, settlements, or payments, and audits trails to maximize transparency. Without laws or common norms governing our commonly held beliefs about issues of justice regarding genomic data, however, blockchains will only solve these problems voluntarily, and there will be no legal recourse for those who think they have been harmed. Under any reasonable legal definition of ownership, no one can own genomic data, not without severely challenging our historically entrenched notions of ownership in general. Nevertheless, we can work to create laws, norms, rules, and regulations that enhance the application of just principles to our shared genetic heritage and our individual roles as part of that heritage, to better encourage and make safe our participation in scientific discovery and share the wealth generated not only by money that is spent on that data but the benefit we all receive through medical science conducted with genomic data.

Researchers need lots of data, and the consumer testing market is just a small sample of the data required to fully realize the promises of genomic sciences and industry. Blockchains can help secure more participation by subjects, fulfill better-informed consent over the uses of data, provide greater access to a broader research community, and potentially unlock new, unanticipated means of delivering the science itself from current silos into a connected, interplanetary grid of ethical inquiry. Life sciences research and industry continue to explore and build upon the promises of blockchains in genetic data science, and opportunities abound for revolutionizing genomic science in general if the law, culture, and regulations necessary to enable it and protect individuals keep pace.

References

1. Hartl DL (2000) Fly meets shotgun: shotgun wins. Nat Genet 24(4):327–328. https://doi.org/10.1038/74125
2. Harris RF (2000) Patenting genes: is it necessary and is it evil? Cur Biol 10(5):R174–R175. https://doi.org/10.1016/s0960-9822(00)00369-9
3. Kaiser J (2018) There are about 20,000 human genes. So why do scientists only study a small fraction of them? Science. https://doi.org/10.1126/science.aav4546

 4. Lai JC (2015) Myriad genetics and the BRCA patents in Europe: the implications of the U.S. supreme court decision. UC Irvine Law Rev, 5, 1041. https://scholarship.law.uci.edu/ucilr/vol5/iss5/5
 5. Koepsell D (2009) Who owns you?: the corporate gold rush to patent your genes. Wiley. https://doi.org/10.1002/9781444308587
 6. Shabani M, Borry P (2015) Challenges of web-based personal genomic data sharing. Life Sci Soc Policy 11:3. https://doi.org/10.1186/s40504-014-0022-7
 7. Stoeklé HC, Mamzer- MF, Vogt G, Hervé C (2016) 23andMe: a new two-sided data-banking market model. BMC Med Ethics 17:19. https://doi.org/10.1186/s12910-016-0101-9
 8. Kim D, Shin H, Song YS, Kim JH (2012) Synergistic effect of different levels of genomic data for cancer clinical outcome prediction. J Biomed Inform 45(6):1191–1198. https://doi.org/10.1016/j.jbi.2012.07.008
 9. Johnson AD, O'Donnell CJ (2009) An open access database of genome-wide association results. BMC Medical Genet 10:6. https://doi.org/10.1186/1471-2350-10-6
10. Karczewski KJ, Weisburd B, Thomas B, Solomonson M, Ruderfer DM, Kavanagh D, Hamamsy T, Lek M, Samocha KE, Cummings BB, Birnbaum D (2017) The exome aggregation consortium. In: Daly MJ, MacArthur DG (eds) The ExAC browser: displaying reference data information from over 60 000 exomes. Nucleic Acids Res, 45(Database issue), D840. https://doi.org/10.1093/nar/gkw971
11. Beck T, Hastings RK, Gollapudi S, Free RC, Brookes AJ (2014) GWAS central: a comprehensive resource for the comparison and interrogation of genome-wide association studies. Eur J Hum Genet 22(7):949–952. https://doi.org/10.1038/ejhg.2013.274
12. Hogarth S, Saukko P (2017) A market in the making: the past, present and future of direct-to-consumer genomics. New Genet Soc 36(3):197–208. https://doi.org/10.1080/14636778.2017.1354692
13. Grothaus M (2015) How 23 and me is monetizing your DNA. Fast Co. https://www.fastcompany.com/3040356/what-23andme-is-doing-with-all-that-dna
14. DeFrancesco, & (2019) Your DNA broker. Nat Biotechnol 37(8):842–847. https://doi.org/10.1038/s41587-019-0200-5
15. Molla R (2020) Why DNA tests are suddenly unpopular, Vox https://www.vox.com/recode/2020/2/13/21129177/consumer-dna-tests-23andme-ancestry-sales-decline Accessed 23 Nov 2021
16. Zhang S (2018) Big pharma would like your DNA. Atl, Atl Media Co. www.theatlantic.com/science/archive/2018/07/big-pharma-dna/566240/
17. Hummel P, Braun M, Dabrock P (2021) Own data? Ethical reflections on data ownership. Philos Technol 34(3):545–572. https://doi.org/10.1007/s13347-020-00404-9
18. Mill JS (1989) On Liberty' and Other Writings. Cambridge University Press. https://www.cambridge.org/us/academic/subjects/politics-international-relations/texts-political-thought/j-s-mill-liberty-and-other-writings
19. Skloot R (2010) The immortal life of Henrietta Lacks. Penguin Random House. https://www.penguinrandomhouse.com/books/168191/the-immortal-life-of-henrietta-lacks-by-rebecca-skloot/
20. Zimmer C (2013) A family consents to a medical gift, 62 years later. N Y Times 7. https://www.nytimes.com/2013/08/08/science/after-decades-of-research-henrietta-lacks-family-is-asked-for-consent.html
21. Kubát M (2015) Virtual currency bitcoin in the scope of money definition and store of value. Procedia Econ Financ 30:409–416. https://doi.org/10.1016/S2212-5671(15)01308-8
22. Garzik J, Donnelly JC (2018) Blockchain 101: an introduction to the future. In: Handbook of blockchain, digital finance, and inclusion, vol 2. Academic Press, pp 179–186. https://doi.org/10.1016/b978-0-12-812282-2.00008-5
23. Yoon HJ (2019) Blockchain technology and healthcare. Health Inform Res 25(2):59–60. https://doi.org/10.4258/hir.2019.25.2.59

24. Leeming G, Cunningham J, Ainsworth J (2019) A ledger of me: Personalizing healthcare using blockchain technology. Front Med (Lausanne) 6:171. https://doi.org/10.3389/fmed.2019.00171

25. Erlich Y, Shor T, Carmi S, Pe'er I (2018) Re-identification of genomic data using long range familial searches. BioRXIV, 350231. https://www.biorxiv.org/content/https://doi.org/10.1101/350231v2

26. Raisaro JL, Tramer F, Ji Z, Bu D, Zhao Y, Carey K, Hubaux JP (2017) Addressing beacon re-identification attacks: quantification and mitigation of privacy risks. J Am Med Inf Assoc 24(4):799–805. https://doi.org/10.1093/jamia/ocw167

27. MacDonald A, Marcus A (2021) 23 and me to go public with Richard Branson-backed SPAC. Wall Str J. https://www.wsj.com/articles/23andme-to-go-public-with-richard-branson-backed-spac-11612450687 Accessed Nov 23 2021

28. Katuwal GJ, Pandey S, Hennessey M, Lamichhane B (2018) Applications of blockchain in healthcare: current landscape and challenges. arXiv preprint arXiv:1812.02776. https://arxiv.org/abs/1812.02776

29. Racine V (2021) Can blockchain solve the dilemma in the ethics of genomic biobanks? Sci Eng Ethics 27:35. https://doi.org/10.1007/s11948-021-00311-y

30. Crow D (2019) A new wave of genomics for all. Cell 177(1):5–7. https://doi.org/10.1016/j.cell.2019.02.041

31. Zarpala L, Casino F (2021) A blockchain-based forensic model for financial crime investigation: the embezzlement scenario. Digit Financ. https://doi.org/10.1007/s42521-021-00035-5

32. Koepsell D (2014) It is not ethical to patent or copyright genes, embryos, or their parts. In: Caplan AL, Arp R (eds) Contemporary debates in bioethics. Wiley-Blackwell, pp 152–161. https://www.wiley.com/en-us/Contemporary+Debates+in+Bioethics-p-9781444337143

33. Staunton C, Slokenberga S, Mascalzoni D (2019) The GDPR and the research exemption: considerations on the necessary safeguards for research biobanks. Eur J Hum Genet 27(8):1159–1167. https://doi.org/10.1038/s41431-019-0386-5

34. Kasperbauer TJ, Schwartz PH (2020) Genetic data aren't so special: causes and implications of reidentification. Hastings Cent Rep 50(5):30–39. https://doi.org/10.1002/hast.1183

35. McKeon H (2021) Millions of patients receive healthcare data breach notifications. Health IT Secur. https://healthitsecurity.com/news/millions-of-patients-receive-healthcare-data-breach-notifications Accessed 24 Nov 2021

36. Koepsell D (2007) Individual and collective rights in genomic data: preliminary questions. J Evol Technol 16(1):151–159. https://www.jetpress.org/v16/koepsell.html

37. Visser A (2021) NFTs: How do they work? Finweek (9):44–45. https://hdl.handle.net/10520/ejc-nm_finweek_v2021_n9_a16

38. Fairfield J (2021) Tokenized: The law of non-fungible tokens and unique digital property. Indiana Law J. Forthcom. https://papers.ssrn.com/sol3/papers.cfm?abstract_id=3821102

39. Taylor M (2012) Genetic data and the law: a critical perspective on privacy protection vol 16, Cambridge University Press. https://doi.org/10.1017/CBO9780511910128

Dr. David Koepsell is a philosopher and lawyer who has authored numerous books, chapters, and articles primarily on the subjects of ontology, ethics, and technology, and been a tenured Associate Professor of Philosophy at the Delft University of Technology, Faculty of Technology, Policy, and Management in the Netherlands, Visiting Professor at UNAM, Instituto de Filosoficas and the Unidad Posgrado, Mexico, Director of Research and Strategic Initiatives at COMISION NACIONAL DE BIOETICA in Mexico, and Asesor de Rector at UAM Xochimilco. He is the co-founder of EncrypGen, Inc. http://davidkoepsell.com and Lecturer at Texas A&M Univeristy, Dept. of Philosophy.

Dr. Mirelle Vanessa Gonzalez Covarrubias completed a degree in Biological Pharmaceutical Chemistry from the Faculty of Chemistry of the Universidad Nacional Autónoma de México (UNAM), obtained a Master of Science degree from the Faculty of Chemistry at UNAM and a Doctorate in Pharmaceutical Sciences from the Faculty of Pharmacy and Pharmaceutical Sciences at the State University of New York at Buffalo, as well as a post-doctorate from the University of Leiden in the Netherlands. She is currently a researcher in Medical Sciences "C", attached to the Pharmacogenomics laboratory of INMEGEN and a member of the National System of Researchers (SNI) Level I.

Convergence of Blockchain and AI for IoT in Connected Life Sciences

Orlando Lopez⬤, **Frederic de Vaulx, and William Harding**⬤

Abstract The internet of things (IoT) has emerged at the forefront of many indus-tries, exemplified by the connected home, heterogeneous manufacturing environ-ments, and even interconnected wearable technologies. However, when IoT-enabled solutions are integrated with blockchain and artificial intelligence (AI) approaches, it is possible to mitigate the uncertainty of ongoing challenges and existing bottle-necks to streamline the clinical translation of innovation in life sciences into clinical practice. It is then that humanity can experience growth that exponentially improves an individual's quality of life. With that growth in mind, this chapter discusses how blockchain and AI can help govern inherent and residual risks associated with (1) IoT-enabled technologies that need to enable secure aggregation, analysis, and acces-sibility of data across standard models of interoperability, and (2) how these emerging technology platforms can help catalyze the transfer of scientific discovery into new biomedical products and services to improve the delivery of healthcare and patient outcomes.

Keywords Blockchain · Machine learning · Internet of things (IoT) · Digital transformation · Bring your own device (BYOD) · Artificial intelligence (AI)

O. Lopez (✉)
National Institute of Dental and Craniofacial Research (NIDCR), National Institutes of Health (NIH), Bethesda, MD, USA
e-mail: orlopa@gmail.com

F. de Vaulx
Strategy & Innovation, Prometheus Computing LLC, Bethesda, MD, USA
e-mail: f.devaulx@prometheuscomputing.com

W. Harding
Medtronic, Tempe, AZ, USA
e-mail: William.Harding@Medtronic.com

© The Author(s), under exclusive license to Springer Nature Singapore Pte Ltd. 2022   85
W. Charles (ed.), *Blockchain in Life Sciences*, Blockchain Technologies,
https://doi.org/10.1007/978-981-19-2976-2_5

1   Introduction

1.1   Fueling the Digital Transformation in Health and Life Sciences

A variety of scientific domains associated with the study of living organisms, including biology, botany, zoology, microbiology, physiology, biochemistry, and related subjects make up the field of Life Sciences. The life sciences industry is on the cusp of transformational change and a digital revolution. While challenging, digital transformation creates new opportunities. Novel tools and approaches are being developed to capitalize on the continuous advancement of new techniques.

Creating positive change in healthcare requires coordinated efforts and the collaboration of a community of multi-domain experts and innovators to share and shape ideas that will transform the healthcare ecosystem. The digital transformation in healthcare has experienced a very dynamic year resulting from the coronavirus pandemic and all the implications that it represents. Blockchain, artificial intelligence (AI)/machine learning (ML), and clinical Internet of Things (IoT) are rapidly becoming core tools in the health and life sciences to drive connectedness across multiple applications, communities, and key stakeholders in a manner with potential implications for social change.

As a result of the pandemic, many sectors of the economy were forced to go through a digital transformation to keep working as people had to stay separated and physical connections were becoming digital connections. Many sectors saw programs and projects become more distributed and decentralized to accommodate this new reality. Concepts like Bring Your Own Device (BYOD) are part of the solution to bridge the gap between our personal environment and enterprise needs. In the context of life sciences research, this can translate into interesting opportunities for decentralized clinical trials, data management, and governance or processes management. A digital transformation also comes with risks explored later in the chapter. This chapter explores how emerging technologies like blockchain and AI can be leveraged to enable opportunities and mitigate risks around the use of BYOD. Figure 1 shows that BYOD can be associated with technologies like blockchain, AI, and wearables to support various use cases for life sciences.

1.2   Technology Unification

Blockchain involves many blocks of data chained together across divergent systems, where each block of data might be owned by a different person or organization [1]. Similarly, BYOD technology embraces a model where each person has brought a unique and personalized device [2]. Then, it is possible to conceive of data blocks on individually-owned devices existing within a decentralized scheme of divergent technologies. Combining those elements to form a decentralized scheme results in

Fig. 1 BYOD and life
sciences relationship

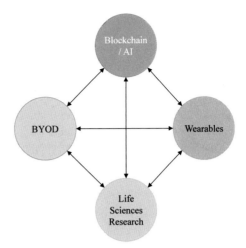

a federated model perfectly positioned for use by an AI engine seeking to use the power of many decentralized BYOD technologies to crunch volumes of data [3]. Thus, with the data (or blocks) being owned by many users across many devices, the elements associated with blockchain, BYOD, federated/decentralized data, and AI harmonize to form a powerful, secure, and cohesive system. Lastly, though it was not mentioned as an element of this unifying theory, an IoT/transformational engine plays a critical role in standardized communications and data aggregation methods. That IoT connection model ties all discussed elements together to form an intraoperative solution that can securely integrate divergent technologies to exchange data using standardized protocols [4].

2 Harnessing the Power of Data-Driven Technologies in Life Sciences

2.1 Data-Driven Technologies in Life Sciences

Data-driven technologies such as AI and ML are catalyzing transformational advancements in scientific, clinical, and industrial domains through process automation, extraction of patterns, and inferences from complex datasets to enhance decision-making and problem-solving. The existing implementations of data-driven tools in biomedical sciences can range in scope, technical approaches, and context-of-use, such as assisting identification of biomarkers in drug development, genomics, and protein structure prediction, powering clinical treatment planning with decision support systems, enhancing the performance of digital pathology with intelligent image processing algorithms, automating data acquisition, aggregation and processing from multi-sensor arrays, and many others. Among the many new

AI concepts and techniques launching almost daily, emerging trends are rapidly gaining adoption by data scientists and biomedical researchers alike to help guide experimental designs that in turn lead to better science and speed up innovation cycles.

A blockchain is a digital ledger of transactions that makes it difficult or impossible to change, hack, or cheat the system. This system of recording information offers a unique set of capabilities that enables individuals to achieve digital endpoints in new innovative ways that minimize the need for intermediaries. It is a capability that can do a wonderful job of empowering people by facilitating information exchange while instilling trust at transactional layers [3]. As shown in Fig. 2, this chapter seeks to illustrate the value of using an AI-powered digital mesh (e.g., distributed ledger technologies (DLTs)), which connects wearable technologies with an individual as a way to enhance the value of data for clinical decision making. For example, a blockchain can be used to create an identity for a proteomic genetic makeup: a metabolic makeup combined with a comprehensive physiologic profile that allows stress testing to gain insights into those differentiators. This information can determine health status and support clinical diagnosis, treatment planning, and management. Beyond automating back-office operations, empowering people and giving people the ability to consent to share data, allowing people to manage data on themselves, it is possible to build digital endpoint capability to establish digital twin infrastructure to enable simulation of environments based on genomic makeup, phenotypic profiles, personality traits and a wide range of physiologic and Social Determinants of Health (SDoH). This capability has the potential to disrupt entire value chains across the health and life sciences ecosystem and change the way consumers consume. Further, this capability allows patients, researchers, and clinicians to exchange information for data-driven solutions [5].

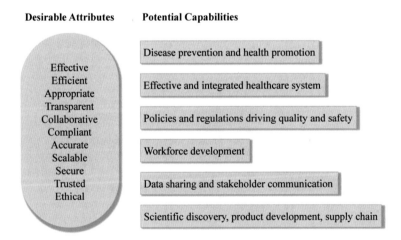

Fig. 2 The value in the convergence of emerging technologies in the life sciences ecosystem

Emerging technologies are fueling the development and adoption of next-generation tools to gather and process multiple data streams. This creates the potential to enhance the value of the scientific pursuit through data-led insights to facilitate a deeper understanding of health and disease states through data-driven infrastructures with the promise to advance the efficiency further, cost, and quality of basic, applied, and clinical research. The integration of IoT, blockchain and AI is demonstrating unique value in supporting the advancement of innovation in one of the most complex and regulated industries.

Digital twins are virtual representations of a physical object, process, or service, with dynamic, bi-directional links between the physical entity and its corresponding twin in the digital domain [6]. Digital twin technology can characterize systems and replicate processes to collect data to predict how they will perform. Digital twin platforms are increasingly finding applicability in medicine and public health and enabling a new era of precision (and accuracy) medicine and public health. Complementing traditional electronic health record information or experimental data throughout the life sciences pipeline, with streams of real-world data from BYOD technologies, empowered by blockchain-AI-based infrastructure, have the potential to enable a new generation of intelligent digital twins for learning and discovering new knowledge, new hypothesis generation, and testing, and in silico experiments and comparisons to accelerate life sciences discovery and improve patient outcomes.

3 Innovating in a Highly Regulated Industry

While many life sciences originate in biology, new biotechnological specializations have emerged that seek to translate promising new scientific discoveries into life-changing biomedical innovations. Life sciences organizations operate in a highly regulated industry where laws, regulations, and statutes evolve in an increasingly global marketplace with heightened transparency expectations. Considering that there are high research and development (R&D) costs with minimal revenue in the initial years, product development in the life sciences demands continuous attention to a rapidly evolving landscape and the adoption of robust risk management strategies at every phase of product development and clinical validation cycles [7].

Innovation is rarely easy in any industry, and setbacks are often expected. Introducing innovation can be especially difficult in highly complex and regulated industries like life sciences and healthcare. Nonetheless, it is possible to accelerate the translation of new scientific discoveries into practical and functional innovations that are safe and effective. These innovations require proactive strategic planning by integrating relevant legal, regulatory, and compliance milestones into a robust risk management framework at every phase of preclinical product development and clinical testing. Accounting for legal, regulatory, and compliance requirements early in the R&D process makes it possible to gain a sustained competitive advantage in life sciences innovation through shifting clinical and regulatory requirements without compromising capital, resources, quality, or patient safety.

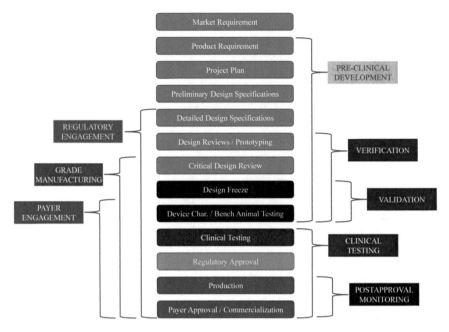

Fig. 3 Lifecycle of life sciences product development

Maintaining regulatory compliance and demonstrating thorough preclinical characterization of a new medical product before entering clinical testing are some of the biggest challenges for organizations in life sciences, especially due to globalization, heightened transparency expectations, and increased emphasis on innovative technologies. The general process of translating new life sciences research from the lab to human use according to specific claims on the performance of the technology or new clinical approach is summarized in Fig. 3.

The capabilities of IoT and BYOD technologies to capture multiple data streams combined with an intelligent governance infrastructure provided by AI-enabled blockchain platforms have the potential to impact all stages of life sciences product development, from early conception, design, and prototyping to clinical evidence generation, regulatory review, and finally commercialization [8]. The integration of emerging technologies from the early stages of life sciences R&D into a 'Quality Risk Management' framework (21 CFR § 820, Quality system regulations, ISO 13485:2016, Medical devices—quality management systems; and ISO 14971:2019, Medical devices—Application of risk management to medical devices) has the potential to offer a unique advantage to enhance the timeliness of evidence available to assess, control, identify, and monitor the risks that could reduce the safety, efficacy or performance of new life sciences products [9, 10]. For example, integrating IoT/BYOD with blockchain and AI allows the establishment of valuable tools and infrastructure that promotes data reliability, transparency, and traceability that gets captured and flows across the entire life cycle of life sciences research. The content in this chapter promotes taking a proactive mindset to improve workflows and value-chains in the life sciences ecosystem by risk management principles and leveraging the enhanced features an integrated IoT/BYOD-blockchain and AI platform affords.

To create successful solutions, it is important to have a team with critical skills in analytical thinking, data analysis, working collaboratively to take innovations from 'bench-to-bedside.' However, before work begins, it is important to understand how managing uncertainty and incorporating strategic planning throughout key milestones in biomedical product development can dramatically improve the overall efficiency, quality, and safety of new scientific endeavors. In turn, it is possible to reduce overall risk across the life cycle of biomedical product development and ensure critical milestones are adequately met on time and in alignment with relevant clinical, regulatory, and market requirements. The adoption of a robust risk management framework is recommended early. A framework allows an organization to determine quality-based strategies to identify, analyze, mitigate, evaluate, and treat different degrees of uncertainty involved in the process to translate scientific discoveries into innovations that address specific clinical and market needs.

Additionally, the pressure to reduce product development costs is intense across the pharmaceutical and medical device industries. Companies have been reducing R&D budgets and extracting the maximum value from available funding. Because clinical trials involve the greatest expense, they are under even greater scrutiny to reduce R&D costs. Many pharmaceutical and device companies are cutting budgets and scrutinizing expenses for clinical trials. Therefore, Contract Research Organizations need to better inform their business decision-making and confidence to achieve milestones or risk walking into bids blind. These pressures create a challenging operating environment.

4 Essential Elements for Data Strategy in Life Sciences

Data strategies are necessary to create long-term organizational objectives through the product lifecycle—not only consist of technical processes for data management and analytics—but the human element in managing and understanding data. The following are data strategies adapted from [11] to be given significant consideration:

1. Types of data.
2. Data acquisition tools (e.g., wearables, IoTs, BYODs).
3. Data aggregation streams.
4. Analytical techniques.
5. Collaboration methods.
6. Documentation and auditing (e.g., blockchain-AI-enabled).

Data integration powered by blockchain and AI can turn vulnerabilities, such as data size and complexity, into an advantage by enabling the transformation of disparate devices and connections into an integrated network of sensors (IoTs/BYODs). This intersection of increasingly mature AI with platforms is a potentially transformational element. The platform IoT sensors are the source of data, and AI is the engine that can make sense of that data. The platform devices are also controls that can take action at scale and in real time by implementing a data governance framework as a key infrastructural component to analytics success. It is important to keep in mind that successful data strategies require a strong and thoughtful data governance plan. A data governance plan can balance the competing needs for protecting access and creative data exploration.

4.1 Data Building Blocks

With consideration for the unification of solutions and technologies associated with blockchain, IoT, and AI, intersects and commonalities have been established such that there is clear alignment across those elements with an emphasis on interoperability, data security, and data analysis. Additionally, as suggested within a unification that embodies blockchain, IoT, and AI, there is clear support for the integration of BYOD technology. However, when considering where to start constructing a solution that integrates all four of those technological advancements, it might not be clear where it might be perceived as a chicken or egg thing. Thus, one could wonder which comes first or, more precisely, is it preferable to start designing around the data or the technology. Accordingly, the first inclination might be to say, 'collect the data,' 'distribute the data,' or maybe 'protect the data.' However, a data framework must exist first, in which data can be collected, distributed, analyzed, and reported. Nonetheless, a framework cannot be built without a vision of (1) what data will look like, (2) how data will be gathered, (3) how data will be transformed across platforms, (4) how data will be protected, (5) how data will be stored, (6) how data will be analyzed, and (7) how will the results inform and influence future decisions?

Recommended steps:

1. Establish the need (i.e., what problem requires solving?).
2. Construct a framework (i.e., architect a design that enables visualization of all the phases).
3. Assess the risks, barriers, critical line elements, and potential roadblocks.
4. Identify all the relevant stakeholders.
5. Execute the plan.

With those five steps in mind, it is easier to see that the next step is to construct a prototype from which some actual or simulated data can be generated. The point of the prototype makes it easier to see any potential pitfalls while testing the interoperability of any needed technology. Further, interoperability represents one of the most important elements that are not fully explored nor understood. Respectively, when attempting to deploy a solution into any life sciences environment, there is a desire to aggregate data associated with blockchain, IoT, AI, and BYOD technology. Specifically, if technologists neglect to conduct a complete examination of the need for effective interoperability, then the results of their efforts will be a standalone environment where data move containing manually entered/transcribed data filled with human errors.

Additionally, to improve the vocabularies and infrastructure supporting the reuse of scholarly data in the life sciences enterprise, it is valuable to adopt the FAIR Guiding Principles (Findable—Accessible—Interoperable—Reusable) across academia, industry, funding agencies, and scholarly publishers. FAIR principles promote scientific data management and stewardship in a concise and measurable framework to be implemented by life sciences practitioners across human-driven and machine-driven activities [12].

5 Prioritizing Risk Management in Life Sciences

Establishing a robust risk management framework is not only critical but required to the success and efficient commercialization of new life sciences research. To enable timely identification, analysis, mitigation, evaluation, and treatment of risks, managing potential risks in a life sciences project from an early stage will help manage uncertainties that could lead to irreparable outcomes in the future. Taking a proactive stand to risk management at the early stages of a new technology development project in life sciences will help guide preclinical research activities to align with strict performance requirements regarding the safety and effectiveness of new candidate technologies. The ability to monitor, analyze, secure, and learn from data flowing across multiple streams of connected IoT and BYOD technologies and from other data sources can be used to formulate digital representations or digital twins [13]. This information leads to new candidate life sciences technologies to assess technical readiness and establish a data-driven framework to demonstrate usefulness and compliance with regulatory and market expectations in an ever-increasing patient/user-centric translational science spectrum.

6 Opportunities and Challenges for Emerging Digital Technologies in Life Sciences

The convergence of IoT and BYOD with blockchain and AI offers a unique inflection point for technical readiness that enables enhanced capabilities to address given functional, operational, and scientific needs using data-led approaches. The ability to obtain meaningful insights from multiple data streams provides an opportunity to shed light on complex and unexpected processes and conditions that help establish associations against factors with the potential to guide transformative strategies for complex scenarios in biomedical research and product development.

Translating new scientific discoveries into new clinical tools has the potential to transform the practice of health care and dramatically improve patient outcomes.

Adopting a systematic approach and adopting quality principles, especially at the early stages of product development, can help establish a significant degree of reliability, trust, and acceptability for the new product.

6.1 Major Milestones in Life Sciences Product Development

Several major milestones are common in the R&D of drug, biologic, medical devices, and combination products. Emerging technologies are fueling the development and adoption of a new generation of tools to gather and process multiple data streams. These technologies create the potential to enhance the way basic, applied, and clinical research is conducted when experimental workflows can be optimized through evolving data-led insights. The integration and adoption of IoT networks, blockchain, and AI capabilities demonstrate the potential to serve as a new infrastructure that helps advance the development of innovation in life sciences, one of the most complex and regulated industries.

6.1.1 Challenges and Opportunities for BYODs—Blockchain-AI in Life Sciences

'Bring Your Own Device (BYOD)' emerged as a common term in the early 2000s, as a direct result of Voice Over IP (VOIP) solution providers proposing services that could be supported on a variety of user-provided devices. The term became even more prevalent a few years later as companies recognized that many of their new employees looked to continue using the devices that they brought with them. Accordingly, the basic options that align well with the concept of IoT compatible BYOD are as follows:

- BYOD—Bring your own device.
- CYOD—Choose your own device.
- COPE—Corporate owned, personally enabled.
- POCE—Personally owned, company enabled.

At that point, companies realized that they could reduce costs usually associated with building/buying technology that was both restrictive and costly to maintain [14]. Additionally, when referring to BYOD solutions, it is important to recognize that those solutions are a composite of both hardware and software elements.

The acceptance of the BYOD model enabled companies to narrow their focus on just controlling device connectivity and data at the application level, versus controlling both applications and hardware. That acceptance also exposed that no technology can be 100% secured. A hybrid model of application layer security and procedural/policy controls enabled companies to maintain control over how devices connect to their networks and how their data and intellectual property (IP) are protected. That control over the data and connectivity of user-supplied devices enables companies to reduce risks associated with their brand or market share if a user-supplied device is stolen or lost.

Leveraging BYOD in Life Sciences

Mainstream use of BYOD devices in life sciences lags that of other industries, where many organizations within the life sciences industry could not instantly move to the BYOD model. For example, with most life sciences technologies requiring government acceptance and validation, the ability to switch to an unregulated technology was nearly impossible. That point becomes clearer by examining the time and money it takes to design, validate, deploy, and sustain technology that meets regulatory requirements. However, as more life sciences companies embrace a BYOD model, they are starting to realize greater user satisfaction and reduced costs, generally associated with the design and support of unique technology [15]. At the same time, the ability of the life sciences industry to accept the BYOD model reduces a dependency on technology suppliers who might have cornered the life sciences technology market. That dependency on specialized technology suppliers has not been entirely removed across all three classes of life sciences technology, but it is becoming more common for class 1 technology. Specifically, many class 1 technologies are starting to emerge in readily available wearable devices such as EKG, pulse oximeter, blood pressure, and heart rate (Fig. 4).

Some life sciences organizations have embraced a hybrid BYOD model where users can select technology from a growing list of compliant devices. From that hybrid model, some control over what users can use is realized. Additionally, there is a growing trend among technology developers to build technology that can be easily secured and maintained by life sciences IT organizations [16]. That movement is being driven by the recognition that data are a commodity and that more users are seeking devices that can assist with producing healthy outcomes. Further, with the integration of fitness devices into multi-purpose mobile devices such as phones, tablets, glasses, and watches, there is a need to ensure that user data are protected. The user data, which might exist on technology as found within life sciences environments, represent an element that needs to be protected even when user-supplied technology is not being used for work. For example, individuals might use their company-regulated device to track their health while being connected to their company network, representing a real-world scenario requiring a company to apply an equal level of protection for both company and user data.

A common question that arises during any conversation related to BYOD technology within life sciences is, 'Can BYOD be embraced in hospitals/clinics?' The question immediately stimulates thoughts around how a hospital might promote secure connections between BYOD technology and existing hospital systems. More importantly, any organization operating within the scope of life sciences and healthcare has to be cognizant of how data are exchanged between BYOD and healthcare systems can be secured. Those points are obviously important, but currently evolving global conditions have forced many life sciences organizations to reassess BYOD

Fig. 4 Typical wearable
devices

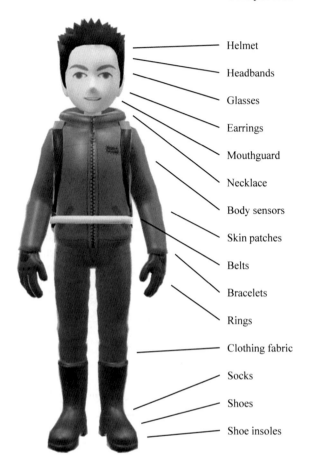

Helmet

Headbands

Glasses

Earrings

Mouthguard

Necklace

Body sensors

Skin patches

Belts

Bracelets

Rings

Clothing fabric

Socks

Shoes

Shoe insoles

technology with a greater emphasis on an individual's needs versus on security [17]. Additionally, as BYOD technology and digital data transformation in life sciences are assessed, medical use cases represent a real-world scenario where data collected from a personally-supplied device would assist life sciences professionals in performing preemptive data analysis.

For example, the emergence of COVID-19 has impacted the ability of healthcare individuals to perform in-person diagnostics, consultations, health monitoring, etc. For that reason, telepresence has replaced clinic visits where many healthcare professionals have had to rely on patients using their own devices to collect health data and to convey health-related information [17]. Those methods of collecting and sharing data from BYOD technology to healthcare systems can be accomplished through direct data sharing or cloud-based solutions to reduce security risks. Subsequently, some healthcare/life sciences organizations have started to better understand the benefits of BYOD technology and have started adapting to the need for change

[18]. Thus, some of the reasons to embrace a BYOD model are (1) improved productivity or efficiency, (2) improved employee satisfaction, (3) easier to telecommute, (4) increased money savings, and (5) reduced workload on IT organizations [19]. Those positive results and perceptions can be combined with the apparent benefits associated with multiple life sciences use cases, such as the timely application of medical skills and the collection of archived data that can be used to improve outcomes. Specifically, through the lens of responsiveness, data streamed from individually supplied devices will enable in-field scientists, healthcare professionals, and researchers to assess current data states/conditions more easily while simultaneously transmitting that data securely to targeted institutions.

From BYOD to Meaningful Patient Outcomes

Harnessing new consumer data sources through BYODs has the potential to drive better health outcomes. Wearable devices for remote patient monitoring are becoming essential tools in clinical research and are rapidly gaining increased adoption in mainstream clinical practice. These technologies allow clinical researchers and providers to see what happens when patients go home and throughout a treatment regimen. Consumer-grade wearable devices enable clinical providers with a better view of patients' SDoH. The wearables track a wide range of factors associated with an individual's health status, such as activity levels, geolocation, heart rate, and changes in behavior indicative of treatment, compliance, effectiveness, and safety. Combined with computational and analytical technologies, BYODs are beginning to empower the next generation of clinical researchers through essential infrastructure that facilitates a wide spectrum of patient monitoring for indicators associated with treatment safety and effectiveness. In turn, these capabilities are poised to lead to new and better treatment modalities with reduced potential readmissions or complications and ultimately improving overall patient outcomes.

SDoH is also being addressed to connect patients with resources outside of typical healthcare resources, like appropriate shelter, clothing, food, and nutrition. In a value-based care world, those things matter more because they impact a patient's overall health. Focusing on overall health is rapidly becoming a more appealing model to health care than whatever the procedure or medication a patient might be taking. Therefore, technology can track and guide patients into changing behavior to promote overall health. Figure 5 shows the key SDoH associated with behavioral and socioeconomic factors known to contribute up to 80% of an individual's overall health status [20].

Risks and Challenges with IoT and BYOD

The use of BYOD and IoT technology within life sciences and the data binding elements such as blockchain and AI/ML (i.e., through the lens of distributed data models like federated learning) holds considerable promise in a world of diverse

Fig. 5 Key social determinants of health

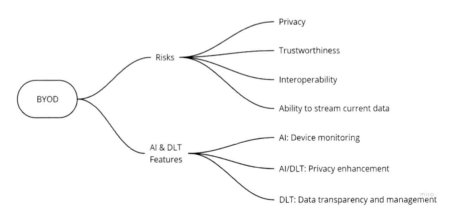

Fig. 6 Risks associated with BYODs and possible mitigations with DLT-AI features

technologies. However, as shown in Fig. 6, the promising evolution of technology integration is not without its own risks. Some of those risks are associated with data leakages or compromised data due to security issues. That point is not just specific to one emerging technology model, whereas previously discussed the unification across blockchain, AI/ML, BYOD, and IoT technology, the focus on data security is a central theme for any technology deployed. Additionally, it is important to recognize that data security must be addressed regardless of status in the lifecycle of technology deployed within life sciences (e.g., public health, basic research, preclinical research, clinical research, and clinical implementation).

Building on the need for data security across BYOD, IoT, AI/ML, and blockchain technologies, data must remain protected and secure. Options include a zero-trust solution, layers of multiple authentication methods, or restrict BYOD or IoT technologies. Regardless of how those options are considered, it is valuable to explore smart contracts. More complex smart contracts might be built around elements associated with government and regulatory frameworks. That point then leads down the path of sophisticated smart contracts that would utilize machine-learning and AI functionality/logic and can help reduce the injection of human errors and might ultimately compromise data security.

Summarizing the need for data security within any Life Sciences that seeks to implement a BYOD, IoT, AI/ML, or blockchain solution and the suggestion that smart contracts be implemented, there are both complex and simple ways to accomplish that goal. That said, it is suggested that technology integrators, technologists, and scientists approach smart contracts with the idea of using standard practices. Accordingly, to reduce the potential risks of data loss or compromise, it is also important that all relevant stakeholders must agree with any and all smart contracts and data security measures put in place.

The proliferation of laptops, tablets, phablets, and especially smartphones and wearables in health care settings is giving rise to the need to structure robust BYOD strategies that address policy, legal, and safety concerns while also satisfying functional and technical requirements that meet user demands. Ensuring that adequate measures are in place to protect human subject information is paramount to the success of any digital technology intended to support health and life sciences research. For example, clinical researchers involved in human subjects research need to ensure all research team members who can access identifiable human subject data are listed on a protocol approved by the respective Institutional Review Board and have completed specialized clinical research training about human research protections.

Table 1 describes an assessment of the potential advantages, disadvantages, and risks of using BYOD in life sciences. Since the device is potentially directly owned and managed by an individual, this can increase the risks that the device could be lost, misconfigured, or not updated.

However, this also can enable attractive benefits like the familiarity of using the device, cost savings of not owning possibly one-time-use devices, or independence. Technologies like AI and blockchain can enter the game to help mitigate some of these risks.

6.1.2 Challenges and Opportunities for Blockchain-AI-Enabled IoT/BYOD Platforms in Regenerative Medicine

Certain operational constraints and vulnerabilities in life sciences could be mitigated through robust implementations of integrated BYOD/IoT-blockchain-AI infrastructure. As another use case example, bioprinting of therapeutic tissue products is an emerging domain in regenerative medicines that offers significant promise to solve

Table 1 Assessment of BYODs in life sciences

Advantages	Disadvantages
• Using familiar devices	• Devices need maintenance and updating
• Institutional cost savings	• Ensuring all devices have the required apps
• Not having to purchase technologies that can rapidly become obsolete	• A BYOD device can be considered a very personal space with people unwilling or
• Familiarity with devices leads to enhanced learning	uncomfortable sharing their data
• Speeds up engagement and ability to achieve goals	• Support of unknown devices
• Allows individuals to be more independent during fieldwork	• Lack of time to set up BYOD device
• Liberates individuals to work independently	• Difficulty supporting a range of devices/apps
• Improves individual engagement	
• Learning is accelerated	
• Device management time savings	

Risks

- Device loss (loss of a device, lost IP, access compromised)
- Damage to device
- Screen visibility (unattended devices and exposed information)
- Employees' negative action (e.g., disgruntled employees)
- Setting up and maintaining privacy requirements
- Unauthorized access and/or threats to data integrity
- Lack of standardized data security
- Maintaining interoperability
- Maximizing full capabilities of mobile devices
- Increased risks for data loss through theft, intruders, and unintentional leaks
- Unpatched vulnerabilities in the operating systems
- End-user anonymity and leakage of private information
- Sharing devices between pairs and groups can be problematic
- Incompatibility/Incorrect software

current challenges with sourcing a variety of biological tissue materials to satisfy a wide range of clinical needs. Many regenerative medicine products offer the potential to mimic sophisticated tissue and grafts, including connecting with vasculature [21].

Tissue-engineered constructs produced using 3D bioprinting technology are progressively improving to simulate the complexity of tissue microenvironments. In the future, the bioprinting process may allow for organ and tissue manufacturing to meet the needs of organ shortages [22]. That said, bioprinting is a relatively new yet evolving technique predominantly used in regenerative medicine and tissue engineering. 3D bioprinting techniques combine the advantages of creating extracellular matrices like environments for cells and computer-aided tailoring of predetermined tissue shapes and structures [23, 24]. Table 2 provides considerations, challenges, and opportunities for the use of therapeutic tissue products.

Table 2 Opportunities for IoT-blockchain-AI in bioprinting of therapeutic tissue products

Key Considerations for Successful Clinical Translation of Tissues Products
1. Demonstrate scientific feasibility
2. Establish preclinical characterization
3. Verification and validation
4. Quality management systems
   a. Quality by design
   b. Quality by testing
5. Supply chain management and optimization
   a. Sourcing and distribution
     i. Acquisition and processing of raw materials
     ii. Storing, packaging, and distribution capacity
6. Assurance and compliance
   a. Achieving practical, scalable, consistent, and cost-effective output
   b. Auditability and transparency
   c. Trust and validation
     i. Test reports, certifications, conformities, expert adjudication, etc.

Challenges with Therapeutic Tissue Products
1. Lack of adequate characterization of therapeutic product
   a. Inadequate quality attributes
2. Proprietary test methods
   a. Lack of transparency
   b. Lack of control of manufacturing process
   c. Limited use of quality by design and process analytics
   d. Reliance on quality by testing manufacturing
   e. Need for each manufacturer to independently validate test methods

Opportunities in Therapeutic Tissue Products
1. Possibility to quickly translate laboratory discoveries into clinical trials
2. Provide better, innovative treatments and cures for indications without effective treatment
3. Adding process analytics and automation to quality by design
4. Adoption of alternative manufacturing models
   a. Centralize—conventional approach
   b. Distributed—considers non-traditional management systems

7 Strategic Planning Frameworks

7.1 Blockchain and AI to Mitigate Risks of IoT/BYOD

The key to incorporating this emerging digital technology is moderation. Leaders must identify specific areas of improvement and assess whether corresponding solutions align with a business, R&D, or clinical solution that addresses the needs of end-users. Other relevant considerations transact operational, policy, ethical, and

Table 3 Opportunities for blockchain and AI solutions in risk management of IoTs and BYODs in health and life sciences

Risks with IoT and BYOD	Use of Blockchain in Risk Mitigation	Use of AI in Risk Mitigation
Privacy	Validation	Automation
Security	Authentication	Governance
	Auditability	Engine
Identity	Traceability	Trend Recognition
Availability	Transparency	Compliance
Interoperability	Tamper resistance	Assurance
	Smart contracts	Surveillance
Quality	Sovereign Identity	Monitoring
Accessibility	Access control	Permissioning
Trust		Risk identification
		Risk predictability
Ownership		Dynamic consent
Safety		

scientific principles. Sometimes, selecting an IoT/BYOD strategy without caution can result in massive volumes of perplexing data and increased inefficiency.

IoTs and BYODs can enable a wide range of real-time insights and edge analytics, and the appropriate blockchain-AI-enabled infrastructure can help maximize their value by allowing for deeper analytics and insights to:

- Make better decisions faster through the translational research spectrum.
- Acquire real-world evidence about product safety and effectiveness.
- Recognize potential quality and manufacturing performance issues.
- Increase ability to forecast demand utilizing real-time supply chain data.

Utilizing blockchain and AI, incorporating BYOD solutions, though, creates a difficult balance of accessibility and cybersecurity. While devices are increasingly available, the number of cybersecurity breaches is increasing. Organizations must develop more robust security for electronic records and cloud-based systems. Mitigation solutions are provided in Table 3.

For instance, it is important to be confident about the identity of a BYOD device even though an organization does not manage it. Using a blockchain-based solution that enables a decentralized tamper-resistant record of information may enable a user to have a verifiable credential issued for his device that life sciences researchers can easily verify. Furthermore, AI/ML models could be developed to monitor the behavior of the BYOD device and increase the confidence that the device is operating as expected.

7.2 Blockchain-AI Platforms and Infrastructures

As explained earlier in the chapter, multiple types of devices could be used as BYOD, from small wearable devices to mobile phones and computers. They need to be connected to a network to exchange data they generate or receive to trigger an action. Many concerns need to be addressed to accept a BYOD device onto the network and the data or actions it may need to provide or execute. Similar concerns are expressed with the use of IoT devices, and there is ongoing research to explore how the use of blockchain and AI can help mitigate some of these concerns. BYOD devices require added scrutiny as they are not directly in the control of the organization interacting with them, and the quality of the data may not be guaranteed.

Organizations like IBM have been collaborating with their partners and customers to enable trust in IoT [25] using blockchain and AI that could be applied to BYOD. The device and its data could have pieces recorded on a blockchain network to increase trust in the provenance of the device and its data. AI models could then be applied to extra meaningful information and predict state change.

7.2.1   IoT-Enabled BYOD Models within Life Sciences Using Blockchain Data Exchange Methods

When thinking of technological solutions in terms of IoT (and clinical IoT) interoperability and the exchange of data within the framework of decentralized blockchains, the types of technologies that are the foundation of those solutions are often described as BYOD. Consequently, the life sciences industry must accept that users and organizations are not always using the same technological platforms. Thus, it is important to embrace a mindset with several different models for exchanging, analyzing, and storing data. Accordingly, the following section provides the reader with information that guides as the reader considers the technologies that can be used in both centralized and decentralized data models, specifically through the lens of using a blockchain data distribution framework across a variety of BYOD technologies.

Embracing an IoT-compatible BYOD solution can enhance the competitive edge of a life sciences organization, but if the risks are too high, then a CYOD is a suitable alternative [15]. However, BYOD models also have some positives associated with the perception that an organization valued the employee enough to buy them a mobile device. Ultimately, the combination of BYOD technology and a Mobile Device Management (MDM) solution should be built with the idea that those elements should replicate internally controlled technology and possibly move security measures to the next level. That said, the risks associated with embracing a BYOD model are the same as with any technology, either inside or outside of an organization. Some of those risks are as follows:

Fig. 7 BYOD simple
planning

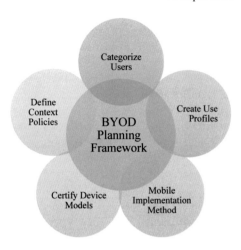

Risks with adopting a BYOD model:

- Device loss (loss of a device, lost IP, access compromised).
- Screen visibility (unattended devices and exposed information).
- Employee action (e.g., disgruntled employee).

When discussing risks associated with using an IoT-compatible BYOD model, it is important to explore options. Options such as having no mobile devices are within the realm of consideration. However, generally, the kinds of technologies and device management methods are the options that most organizations evaluate. Those options should answer basic questions around how an organization describes its desired level of openness and flexibility, specifically as those levels relate to the data stored on the devices and the devices' access (see Fig. 7).

When thinking of managing IoT compatible BYOD technology, there are also a number of management systems that should be considered. Once a device is deployed, many recurring elements still have to be managed [16]. These elements include device support/sustainability, patching and upgrading devices, device communications, data sharing, and storage. So, with the need to consider BYOD management, the systems that require attention are (1) MDM, (2) Mobile Application Management, (3) Mobile Email Management, and (4) Mobile Content Management can help maintain control and reduce risks.

With the risks, options, and management systems in mind, and now thinking of how a distributed ledger system or blockchain solution best fits an organization's needs, there are a few questions that all organizations should first ask themselves. Those questions are listed below, which should be answered before the first device is purchased and deployed.

Questions to ask:

• Can strong passwords be enforced?

• What strategies manage BYOD devices when not within a protected environment (closed-loop versus teleworking)? This may occur when employees work in the field (versus their office within a controlled environment)

• Can devices that have been jailbroken be stopped from connecting with the network?

• Can there be better management of data stored in a device by deploying a BYOD solution? Specifically, people will store information on their personal devices regardless of an organization's attempt to control off-site devices. So, the question pertains to, if by embracing an enterprise BYOD solution, can there be better data protection?

• How should data downloads and uploads be managed from organizational storage locations?

• How should patches, updates, new app deployments, and remote wiping be managed?

• How should malware risks be managed (e.g., email, internet, and WiFi)? Malware examples include trojans, cross-scripting (code injection), unsecured WiFi systems (e.g., fake DNS), key-loggers, and viruses

• Should a zero-trust policy be put in place? Note: Recursive validation should be implemented to ensure that the device or person is still the same person or thing authenticated/validated

• How devices be registered, provisioned, operated, and eventually de-provisioned?

As determined by the focus of each question above, the reader will recognize that organizations are moving from managing a physical device into areas associated with managing how data are stored, analyzed, and transformed. Of course, organizations are still looking at the various platforms from an IoT-compatible BYOD model that a life sciences entity might embrace. This also includes thinking about how trust is established. Specifically, trust must be considered when thinking about enabling or restricting a technology's access as well as how user and device activity is used to determine trust [26]. For example, activities that might not be normal for a particular user or device might set off red flags, and an organization's system should take appropriate actions. However, that is just an example of the many things that should be considered before allowing users to exchange data within an IoT-compatible BYOD model. That said, the reader should consider the following BYOD model and answer the above questions as they define and design their BYOD system using blockchain as their method of secure data exchange.

Possible Traps and Pitfalls

When designing and building a solution that involves blockchain, AI, IoT, and BYOD technology, additional pitfalls or potential traps require thoughtfulness and a deeper understanding of designing and constructing a sustainable solution. Three such traps are:

1. **Thinking tactically versus strategically**—where a series of knee-jerk reactions will spell disaster. In other words, define inputs and outputs and what is in and out of scope before diving into building a solution [27].
2. **Lack of awareness and engagement of relevant stakeholders**—specifically, not including all relevant stakeholders in technology integration phases is the shortest path to long lead times, cost overruns, and failure [28, 29, 30, 31, 32, 33, 34].
3. **Not considering all phases of technology integration**: Exploration, assessment, decision-making, design, development, adoption, deployment, testing, sustainability, and end-of-life—lack of awareness of the phases for successfully integrating a solution will result in missed steps and milestones that can also result in cost overruns and potential failure to deploy as well as support the solution [35, 36].

Note: As previously mentioned during the discussion of technology integration phases, it is necessary to emphasize the importance of 'adoption,' 'sustainability,' and 'end-of-life.' Specifically, adoption occurs when the solution integrates with existing systems/solutions. During the sustainability phase, organizations must measure solution results to ensure that the solution meets objectives and continues to satisfy customer needs. Lastly, the end-of-life phase is often overlooked and underfunded, where the entire lifecycle must be understood from start to finish.

8   Future Directions

8.1   Human as a Platform

When considering how technology such as blockchain, AI, IoT, and BYOD might be integrated into the life sciences industry, do not search for a platform that will be the interconnector that binds humans to technologies, but instead embrace the concept that the human is the platform, data pathway, data collector, transmitter, storage medium, and potentially power source for those technologies [37]. That point emphasizes that our vision of the future cannot be limited to what is known today. Instead, it is necessary to consider where technology might naturally evolve, regardless of present-day limitations. Accordingly, the human body's potential is still emerging as organizations learn to store image data of bacterial DNA or establish sensing across neural pathways, as can be done today when transmitting patient alert signals from medical devices to the auditory cortex.

8.2 Thinking Beyond the Adoption of Technology

The need exists to step back from a myopic view of technology adoption or deployment and examine the blockchain, AI/ML, IoT, and BYOD technologies discussed through the lens of all phases of technology integration. Specifically, the phases that must be examined are technology exploration, assessment, decision-making, adoption, design, development, deployment, testing/validating, sustainability, and end-of-life. That point is supported by the proposal that uses of theoretical models such as Integrated Acceptance and Sustainability Assessment Methodology. When used as a guide for qualitative research, technologists have a great potential for successfully integrating technology such as BYOD into the life sciences industry [38, 39].

9 Conclusions

The integration of IoT and BYOD with blockchain-AI-enabled infrastructures is poised to set a new paradigm for life sciences research by providing new tools for quality risk management across all phases of the product development spectrum, from early conception, design, and prototyping to clinical evidence generation, regulatory review, and through commercialization. Enhancing the timeliness in which scientific evidence becomes available to identify, assess, control, and monitor the risks that could reduce the safety, efficacy, or performance of new life sciences products offers a unique advantage with the potential to accelerate the scientific discovery process and the translation of innovations in a highly regulated environment. The ability to conduct more comprehensive and efficient preclinical and clinical studies by leveraging intelligent digital twins of products, services, patients, or end-users offers the possibility to reduce time-to-market and the overall cost of new therapies and clinical procedures while not compromising product safety, effectiveness, or quality. The reliance on model organisms and animal models could also be reduced via computational platforms. These platforms help bridge gaps in translational sciences between preclinical endpoints and long-term clinical performance by increasing transparency, promoting multi-domain collaborations, and improving the overall quality of life sciences research-related activities through enhanced organizational and clinical workflows. On the other hand, it is also important to recognize that these emerging digital technologies have their own inherent risks to be assessed, evaluated, and mitigated. Several use cases were highlighted to illustrate frameworks to support strategic planning for integrating IoT and BYOD technologies with blockchain-AI-enabled platforms, as well as considerations for data governance and the adoption of FAIR principles to ensure the value of data is maximized across the life sciences enterprise.

References

1. Khalil AA, Franco J, Parvez I, Uluagac S, Rahman MA (2021) A literature review on blockchain-enabled security and operation of cyber-physical systems. (Preprint arXiv:2107. 07916). Florida International University. http://arxiv.org/abs/2107.07916
2. Alirezabeigi S, Masschelein J, Decuypere M (2020) The agencement of taskification: On new forms of reading and writing in BYOD schools. Educ Philos Theory 52(14):1514–1525. https://doi.org/10.1080/00131857.2020.1716335
3. Alevizos L, Ta VT, Eiza MH (2021) Augmenting zero trust architecture to endpoints using blockchain: a state-of-the-art review. Secur Priv, e191. https://doi.org/10.1002/spy2.191
4. Wager KA, Lee FW, Glaser JP (2017) Health care information systems: a practical approach for health care management, 4th edn. Wiley. https://www.wiley.com/en-us/exportProduct/pdf/9781119337188
5. Corral-Acero J, Margara F, Marciniak M, Rodero C, Loncaric F, Feng Y, Fernandes J, Bukhari H, Wajdan A, Martinez MV, Santos MS, Shamohammdi M, Luo H, Westphal P, Leeson P, DiAchille P, Gurev V, Mayr M, … Lamata P (2020) The 'Digital Twin' to enable the vision of precision cardiology. Eur Heart J, 41(48):4556–4564.https://doi.org/10.1093/eurheartj/eha a159
6. Kamel Boulos MN, Zhang P (2021) Digital twins: From personalised medicine to precision public health. J Pers Med 11(8):745. https://doi.org/10.3390/jpm11080745
7. DePamphilis D (2019) Financial modeling basics. In: Mergers, acquisitions, and other restructuring activities: an integrated approach to process, tools, cases, and solutions, 10th edn. Academic Press, pp 233–261. https://doi.org/10.1016/B978-0-12-815075-7.00009-7
8. Ibrahim N, Gillette N, Patel H, Peiris V (2020) Regulatory science, and how device regulation will shape our future. Pediatr Cardiol 41(41):469–474. https://doi.org/10.1007/s00246-020-02296-0
9. Mittal B (2017) Process scale-up, tech-transfer, and optimization. In: Mittal B, Levin M (eds) How to develop robust solid oral dosage forms: From conception to post-approval. Elsevier, pp 137–153. https://doi.org/10.1016/B978-0-12-804731-6.00007-8
10. Robertson J (2013) Risk Management. In: Siegel JA, Saukko PJ, Houck MM (eds) Encyclopedia of forensic sciences. Elsevier. https://doi.org/10.1016/B978-0-12-382165-2.00238-5
11. Farmer D (2021, July 22) 6 key components of a successful data strategy. Search data management. Retrieved December 5, 2021 from https://searchdatamanagement.techtarget.com/tip/6-key-components-of-a-successful-data-strategy
12. Wilkinson M, Dumontier M, Aalbersberg IJJ, Appleton G, Axton M, Baak A, Blomberg N, Boiten J-W, da Silva Santos LB, Bourne PE, Bouwman J, Brookes AJ, Clark T, Crosas M, Dillo I, Dumon O, Edmunds S, Evelo CT, Finkers R, ... Mons B (2016) The FAIR guiding principles for scientific data management and stewardship. Sci Data, 3:160018. https://doi.org/10.1038/sdata.2016.18
13. Huang Z, Shen Y, Li J, Fey M, Brecher C (2021) A survey on AI-driven digital twins in Industry 4.0: Smart manufacturing and advanced robotics. Sensors (Basel) 21(19):6340. https://doi.org/10.3390/s21196340
14. Akande AO, Tran VN (2021) Predicting security program effectiveness in bring-your-own-device deployment in organizations. In: Mori P, Lenzini G, Furnell S (eds) Proceedings of the 7th international conference on information systems security and privacy—ICISSP. SCITEPRESS—Science and Technology Publications, pp 55–65. https://doi.org/10.5220/001 0195800550065

15. Barlette Y, Jaouen A, Baillette P (2021) Bring Your Own Device (BYOD) as reversed IT adoption: Insights into managers' coping strategies. Int J Inf Manage 56:102212. https://doi.org/10.1016/j.ijinfomgt.2020.102212

16. Wani TA, Mendoza A, Gray K (2020) Hospital bring-your-own-device security challenges and solutions: Systematic review of gray literature. JMIR Mhealth Uhealth 8(6):e18175. https://doi.org/10.2196/18175

17. Alashhab ZR, Anbar M, Singh MM, Leau YB, Al-Sai ZA, Alhayja'a, S. A. (2021) Impact of coronavirus pandemic crisis on technologies and cloud computing applications. J Electron Sci Technol 19(1):100059. https://doi.org/10.1016/j.jnlest.2020.100059

18. Chate R, Dhote T (2021) Market estimation of cloud migration services and its security measures. Int J Mod Agric 10(2):162–172. https://www.modern-journals.com/index.php/ijma/article/view/737/

19. Wani TA, Mendoza A, Gray K (2021) Bring-your-own-device usage trends in Australian hospitals–A national survey. In: Merolli M, Bain C, Schaper LK (eds) Healthier lives, digitally enabled. IOS Press, pp 1–6. https://doi.org/10.3233/SHTI210002

20. The Lancet Digital Health (2021) Digital technologies: a new determinant of health. Lancet Digital Health 3(11):e684. https://doi.org/10.1016/S2589-7500(21)00238-7

21. Mao A, Mooney D (2015) Regenerative medicine: current therapies and future directions. Proc Natl Acad Sci USA 112(47):14452–14459. https://doi.org/10.1073/pnas.1508520112

22. Ji S, Guvendiren M (2021) Complex 3D bioprinting methods. APL Bioeng 5(1):011508. https://doi.org/10.1063/5.0034901

23. Jamee R, Araf Y, Naser IB, Promon SK (2021) The promising rise of bioprinting in revolutionalizing medical science: Advances and possibilities. Regen Ther 18(18):133–145. https://doi.org/10.1016/j.reth.2021.05.006

24. Zhang B, Luo Y, Ma L (2018) 3D bioprinting: An emerging technology full of opportunities and challenges. Biodes Manuf 1(1):2–13. https://doi.org/10.1007/s42242-018-0004-3

25. Cuomo, J. (2020, August 5). *How blockchain adds trust to AI and IoT IBM supply chain and blockchain blog.* IBM. Retrieved December 5, 2021, from https://www.ibm.com/blogs/blockchain/2020/08/how-blockchain-adds-trust-to-ai-and-iot/

26. Lian J-W (2020) Understanding cloud-based BYOD information security protection behaviour in smart business: in perspective of perceived value. Enterp Inf Syst 15(9):1216–1237. https://doi.org/10.1080/17517575.2020.1791966

27. Brønn PS (2021) Strategic communication requires strategic thinking. In: Balonas S, Ruão T, Carrillo M-V (eds) Strategic communication in context: theoretical debates and applied research. UMinho Editora/Centro de Estudos de Comunicação e Sociedade, pp 23–43. https://doi.org/10.21814/uminho.ed.46.2

28. Cook EJ, Randhawa G, Sharp C, Ali N, Guppy A, Barton G, Bateman A, Crawford-White J (2016) Exploring the factors that influence the decision to adopt and engage with an integrated assistive telehealth and telecare service in Cambridgeshire, UK: A nested qualitative study of patient 'users' and 'non-users.' BMC Health Serv Res 16(1):137. https://doi.org/10.1186/s12913-016-1379-5

29. Daghfous A, Belkhodja O, Ahmad N (2018) Understanding and managing knowledge transfer for customers in IT adoption. Inf Technol People 31(2):428–454. https://doi.org/10.1108/ITP-10-2016-0222

30. Ingebrigtsen T, Georgiou A, Clay-Williams R, Magrabi F, Hordern A, Prgomet M, Li J, Westbrook J, Braithwaite J (2014) The impact of clinical leadership on health information technology adoption: Systematic review. Int J Med Inform 83(6):393–405. https://doi.org/10.1016/j.ijmedinf.2014.02.005

31. Liebe JD, Hüsers J, Hübner U (2015) Investigating the roots of successful IT adoption processes - An empirical study exploring the shared awareness-knowledge of Directors of Nursing and Chief Information Officers. BMC Med Inform Decis Mak 16:10. https://doi.org/10.1186/s12911-016-0244-0

32. Petersen C (2018) Patient informaticians: Turning patient voice into patient action. JAMIA Open 1(2):130–135. https://doi.org/10.1093/jamiaopen/ooy014

33. van Oorschot JAWH, Hofman E, Halman JI (2018) A bibliometric review of the innovation adoption literature. Technol Forecast Soc Change 134:1–21. https://doi.org/10.1016/j.techfore. 2018.04.032

34. Varsi C (2016) Implementation of eHealth patient–provider communication tools into routine practice: Facilitators and barriers from the perspectives of patients, middle managers and health care providers. (C. M. Ruland, D. Gammon, & M. Ekstedt (eds.)) [Doctoral dissertation, University of Oslo]. https://www.duo.uio.no/bitstream/handle/10852/53265/1/Cecilie-Varsi-2016-PhD.pdf

35. Li W, Long R, Chen H, Geng J (2017) A review of factors influencing consumer intentions to adopt battery electric vehicles. Renew Sust Energ Rev 78:318–328. https://doi.org/10.1016/j. rser.2017.04.076

36. Rezvani Z, Jansson J, Bodin J (2015) Advances in consumer electric vehicle adoption research: A review and research agenda. Transp Res Part D: Trans Environ 34:122–136. https://doi.org/ 10.1016/j.trd.2014.10.010

37. Harding WC, Petroff N, Partridge B (2021) Wearable technology and robotics for a mobile world. In: Douville S (ed) Mobile medicine: Overcoming people, culture, and governance (pp. 13–37). Productivity Press. https://doi.org/10.4324/9781003220473-3

38. Aizstrauta D, Ginters E (2017) Using market data of technologies to build a dynamic integrated acceptance and sustainability assessment model. Procedia Comput. Sci. 104:501–508. https:// doi.org/10.1016/j.procs.2017.01.165

39. Ginters E, Mezitis M, Aizstrauta D (2018, December) Sustainability simulation and assessment of bicycle network design and maintenance environment. In: 2018 international conference on intelligent and innovative computing applications (ICONIC). IEEE, pp 1–6. https://doi.org/10. 1109/ICONIC.2018.8601225

Orlando Lopez, Ph.D. is Director of the Dental Materials and Biomaterials Program and Coordinator of the Small Business (SBIR/STTR) Program at the National Institute of Dental and Craniofacial Research (NIDCR) of the National Institutes of Health (NIH). He leads efforts towards strengthening NIDCR's national agenda supporting basic and translational research on biomaterials, oral biodevices and digital health technologies intended to address clinical needs in oral health and overall health. He represents NIDCR on several important initiatives within the NIH and across several federal agencies, including: NIH Small Business Education and Entrepreneurial Development (SEED) Working Groups, Rapid Acceleration of Diagnostics (RADx)—Radical for COVID-19 Initiative, Interagency research program on development of clinically relevant standard methods for dental biomaterials and oral biosensors at National Institute of Standards and Technology (NIST) and ongoing collaborations with the Food and Drug Administration (FDA). Orlando is major contributor to multiple professional working groups and industry standards committees, including NIH Interagency Modeling and Analysis Group, DoD's Working Group on Computational Modeling, IEEE 2933 Standard on Clinical IoT Interoperability with TIPPSS, IEEE Healthcare Blockchain and AI, American Society of Mechanical Engineers (ASME) Biomedical Engineering Technical Committee and Thermal Medicine Standard, American Dental Association (ADA) Standards on Dental Products, American Council for Technology and Industry Advisory Council (ACT-IAC) Blockchain and AI Working Groups. Prior to NIDCR, he served as lead regulatory reviewer at the Center for Devices and Radiological Health of the Food and Drug Administration (CDRH-FDA) where he led regulatory approvals of medical devices in the areas of diagnostic medical imaging, cardiovascular and digital health technologies. He is an alumnus of the prestigious FDA Commissioner's Fellowship and the NIH Postdoctoral IRTA Fellowship.

Frederic de Vaulx is the Vice President of Prometheus Computing, a small business supporting Federal Agencies with custom software development and emerging technology strategy, where he leads the development of the company's vision, innovation and strategy. He is also a Senior Software Engineer and Program Manager overseeing the design, architecture and development of custom software applications for Prometheus' federal and commercial clients. In addition, Frederic is The CTO at Value Technology Foundation, an emerging technology 501(3)(c) Think Tank. Frederic has more than 15 years of experience in software engineering and emerging technology innovation. He is a member of ACT-IAC, where he is the industry vice-chair of the Emerging Technology Community of Interest and co-chairs the blockchain and AI working group that published primers and the playbooks to help Federal blockchain and AI adoption. Frederic is a member of the IEEE blockchain initiative and the IEEE P2145 Blockchain Governance Standards WG, where he leads a project on DLT governance design patterns.

William Harding, Ph.D. is a Distinguished Technical Fellow with 41 years of industry experience, including 24+ years at Medtronic in Advanced Technologies & Data Science. William has a Bachelor's degree in Computer Science emphasizing Electrical Engineering, a Master's degree in Information Systems, and a Ph.D. emphasizing technology integration. William has had a very successful career in missile launch/tracking systems, drug interdiction, rechargeable cell manufacturing, and medical device manufacturing. William has initiated and championed innovative medical device manufacturing solutions and patented medical product designs that continue to have major impacts across business units in the areas of process improvement, manufacturing automation, product development/traceability, and FDA validation. With more than 90 technical conferences, symposium presentations, lectures, seminars, and workshops under his belt, William continues to establish the highest level of standards through his professionalism, ethical behavior, mentoring, and guidance both internal and external to Medtronic.

A Blockchain-Empowered Federated Learning System and the Promising Use in Drug Discovery

Xueping Liang, Eranga Bandara, Juan Zhao, and Sachin Shetty

Abstract Federated learning is a collaborative and distributed machine learning model that addresses the privacy issues in centralized machine learning models. It emerges as a promising technique that addresses the data sharing concerns for data-private multi-institutional collaborations. However, most existing federated learning systems deal with centralized coordinators and are vulnerable to attacks and privacy breaches. We propose a blockchain-empowered coordinator-less decentralized, federated learning platform "Rahasak-ML" to solve issues in centralized coordinator-based federated learning systems by providing better transparency and trust. It uses an incremental learning approach to train the model by multiple peers in the blockchain network. Rahasak-ML is integrated into the Rahasak blockchain as its data analytics and machine learning platform. Each peer in the blockchain can establish supervised or unsupervised machine learning models with the existing data on its own off-chain storage. Once a peer generates a model, it can be incrementally/continuously trained and aggregated by other peers through the blockchain using the federated learning approach without requiring a centralized coordinator. The model parameters sharing, local model generation, incremental model training, and model sharing functions are implemented in the Rahasak-ML platform. We discussed the promise of Rahasak-ML machine learning in medicine.

Keywords Federated learning · Blockchain · Medicine · Drug discovery · Big data

---

X. Liang (✉)
Department of Information Systems and Supply Chain Management, University of North Carolina at Greensboro, 488 Bryan Building, Greensboro, NC 27402, USA
e-mail: x_liang@uncg.edu

E. Bandara · S. Shetty
Virginia Modeling, Analysis, and Simulation Center, Old Dominion University, Norfolk, VA, USA
e-mail: cmedawer@odu.edu

S. Shetty
e-mail: sshetty@odu.edu

J. Zhao
Department of Biomedical Informatics, Vanderbilt University Medical Center, Nashville, Tennessee, USA
e-mail: juan.zhao@vumc.org

W. Charles (ed.), *Blockchain in Life Sciences*, Blockchain Technologies,
https://doi.org/10.1007/978-981-19-2976-2_6

1 Introduction

Federated learning is a new technique for training machine learning models across decentralized participants without accessing any party's private data [1, 2, 3]. It emerges as a promising paradigm for data-private multi-institutional collaborations by distributing the model training to the data owners and aggregating their results, solving the concerns of sharing data [4]. In a federated learning system, the central server (centralized coordinator) coordinates the learning process and aggregates the parameters from local machine learning models trained on each participant's data [5]. Although such a design minimizes the risk of privacy leakage, the centralized coordinator is vulnerable to attacks and privacy breaches, becoming the single point of failure and trust issues.

While blockchain is a technology that offers assurances of reliability and usage transparency in decentralized settings, researchers started to investigate the combinations of the two promising technologies [6, 7]. In this study, we took advantage of blockchain and federated learning and proposed a platform called Rahasak-ML [8]. Rather than using centralized coordinators to aggregate and learn the global model, the Rahasak-ML used an incremental learning technique [9, 10] to continuously train the models by multiple peers in the blockchain network. Each peer in the blockchain manages its local storage and establishes local models [11]. Once a peer generates a model, it can be incrementally trained and aggregated by other peers through the blockchain by using the federated learning approach. Rahasak-ML stores information (e.g., participating clients who generate and aggregate local models, generation times, etc.) into the blockchain ledger that all participating parties can view. It provides a way to audit the system. All actions performed on the model are entirely traceable by each user giving a clear history of all operations and incremental versions that existed. This system adds more transparency to the federated learning system by providing a traceable record of the model development, potentially alerting to adversarial machine learning attempts or fraudulent actions. Rahasak-ML makes the following contributions:

- Integrates federated learning with blockchain to enable model sharing and aggregations without having centralized authority, increasing the transparency, trust, and provenance of the model generation;
- Adds the ability to audit the federated learning system by storing task details (e.g., who generates local models and aggregates them, model generation times, etc.) in the blockchain;
- Offers different functions in the platform that are implemented as independent services (microservices) that are easy to scale and deploy; and
- Introduces a way to integrate the models in smart contracts to predict the output of real-time data.

The chapter is organized as follows. In Sect. 2, we briefly introduce federated learning, blockchain, and the role of these two technologies in drug discovery. In

Sect. 3, we introduce the architecture of federated learning in the Rahasak-ML platform. In Sect. 4, we further explain the training process and the implementation in a medical use case and offer insights into related work. Finally, in Sect. 5, we discuss the future directions and open questions.

2 Overview of Federated Learning and Blockchain

2.1 Federated Learning

Machine learning represents a set of methods that can automatically uncover patterns in data and then use detected patterns to predict future data. Machine learning models show promise in aiding decision-making in healthcare [12, 13] and finance [14]. However, a large, diverse labeled dataset is the key to making a supervised machine learning model broadly effective. Collaborative learning is an efficient way to increase the data size and diversity, via multi-institutional data sharing for the training of a single model [4]. The current approach to achieving collaborative learning requires sharing the data with a third party to train a global model, such as using data repositories for different purposes (*Fiscal Service Data Registry*, [15]). However, this centralized approach presents many issues, such as high costs for data transmission and storage, security and privacy at high risk, lack of auditing, data ownership, and restrictions of data sharing, e.g., the Health Insurance Portability and Accountability Act (HIPAA) regulations in healthcare [16].

To address these security and privacy issues, a decentralized machine learning approach, i.e., federated learning [17, 18], has been proposed to build a shared machine learning model without storing or having access to any party's private data. In federated learning, the central server coordinates the learning process and aggregates the information from multiple participants (i.e., referred to here as "parties") in a decentralized manner while keeping each participant's raw data private. Each party downloads the global model parameters from the central server at each iteration, locally trains it with their private/local dataset, and sends each of their local model parameters to the central server for aggregation. Then, the central server gathers all the local model parameters, aggregates them, and updates the global model parameters for the next iteration. This learning process continues until pre-defined termination criteria are met. For example, if the maximum number of iterations is reached, or if the model accuracy is greater than a threshold, the learning process is finished and will exit automatically.

2.2 Barriers and Challenges in Drug Discovery

Drug discovery involves identifying potential new medicines, which involves and requires the knowledge of a wide range of scientific disciplines, such as biology, chemistry, and pharmacology. Developing a new drug is a complex, lengthy, and costly process, entrenched with a high risk of uncertainty that a drug will succeed. The drug development pipeline included multiple stages, from identifying targeted therapeutic agents to clinical trial designs, including Phases I, II, and III. Each stage is critical but faces challenges, such as insufficient knowledge about the underlying mechanisms of disease, the heterogeneity of patients who have diverse clinical phenotyping and endotyping, a lack of targets and biomarkers, small or biased samples in clinical trials, and regulatory challenges [19]. These hurdles create barriers to the development of the drugs, leading to increased costs and time, thus increasing the risk of failure. To minimize these challenges, researchers moved toward computational approaches to accelerate pipeline, such as using high-throughput virtual screening and molecular docking to reduce the number of compounds that need to be screened experimentally [20]. However, these approaches have inaccuracy and inefficiency problems. Therefore, new methods and computing technologies to automate analytical model building for pharmaceuticals are needed and could transform drug discovery.

Today, the advances in high-throughput approaches to biology and disease present opportunities to pharmaceutical research and industry [21]. For example, multi-omics ranging from genome, proteome, transcriptome, metabolome, and epigenome are generated at unprecedented speed, improving the capabilities of systematically measuring and mining biological information. In addition, widely adopted electronic health records (EHR) and smart technologies capture detailed phenotypic patterns, allowing researchers to monitor patient outcomes and study medication treatments. The booming of such "big data," including omics, images, clinical characteristics, social/environmental information, and literature, has driven much of researchers' interest in harnessing machine learning to analyze and uncover novel findings and hidden patterns from the massive data [22, 23, 24].

Machine learning and deep learning are fundamental branches of artificial intelligence (AI), which refer to computer systems' ability to learn from input or past data. AI has achieved successful applications in many domains, such as imaging detection and natural language processing. Recently, AI algorithms have been increasingly being applied in all stages of drug discovery, including screening chemical compounds, identifying novel targets [25], examining target–disease associations [26], improving the small-molecule compound design and optimization, studying disease mechanisms [27], evaluating drug toxicity and physicochemical properties [28], predicting and monitoring the drug response [29], and identifying new indications for an existing drug, known as drug repositioning. Moreover, researchers utilized machine learning models to optimize the clinical trials, such as estimating the risks of clinical trials more accurately [30] and improving the patient pre-screening process, as well as approaches to feasibility, site selection, and trial selection [31].

From a machine learning viewpoint, it is desirable to have large and diverse data to inform model training, but access to data remains a challenge in drug discovery. Several public databases contain millions of biological assay results, such as ChEMBL [32] and PubChem [33], which can provide input for machine learning models to retrieve training models and then predict biological activities or physical properties for drug-like molecules. However, the data only represents a small fraction of what has been measured, which might bias the machine learning models and affect the model reliability and reproducibility [34]. Furthermore, many larger datasets are proprietary to pharmaceutical companies or publishers and are not publicly and freely available. To overcome the barriers, researchers seek federated learning to solve data acquisition and data bias problems faced by AI drug discovery by keeping confidentiality and customizing models for users [35].

Federated learning is a new machine learning paradigm where multiple sites collaboratively learn a shared machine learning model while keeping all the training data on a single site [2]. Developing federated health AI technologies are essential and highly demanding in medicine [13]. Examples include the European Union Innovative Medicines Initiative's (https://www.imi.europa.eu/) projects for privacy-preserving federated machine learning. Chang et al. explored data-private collaborative learning methods for medical models for image classification [36]. Xiong et al. [37] proposed using a federated learning work in predicting drug-related properties. The architecture of federated learning is that each participating pharma company (peer) will locally train the model without sharing the training data. Each peer only encrypts and uploads the model updates, and a coordinator server aggregates all the updates from the local client and broadcasts the latest shared global model to them. Thus, individual pharma companies will be able to fine-tune the machine learning model and effectively tailor it to their specific field of inquiry, with the individual research data remaining confidential.

2.3 Challenges in Federated Learning

While the federated learning process has significant improvements to minimize the risk of privacy leakage by avoiding storing raw datasets to a third party, it still presents some major vulnerability issues in the model architecture and the training process.

- First, the central server for coordinating a shared and trained global model presents the single point of failure and trust issues. A malicious behavior or malfunction from the central server could bring inaccurate global model parameters updates, which would misrepresent the local model parameters update sent by the parties. Therefore, decentralization of the entire federated learning process was necessary.
- Second, during the learning process, malicious parties could send manipulated local model parameters to the central server, affecting the global model parameters. If such malicious local parameters are not detected or removed before aggregation, they will compromise the global model and lower the overall model accuracy [1, 38]. Some studies [39] have proposed approaches to verify model parameters, but they mainly rely on the data sample size and the computation time, which could be easily altered by malicious participants to avoid detection.

- In addition, these studies do not address the quality of the data sample that would affect the accuracy and the convergence analysis of the federated learning process. A more difficult malicious behavior, colluding attack, has shown the vulnerabilities of existing defenses based on Sybil [40]. Thus, it is essential to note that verifiable local model parameters update is important for the accuracy of the global model parameters.

2.4 Blockchain Benefit for Federated Learning

Blockchain provides a shared digital ledger that records data in a public or private peer-to-peer network. It guarantees a decentralized trust system without involving trusted third parties. Multiple partners (nodes) can exist in the blockchain network, and each partner (node) has a copy of the data being maintained [41]. The data on the blockchain are organized into blocks. A block contains a set of records (transactions). Each block is linked to its previous block by containing the previous block's hash in its header. If someone was to tamper with the contents of one block, then all blocks in the blockchain following that block would be invalidated.

Depending on the type of access and from where the nodes that support the blockchain are selected, there are two primary types of blockchains: permissionless and permissioned. Permissionless blockchains deal with entirely untrusted/byzantine parties; examples are Bitcoin, Ethereum, and Rapidchain. Permissioned blockchains deal with trusted/known parties; examples are BigchainDB, Hyperledger, and HbasechainDB. Many blockchains, such as Bitcoin, are used for cryptocurrencies. For example, Ethereum and Hyperledger support different transaction storage models related to other business or e-commerce activities. Recently, blockchain has quickly been applied to other areas, including the healthcare and drug industry [42, 43]. For example, studies have integrated blockchain with EHRs, to allow the different stakeholders to manage EHR transparently while guaranteeing fairness and usage (records access) consent [44].

To address the challenges of federated learning, we propose integrating blockchain with federate learning to replace the centralized coordinator. The blockchain network can be deployed among different peers, and the peers can train machine learning modes with the data on their own local storages (e.g., off-chain storage). Then the local models generated by different peers can be aggregated/averaged into a global model using the federated learning approach without using a centralized coordinator. In blockchain-enabled federated learning systems, the model parameter sharing, local model generation, incremental model training, and model sharing functions can be implemented with smart contracts. All federated learning tasks happening in the system (e.g., generate local models and aggregate them) and stored in the blockchain ledger are viewed by all participating parties. It provides a means to audit the system and adds more transparency to the federated learning process. Once local models are generated, these models can be integrated into blockchain smart contracts (e.g., a program that directs client requests to the blockchain) to predict real-time data.

This system adds more transparency to the federated learning system by providing a traceable history of the model development, potentially alerting to adversarial machine learning attempts or fraudulent actions.

2.5 The Benefits of Blockchain-Empowered Federated Learning for Drug Discovery

The blockchain-enabled federated learning enhanced such infrastructure by decentralizing the architecture further and making the training process and model sharing more transparent and traceable. As a result, hospitals, institutions, and drug companies can achieve an accurate and generalizable model; more sites contribute their local insights while remaining in full control and possession of their data. This approach allows complete traceability of data access, limiting the risk of misuse by third parties. There is a consortium of pharmaceutical, technology, and academic partners, the Machine Learning Ledger Orchestration for Drug Discovery (MELLODDY, https://www.melloddy.eu/), that uses deep learning methods on the chemical libraries of ten pharma companies to create a modeling platform that can more quickly and accurately predict promising compounds for development, all without sacrificing the data privacy of the participating companies. Specifically, the benefits of a blockchain-empowered federated learning system are as follows:

- Entails training algorithms across decentralized sites or devices holding data samples without exchanging those samples.
- Small pharmaceutical companies and research institutions would achieve accurate, less biased models by gaining insights from other sites containing diverse data.
- Provides a platform with more transparency, trust, and provenance for model training and sharing.
- Provides the ability to audit the system and make sure local data and models are traceable. For example, the task information related to who generates models, aggregate parameters, and model generation time would be recorded in the blockchain.
- Offers flexibility with connecting more participating sites and devices.
- Provides the ability to process real-time data.

3 The Rahasak-ML Platform

3.1 Overview

The Rahasak-ML platform integrates federated learning with blockchain to enable model sharing and model training without having a centralized coordinator, which keeps the data private [45, 46]. The proposed platform has been implemented on

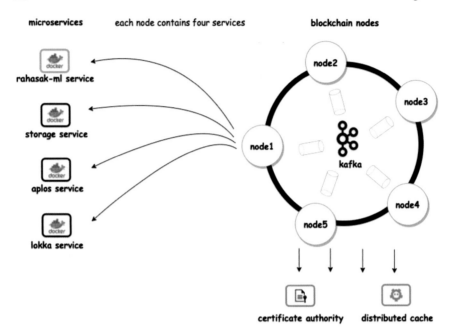

Fig. 1 Rahasak-ML platform's microservices-based architecture. Each blockchain node contains four services: Rahasak-ML service, Storage service, Aplos service, and Lokka service

top of the Rahasak blockchain [5], a highly scalable blockchain system for big data. The architecture of the Rahasak-ML federated learning environment is discussed in Fig. 1.

Its proposed platform has been designed with microservice-based distributed system architecture [47]. In Rahasak-ML, all the functionalities are implemented as independent microservices. These services are Dockerized [48] and available for deployment using Kubernetes [49]. The following are the main services/components of the Rahasak-ML platform:

- Storage service: Apache Cassandra-based block, transaction, and asset storage service [50].
- Aplos service: smart contract service implemented using Scala functional programming language and Akka actors [51].
- Lokka service: block creating service implemented using Scala and Akka streams [52].
- Distributed message broker: Apache Kafka-based distributed publisher/subscriber service used as consensus and message broker platform in the blockchain, Rahasak-ML service federated machine learning service.
- Distributed cache: Etcd-based distributed key-value pair storage (open-source distributed key-value storage system).
- Certificate authority: certificate authority that issues certificates for peers and clients.

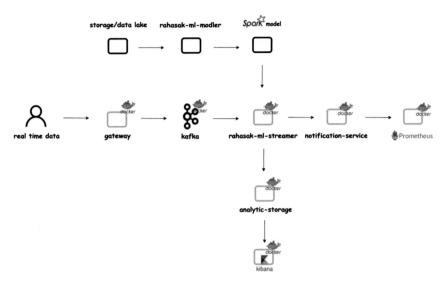

Fig. 2 Rahasak-ML service architecture. Each blockchain peer has its own Rahasak-ML service. Machine learning models will be generated with the data on each peer's off-chain storage

Each peer in the network has its own off-chain storage for storing the raw data. The hash of these data is published to a blockchain ledger and shared with other peers. The blockchain storage on the Rahasak-ML platform keeps all its transactions, blocks, and asset information (hash of the data in off-chain storage) on Cassandra-based Elassandra Storage (https://github.com/strapdata/elassandra). It exposes Elasticsearch application programming interfaces [53] for transactions, blocks, and assets on the blockchain. Each peer in the blockchain can establish supervised or unsupervised machine learning models with the existing data on its own off-chain storage. Once a peer generates a model, it can be incrementally trained and aggregated by other peers through the blockchain by using the federated learning approach. The model parameter sharing, local model generation, incremental model training, and model sharing functions are implemented in the Rahasak-ML platform. Once machine learning models are generated, these models can be integrated into blockchain smart contracts to predict real-time data. Figure 2 shows the architecture of the Rahasak-ML services in a single blockchain peer.

Each peer in the network runs its own Rahasak-ML service. The Rahasak-ML service contains the following components. All these components are Dockerized and deployed via Kubernetes.

- Storage Service.
- Rahasak-ML Modeler Service.
- Rahasak-ML Streamer Service.
- Gateway Service.
- Apache Kafka.

3.2 Key Components

3.2.1 Storage Service

Each peer in the Rahasak-ML platform has two storage mechanisms: off-chain and on-chain storage. Both are built with Apache Cassandra-based Elassandra storage. Off-chain storage stores the data generated by the peers. The hash of these data is published to on-chain storage and shared with other peers. Blockchain keeps all its transactions, blocks, and asset information on this on-chain storage. The on-chain storage in each peer is connected in a ring cluster architecture. The data saved in one node will be replicated with other nodes via this ring cluster. After executing transactions with smart contracts, the state updates in a peer are saved in Cassandra storage and distributed with other peers, Fig. 3.

Blockchain can keep any data structure as blockchain assets since it uses Cassandra as the underlying asset storage. As a use case of Rahasak-ML, the authors built a blockchain-based secure NetFlow network packet storage and network anomaly detection (e.g., network attack) service. It stored actual NetFlow packet data in the blockchain peers' off-chain storage. The hash of the data was stored in the on-chain storage as a blockchain asset. The smart contracts in the blockchain parsed the NetFlow packets coming through the router and stored them in the blockchain storage. Rahasak-ML can build machine learning models with the data saved in the peers' off-chain storage. In federated learning scenarios, the local models are stored in the off-chain storage. The hash of the model and storage Uniform Resource Identifier (URI) of the model are stored in on-chain storage and distributed with other peers.

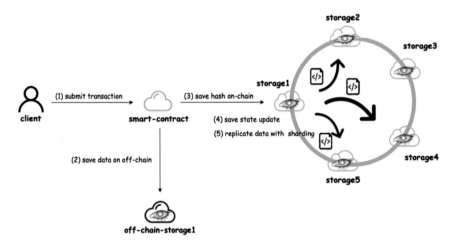

Fig. 3 Rahasak-ML storage service architecture. Each peer comes with two types of storage: on-chain storage and off-chain storage. Off-chain storage stores the actual data generated by the peers. The hash of these data is published to on-chain storage and shared with other peers

3.2.2 Rahasak-ML Modeler Service

Rahasak-ML modeler service is responsible for building the machine learning model by analyzing the peers' off-chain storage data. It supports building both supervised (e.g., Decision Tree, Random Forest, and Logistic Regression) and unsupervised (e.g., K-Means and Isolation Forest) machine learning models. To build a new machine learning model, the first step is training, which uses a dataset as an input and adjusts the model weights for the model accuracy. The second step is testing, which takes in an independent dataset for testing the accuracy.

Figure 4 shows the overall flow of these steps, which is performed by the Rahasak-ML Modeler service. Once the prediction model is built and trained by the Rahasak-ML Modeler service, it can be used to perform tasks on new data. In a federated learning environment, each peer in the network will continuously train the generated model with the data on their off-chain storage using an incremental training approach. The continuous model training can be done with Spark Streams [54], such as real-time training libraries. More information about the continuous model training is discussed in Sect. 4.

Following the model generation, the training models can be used in smart contracts to predict/cluster real-time data. For example, Rahasak-ML Modeler can be used to build the Isolation Forest and K-Means-based models to detect outliers of network traffic data. This model will split network data into two clusters: normal network traffic and suspicious (attacks) network traffic. Once local models are built and aggregated, the models can be integrated into blockchain smart contracts to predict

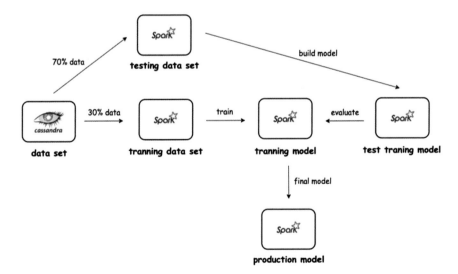

Fig. 4 Rahasak-ML modeler service architecture. Seventy percent of the data is used to train the model, and 30% will be used for testing

the real-time network data. When new network packets come to the blockchain, smart contracts can use the model and predict the category (normal or suspicious) of real-time network traffic.

3.2.3   Rahasak-ML Streamer Service

Rahasak-ML streamer service clusters the real-time data with the machine learning models built by the Rahasak-ML Modeler service. It uses blockchain smart contracts [55, 56] to run the machine learning model with the newly generated data. Smart contract functions are written to use the model and predict the cluster output. This service consumes real-time data via Kafka (e.g., Kafka Streams and Spark Streams). For example, in the previously mentioned network traffic analysis scenario, the Rahasak-ML streamer will consume real-time network packets via Apache Kafka and run through the model built by the Rahasak-ML Modeler service. It will decide the clustering output (normal and suspicious) of the new packets, and if a suspicious packet is found, it will publish the entry to a notification service. Alerts will be generated, notifying experts via notification dashboards (e.g., Prometheus and Grafana), as shown in Fig. 5.

3.2.4   Gateway Service

When analyzing real-time data, the Gateway service is used as the entry point to the Rahasak-ML platform. It fetches (or pushes from other services) real-time data from various data sources, such as log fields, NetFlow, TCP, UDP, and database. For example, the gateway service can receive real-time network traffic data via NetFlow. Once data arrive, they are prepared (by removing noise, parsing the data, etc.) and published to the Rahasak-ML streamer service via Kafka as JSON encoded objects.

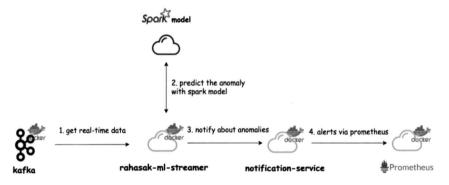

Fig. 5 Rahasak-ML streamer service architecture. Streamer service clusters the real-time data with the machine learning models built by the Rahasak-ML Modeler service

Fig. 6 Gateway service architecture. Gateway service is used as the entry point to the Rahasak-ML platform. It fetches (or pushes from other services) real-time data from various data sources such as log fields, NetFlow, TCP, UDP, and database

When the platform receives NetFlow packets, it extracts relevant fields, aggregates them, constructs a JSON object, and forwards it to the Rahasak-ML streamer service via Kafka, as shown in Fig. 6.

3.2.5 Kafka Message Broker

Apache Kafka is the consensus and message broker service in the Rahasak-ML blockchain environment. The authors use a Reactive Programming and Reactive Streaming model [57] where the services published events/messages with Kafka. The events will be subscribed by relevant services and take corresponding actions. The real-time data that come through the gateway service are published into Kafka first. Then Rahasak-ML streamer service consumes them and runs with the model, which is built by the Rahasak-ML Modeler service, as shown in Fig. 7.

4 Rahasak-ML Federated Learning Process

4.1 Overview

Rahasak-ML proposed a blockchain-based federated learning approach to build and share the models. With this approach, model generation, incremental model training, model aggregation, and sharing can be done without having centralized authority. Federated learning approaches increase privacy but still rely on centralized control

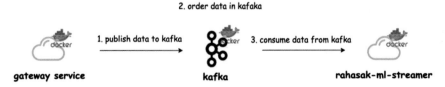

Fig. 7 Rahasak-ML message broker architecture. Apache Kafka is the message broker of the Rahasak-ML platform. Each microservice communicates with other services via Kafka

to manage the process. Centralized control can be compromised, causing a potential weak link in the system and a lack of trust in the authority that owns the centralized server [2]. A blockchain-based decentralized system provides a logical ruleset that all participants are aware of and agree on, allowing participants to audit operations to ensure that all parties follow the rules. It improves the ability to audit and adds more transparency to the federated learning process. Each peer in the blockchain network incrementally trains the machine learning models with the data on its own local off-chain storage. Once all peers (or a majority of peers) are trained, the finalized model details will be integrated into a block and published to the other peers in the network by the block-generating service of Rahasak-ML (Lokka service).

4.2 Incremental Training Flow

Assume a scenario where blockchain nodes are deployed in three companies, Companies A, B, and C. The blockchain is configured to store the data related to network traffic. Each company has its own off-chain storage, which stores the actual network traffic data. The hash of the network traffic data is published into the blockchain ledger. First, the Lokka service (that generates blocks) creates a genesis block with the incremental learning flow and the model parameters, as shown in Algorithm 1. Each peer in the network has its own Lokka service. The Block Creator is determined in a round-robin distributed scheduler. Consider the scenario in Fig. 8, which has three Lokka services, and assume that the first block is created by Lokka A, the second block will be created by Lokka B, and Lokka C creates the third block. This process is repeatedly performed to generate future blocks.

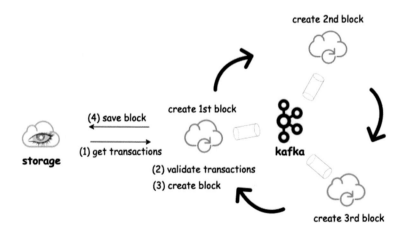

Fig. 8 The block creator is determined in a round-robin distributed scheduler. The block approval process is performed via the federated consensus implemented between Lokka services

Algorithm 1 Training pipeline initialization

1 INITIALIZE TRAINING PIPELINE:
2 Choose Lokka node l_i by the round-robin scheduler to initialize the training pipeline
3 Find available blockchain peers p(1, ..., n) from the distributed cache
4 Define incremental learning flow based on each peer join time (ttl) to the network
5 Define ML model training parameters and algorithm information
6 GENERATE GENESIS BLOCK:
7 Create genesis block b_i with model parameters and incremental training flow
8 Save b_i in ledger and broadcast it to other peers in the network

Incremental learning flow defines the order of the model training process. When defining a learning flow, the Lokka service finds the existing nodes in the network via distributed cache service in the Rahasak-ML. Rahasak-ML uses Etcd distributed key/value pair storage as the distributed cache and service registry. Etcd stores the health information of the blockchain nodes in the network. When a blockchain node is added to the network, it registers a node name (with meta-information) in the Etcd with the time to live (TTL) key. The node will periodically update this TTL key (before TTL reach) to prove it is alive. If a node is dead/exits, the TTL key will automatically be removed from Etcd. By using the TTL keys in Etcd, other nodes can know the available nodes in the network. The order of the incremental learning flow is decided by the TTL key created timestamp in the Etcd. This timestamp defines the blockchain nodes' added time to the network. Assume the Lokka service has the incremental learning flow as A→B→C based on the TTL keys in the Etcd registry. This flow represents that peer A will produce a model, and then this model will be incrementally trained by peer B and then peer C. Once a miner node publishes the genesis block with model parameters and incremental flow to the blockchain ledger, other peers take the block and process it according to the defined flow, as shown in Fig. 9.

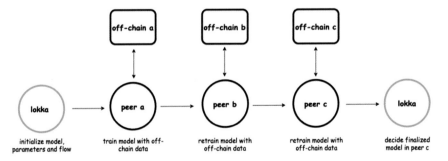

Fig. 9 Rahasak-ML training pipeline. Once a miner node publishes the genesis block with machine learning model parameters and incremental flow to the blockchain ledger, other peers take the block and process it according to the defined flow

According to the incremental learning flow, first, peer A generates the anomaly detection model with the data on the off-chain storage based on the model parameters in the genesis block. Then it saves the model built on its off-chain storage. The actual model is not published onto the blockchain ledger or any central storage. The hash and URI of the built model saved in the off-chain storage are published to the blockchain ledger as a transaction. Then peer B starts to incrementally train the model built by Peer A. To achieve this, peer B fetches the model built by peer A from peer A's off-chain storage using the given URI. Then it trains that model with the data on peer B's off-chain storage. This training model will be saved on peer B's off-chain storage, and peer B will publish the model hash and off-chain storage URI of the model to the blockchain ledger as a transaction. Next, peer C will incrementally train the model trained by peer B and publish the details to the blockchain ledger as a transaction, as shown in Algorithm 2.

Algorithm 2 Incremental training flow

---

1 Wait till publishing genesis block b_i
2 **for** each peer p= 1, ..., n **do**
3   **INCREMENTAL MODEL TRAINING:**
4   **if** p == 1 **then**
5   Fetch genesis block b_i from the ledger and get model training parameters
6   Build initial model with the data in the off-chain storage
7   **else**
8   (assume p=x)
9   Fetch ML model from the peer p=x−1 off-chain storage
10   Incrementally train that model with the data on the peer
         p=xoff-chain storage
11 **end**
12 Save built ML model in off-chain storage
13 **PUBLISH MODEL UPDATES:**
14 Create transaction t_i with ML model hash and off-chain storage URI of the model
15 Publish t_i to the ledger
16 **end**

The flow of the incremental learning process is described in Fig. 10.

4.3 Finalizing Model

Assume all three companies (or a majority of the companies) incrementally train the prediction model and publish the model hash and URI to the blockchain ledger as a transaction. Then Lokka service takes these transactions and creates a block with finalized model details with the final model stored in the peer C's off-chain storage. Currently, the model trained by the last peer (peer C in this scenario) is identified as the finalized model. In future work, there are plans to determine the

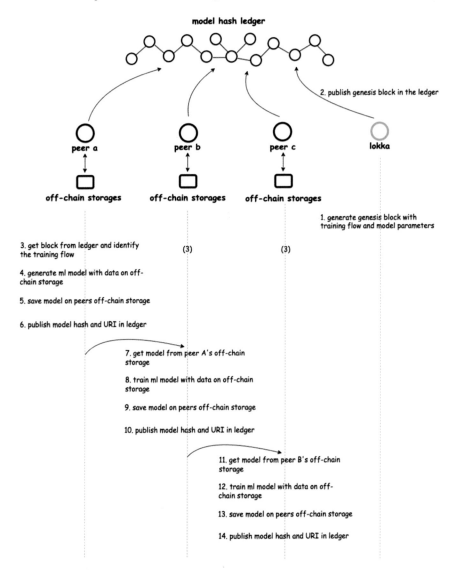

Fig. 10 Rahasak-ML incremental training flow. Each peer trains the model with the data on the off-chain storage. The state update in each training step will be published to the blockchain ledger

finalized model by evaluating the accuracy of each model trained by its peers. Lokka service includes the URI of peer C's off-chain storage (which stores the final model) and model training transaction details into the block. Then Lokka service saves the generated block in the ledger and distributes it to other peers. Once the peers receive the new block, they validate the learning process with the transactions in the block. If the process is valid, peers fetch the final model stored in peer C's off-chain storage

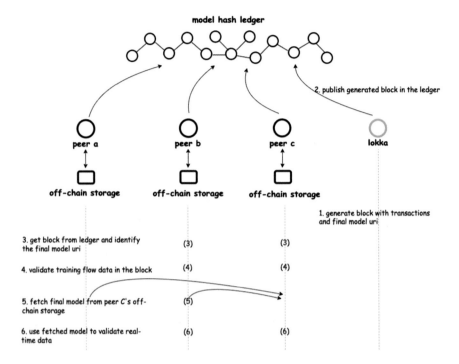

Fig. 11 Rahasak-ML finalizes the machine learning model. The final model will be decided by the Lokka service when generating the final block

via the given URI in the block. The incrementally trained model sharing process is described in Fig. 11. Once the finalized model is fetched, it can be used in smart contracts for prediction.

For the Lokka service to decide the final model, the majority of the nodes in the network need to complete the incremental learning process. If there are five nodes in the federated learning flow, three of these nodes need to finish the incremental learning flow to decide on the finalized model. Once the Lokka service has generated the block with the finalized model details, other Lokka services in the network need to approve that block. When approving, they first validate the transactions in the block. If all transactions in the block are valid, it gives a vote for the block (mark block as valid or invalid), as shown in Algorithm 3. To handle the voting process, the Lokka service digitally signs the block hash and adds the signature to the block header. When the majority of Lokka services submit the vote for a block, that block is considered as a valid/approved block.

Algorithm 3 Choose final model

---

1 Wait till the majority of the peers complete the incremental training process in the training pipeline
2 DEFINE FINAL MODEL:
3 Get transaction list t(1, ..., n) from ledger
4 Find the transaction t_n which submitted by the last peer p_n (model trained by the last peer identified as the finalized model)
5 Create block b_{i+1} with final model URI, model hash and transactions
6 Save blockb_{i+1} in the ledger and broadcast it to other peers
7 UPDATE FINAL MODEL:
8 **for** each peer p= 1, ..., n **do**
9   Fetch blockb_{i+1} from ledger
10   Verify transactions in the block
11   If the block is valid, fetch final ML model from peer p_n
12 **end**

4.4 The Use Case of Blockchain-Empowered Federated Learning in the Medical Field

Blockchain-empowered federated learning provides a secure, transparent, and privacy-preserving computing solution for building accurate and robust predictive models using biomedical data from multiple parties (e.g., institutions, hospitals, and drug companies). It does not need a centralized server to collect data from various parties, which is often difficult to share due to HIPAA. As a proof of concept, the authors built blockchain-empowered federated learning for diagnosing acute inflammation of the bladder. We used inflammation of the bladder health dataset [58] and chose logistic regression as the prediction model. In this use case, a blockchain network is deployed at five peers (five hospitals). Each peer has its own dataset and trains and validates a local logistic regression model. Finally, these local models are averaged. The loss and accuracy of the models were computed, and block generation time was measured in the blockchain-enabled federated learning system. The preliminary study can be extended to more scenarios in medicine and drug discovery use cases.

4.4.1 Federated Model Accuracy and Training Loss

In the federated learning scenario, the model was trained with 1000 iterations. A copy of the shared model is sent to all peers participating in the iteration. Each peer trains its own model with its own dataset locally. Each local model is improved in its own direction. Then total loss and accuracy were computed as shown in Fig. 12. Figure 13 shows how the total training loss varies at different peers in each iteration.

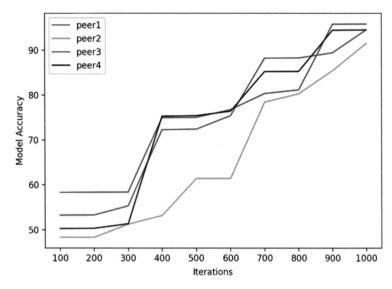

Fig. 12 Federated model accuracy in different peers

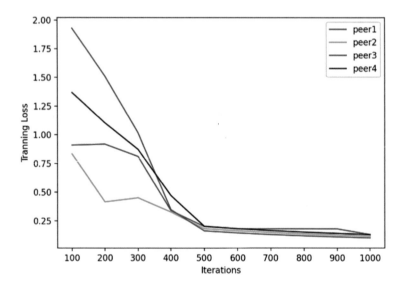

Fig. 13 Federated model training loss in different peers

4.4.2   Block Generation Time

Block generation time was measured in the Bassa-ML federated learning system with a different number of blockchain peers (up to 7). Figure 14 shows the average block generation time when having a different number of blockchain peers in the network.

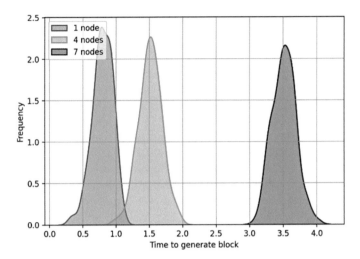

Fig. 14 Average block generation time

Each experiment was repeated 100 times in this evaluation—each with different peer sets—and average values were plotted. When adding peers to a cluster, each peer needs to validate transactions in the block and recalculate the block header. Accordingly, block generation time increases as peers are added.

5 Future Directions

The proposed platform took full advantage of blockchain and AI technologies to provide a more efficient and secure solution with a promise to accelerate the research in medicine. The following is a summary of future work and several open directions.

The proposed system overcomes several key concerns faced in centralized systems. While individual nodes (peers) develop local models based on their local data, the resulting models and parameters are shared through the blockchain platform. The model parameter sharing, local model generation, model averaging, and model sharing functions are implemented with smart contracts implemented on the platform. Most recently, the Rahasak-ML federated learning system was integrated into Rahasak blockchain version 3.0. The following are features of the Rahasak-ML platform that are planned to release in the future:

- Decide the finalized model by evaluating the accuracy of each model trained by the peers.
- Support more supervised/unsupervised machine learning algorithms with Rahasak-ML.
- Automate the deployment of the Spark cluster in Rahasak-ML with Kubernetes.
- Integrate TensorFlow-Federated libraries into Rahasak-ML.

5.1 Data Heterogeneity

Medical data are particularly diverse—in terms of the variety of modalities, dimensionality, and characteristics—even for a specific protocol, there are acquisition differences, a brand of the drugs, or local demographics [59]. Although federated learning can address the data bias issue by collecting more data sources, inhomogeneous data distribution is still challenging, as many assume independently and identically distributed data across their peers. Another challenge is the different data standards and data heterogeneity among peers. For example, hospitals may adopt EHR systems from different vendors, and different countries use different diagnostic and procedure coding systems. For example, health systems in the United Kingdom use the International Classification of Diseases ICD-10 code, but the United States adopted ICD-10-CM. This heterogeneity may lead to a situation where the optimal global solution may not work well for an individual local participant.

5.2 Efficiency and Effectiveness

From the technical view, efficiency and effectiveness are the major concerns of federated learning. Federated learning needs peers to share and update the models, and thus, the communication cost between different peers is an issue. Especially when integrated with blockchain, how to minimize the communication time and improve the efficiency of the training process is important. Studies have focused on improving the framework to jointly improve the federated learning convergence time and the training loss [60], but the tradeoff between accuracy and communication expenditure should also be considered.

5.3 Model Interpretation

Integrating machine learning models is important, particularly for healthcare and medicine. The core question of interpretability is whether humans understand why the model makes such predictions on unseen instances. Many machine learning models, such as deep learning, are a "black-box" to humans, and thus, many studies have explored tools to interpret the models [61, 62, 63, 64]. In a federated learning context, as the model was kept updated through multiple parties, the interpretation would be a challenge.

To summarize, federated learning for life sciences will benefit the process of data sharing among multiple organizations without a central authority. The data sharing process will monitor and track the data operations efficiently to ensure data integrity and provenance. Still, the data ownership problem is the key to adopting Rahasak-ML in FDA- or EMA-regulated research.

6 Conclusions

Federated learning emerges as a new technique that uses collaboration and distribution to train machine learning models without sharing the local raw data. It promises to benefit the medical field and drug industry that require strict data protection. However, most of the existing federated learning systems deal with centralized coordinators that are vulnerable to attacks and privacy breaches. We proposed a blockchain-empowered coordinator-less decentralized federated learning platform, named Rahasak-ML, to solve issues in centralized coordinator-based federated learning systems by providing better transparency and trust. We introduced the architecture and learning process of Rahasak-ML. We introduced a use case of using Rahasak-ML to train a machine learning model for diagnosis, which could be applied to other biomedical data to facilitate decision-making. Still, data standardization, communication efficiency, and model interpretation need to be resolved.

References

1. Kairouz P, McMahan HB, Avent B, Bellet A, Bennis M, Bhagoji AN, Zhao S (2021) Advances and open problems in federated learning. Found Trends Mach Learn 14(1–2):1–210. https://doi.org/10.1561/2200000083
2. Konečný J, McMahan HB, Yu FX, Richtarik P, Suresh AT, Bacon D (2016) Federated learning: strategies for improving communication efficiency. NIPS Workshop on Private Multi-Party Machine Learning. Retrieved from https://arxiv.org/abs/1610.05492
3. McMahan B, Moore E, Ramage D, Hampson S, Arcas BA (2017) Communication-efficient learning of deep networks from decentralized data. In: Proceedings of the 20th international conference on artificial intelligence and statistics; Proc Mach Learn Res 54:1273–1282. https://proceedings.mlr.press/v54/mcmahan17a.html
4. Sheller MJ, Edwards B, Reina GA, Martin J, Pati S, Kotrotsou A, Bakas S (2020) Federated learning in medicine: facilitating multi-institutional collaborations without sharing patient data. Sci Rep 10(1):12598. https://doi.org/10.1038/s41598-020-69250-1
5. Yang T, Andrew G, Eichner H, Sun H, Li W, Kong N, Beaufays F (2018) Applied federated learning: improving google keyboard query suggestions. https://arxiv.org/abs/1812.02903
6. Lu Y, Huang X, Zhang K, Maharjan S, Zhang Y (2020) Blockchain empowered asynchronous federated learning for secure data sharing in the internet of vehicles. IEEE Trans Veh Technol 69(4):2020 4298–4311. https://doi.org/10.1109/TVT.2020.2973651
7. Qu Y, Gao L, Luan TH, Xiang Y, Yu S, Li B, Zheng G (2020) Decentralized privacy using blockchain-enabled federated learning in fog computing. IEEE Internet Things J 7(6):5171–5183. https://doi.org/10.1109/JIOT.2020.2977383
8. Bandara E, Liang X, Foytik P, Shetty S, Ranasinghe N, De Zoysa K (2021) Rahasak-scalable blockchain architecture for enterprise applications. J Syst Archit 102061. https://doi.org/10.1016/j.sysarc.2021.102061
9. Nallaperuma D, Nawaratne R, Bandaragoda T, Adikari A, Nguyen S, Kempitiya T, Pothuhera D (2019) Online incremental machine learning platform for big data-driven smart traffic management. IEEE Trans Intell Transp Syst 20(12):4679–4690. https://doi.org/10.1109/TITS.2019.2924883

10. Shan N, Ziarko W (1994) An incremental learning algorithm for constructing decision rules. In: Ziarko WP (ed) Rough sets, fuzzy sets and knowledge discovery. Springer, pp 326–334. https://doi.org/10.1007/978-1-4471-3238-7_38

11. Sathya R, Abraham A (2013) Comparison of supervised and unsupervised learning algorithms for pattern classification. Int J Adv Res Artif Intell 2(2):34–38. https://doi.org/10.14569/IJARAI.2013.020206

12. Johnson KB, Wei W-Q, Weeraratne D, Frisse ME, Misulis K, Rhee K Snowdon JL (2021) Precision medicine, AI, and the future of personalized health care. Clin Transl Sci 14(1):86–93. https://doi.org/10.1111/cts.12884

13. Wang F, Preininger A (2019) AI in health: state of the art, challenges, and future directions. Yearb Med Inform 28(1):16–26. https://doi.org/10.1055/s-0039-1677908

14. Emerson S, Kennedy R, O'Shea L, O'Brien J (2019) Trends and applications of machine learning in quantitative finance (SSRN Scholarly Paper No. ID 3397005). Rochester, NY, Social Science Research Network. Retrieved from Social Science Research Network, https://papers.ssrn.com/abstract=3397005

15. List of registries (2015) Retrieved August 30, 2021, from National Institutes of Health (NIH). https://www.nih.gov/health-information/nih-clinical-research-trials-you/list-registries

16. Annas GJ (2003) HIPAA regulations—a new era of medical-record privacy? N Engl J Med 348(15):1486–1490. https://doi.org/10.1056/NEJMlim035027

17. Federated learning: Collaborative machine learning without centralized training data (2017) Retrieved August 30, 2021, from Google AI Blog, http://ai.googleblog.com/2017/04/federated-learning-collaborative.html

18. Liao W, Luo C, Salinas S, Li P (2019) Efficient secure outsourcing of large-scale convex separable programming for big data. IEEE Trans Big Data 5(3):368–378. https://doi.org/10.1109/TBDATA.2017.2787198

19. Forum on Neuroscience and Nervous System Disorders, Board on Health Sciences Policy, Institute of Medicine (2014) Drug development challenges. In: Improving and Accelerating Therapeutic Development for Nervous System Disorders: Workshop Summary. National Academies Press (US). Retrieved from https://www.ncbi.nlm.nih.gov/books/NBK195047/

20. Liao C, Peach ML, Yao R, Nicklaus MC (2013) Molecular docking and structure-based virtual screening. In: Future Science Book Series. In Silico Drug Discovery and Design. Future Science Ltd, pp 6–20. https://doi.org/10.4155/ebo.13.181

21. Subramanian I, Verma S, Kumar S, Jere A, Anamika K (2020) Multi-omics data integration, interpretation, and its application. Bioinform Biol Insights 14. https://doi.org/10.1177/1177932219899051

22. Pendergrass SA, Crawford DC (2019) Using electronic health records to generate phenotypes for research. Curr Protoc Hum Genet 100(1):e80. https://doi.org/10.1002/cphg.80

23. Zhao J, Feng Q, Wu P, Lupu RA, Wilke RA, Wells QS, Wei W-Q (2019) Learning from longitudinal data in electronic health record and genetic data to improve cardiovascular event prediction. Sci Rep 9(1):717. https://doi.org/10.1038/s41598-018-36745-x

24. Zhao J, Grabowska ME, Kerchberger VE, Smith JC, Eken HN, Feng Q, Wei W-Q (2021) ConceptWAS: a high-throughput method for early identification of COVID-19 presenting symptoms and characteristics from clinical notes. J Biomed Inform 117:103748. https://doi.org/10.1016/j.jbi.2021.103748

25. Jeon J, Nim S, Teyra J, Datti A, Wrana JL, Sidhu SS, Kim PM (2014) A systematic approach to identify novel cancer drug targets using machine learning, inhibitor design and high-throughput screening. Genome Med 6(7):57. https://doi.org/10.1186/s13073-014-0057-7

26. Ferrero E, Dunham I, Sanseau P (2017) In silico prediction of novel therapeutic targets using gene-disease association data. J Transl Med 15(1):182. https://doi.org/10.1186/s12967-017-1285-6

27. Godinez WJ, Hossain I, Lazic SE, Davies JW, Zhang X (2017) A multi-scale convolutional neural network for phenotyping high-content cellular images. Bioinformatics 33(13):2010–2019. https://doi.org/10.1093/bioinformatics/btx069

28. Brown N, Ertl P, Lewis R, Luksch T, Reker D, Schneider N (2020) Artificial intelligence in chemistry and drug design. J Comput Aided Mol Des 34(7):709–715. https://doi.org/10.1007/s10822-020-00317-x

29. Bibault J-E, Giraud P, Housset M, Durdux C, Taieb J, Berger A, Burgun A (2018).Deep learning and radiomics predict complete response after neo-adjuvant chemoradiation for locally advanced rectal cancer. Sci Rep 8(1):12611. https://doi.org/10.1038/s41598-018-30657-6

30. Harrer S, Shah P, Antony B, Hu J (2019) Artificial intelligence for clinical trial design. Trends Pharmacol Sci 40(8):577–591. https://doi.org/10.1016/j.tips.2019.05.005

31. Calaprice D, Galil K, Salloum W, Zariv A, Jimenez B (2020) Improving clinical trial participant prescreening with artificial intelligence (AI): a comparison of the results of AI-assisted vs standard methods in 3 oncology trials. Ther Innov Regul Sci 54(1):69–74. https://doi.org/10.1007/s43441-019-00030-4

32. Gaulton A, Hersey A, Nowotka M, Bento AP, Chambers J, Mendez D, Leach AR (2017) The ChEMBL database in 2017. Nucleic Acids Res 45(D1):D945–D954. https://doi.org/10.1093/nar/gkw1074

33. Kim S, Chen J, Cheng T, Gindulyte A, He J, He S, Bolton EE (2019) PubChem 2019 update: improved access to chemical data. Nucleic Acids Res 47(D1):D1102–D1109. https://doi.org/10.1093/nar/gky1033

34. Parikh RB, Teeple S, Navathe AS (2019) Addressing bias in artificial intelligence in health care. JAMA 322(24):2377–2378. https://doi.org/10.1001/jama.2019.18058

35. Schneider P, Walters WP, Plowright AT, Sieroka N, Listgarten J, Goodnow RA, Schneider G (2020). Rethinking drug design in the artificial intelligence era. Nat Rev Drug Discov 19(5):353–364. https://doi.org/10.1038/s41573-019-0050-3

36. Chang K, Balachandar N, Lam C, Yi D, Brown J, Beers A, Kalpathy-Cramer J (2018) Distributed deep learning networks among institutions for medical imaging. J Am Med Inform Assoc 25(8):945–954. https://doi.org/10.1093/jamia/ocy017

37. Xiong Z, Cheng Z, Lin X, Xu C, Liu X, Wang D, Zheng M (2021) Facing small and biased data dilemma in drug discovery with enhanced federated learning approaches. Sci China Life Sci https://doi.org/10.1007/s11427-021-1946-0

38. Li L, Xu W, Chen T, Giannakis GB, Ling Q (2019) Rsa: Byzantine-robust stochastic aggregation methods for distributed learning from heterogeneous datasets. [Cs, Math]. Retrieved from http://arxiv.org/abs/1811.03761

39. Kim H, Park J, Bennis M, Kim S-L (2020) Blockchained on-device federated learning. IEEE Commun Lett 24(6):1279–1283. https://doi.org/10.1109/LCOMM.2019.2921755

40. Xu G, Li H, Liu S, Yang K, Lin X (2020) Verifynet: secure and verifiable federated learning. IEEE Trans Inf Forensics Secur 15:911–926. https://doi.org/10.1109/TIFS.2019.2929409

41. Androulaki E, Barger A, Bortnikov V, Cachin C, Christidis K, De Caro A, et al (2018) Hyperledger fabric: A distributed operating system for permissioned blockchains. In: Proceedings of the thirteenth EuroSys conference, vol 30. ACM. https://doi.org/10.1145/3190508.3190538

42. Azaria A, Ekblaw A, Vieira T, Lippman A (2016) Medrec: using blockchain for medical data access and permission management. In: 2016 2nd International conference on open and big data (OBD). pp 25–30. https://doi.org/10.1109/OBD.2016.11

43. Liang X, Shetty S, Zhao J, Bowden D, Li D, Liu J (2018) Towards decentralized accountability and self-sovereignty in healthcare systems. In: Qing S, Mitchell C, Chen L, Liu D (eds) Information and communications security. ICICS 2017. Lecture notes in computer science, vol 10631. pp 387–398. https://doi.org/10.1007/978-3-319-89500-0_34

44. Yang G, Li C (2018) A design of blockchain-based architecture for the security of electronic health record (EHR) systems. IEEE international conference on cloud computing technology and science (CloudCom) 2018:261–265. https://doi.org/10.1109/CloudCom2018.2018.00058

45. Abadi M, Chu A, Goodfellow I, McMahan HB, Mironov I, Talwar K, Zhang L (2016) Deep learning with differential privacy. In: Proceedings of the 2016 ACM SIGSAC conference on computer and communications security. pp 308–318. https://doi.org/10.1145/2976749.2978318

46. Bonawitz K, Ivanov V, Kreuter B, Marcedone A, McMahan HB, Patel S, Seth K (2017) Practical secure aggregation for privacy-preserving machine learning. In: Proceedings of the 2017 ACM SIGSAC conference on computer and communications security. pp 1175–1191. https://doi.org/10.1145/3133956.3133982
47. Thönes J (2015) Microservices. IEEE Softw 32(1):116–116. https://doi.org/10.1109/MS.2015.11
48. Merkel D (2014) Docker: Lightweight Linux containers for consistent develop-ment and deployment. Linux J 2014(239):2. https://doi.org/10.5555/2600239.2600241
49. Burns B, Grant B, Oppenheimer D, Brewer E, Wilkes J (2016) Borg, omega, and kubernetes. Queue 14(1):70–93. https://doi.org/10.1145/2898442.2898444
50. Lakshman A, Malik P (2010) Cassandra: a decentralized structured storage system. Oper Syst Rev (ACM) 44(2):35–40. https://doi.org/10.1145/1773912.1773922
51. Gupta M (2012) Akka essentials. Packt Publishing Ltd. Retrieved from https://www.packtpub.com/product/akka-essentials/9781849518284
52. Davis AL (2019) Akka streams. In: Reactive streams in java. Apress, pp 57–70 https://doi.org/10.1007/978-1-4842-4176-9_6
53. Gormley C, Tong Z (2015) Elasticsearch: the definitive guide: a distributed real-time search and analytics engine. O'Reilly Media, Inc. Retrieved from https://www.oreilly.com/library/view/elasticsearch-the-definitive/9781449358532/
54. Beneventi F, Bartolini A, Cavazzoni C, Benini L (2017) Continuous learning of hpc infrastructure models using big data analytics and in-memory processing tools. In: Design, automation and test in Europe conference and exhibition (DATE). IEEE, 1038–1043. https://doi.org/10.23919/DATE.2017.7927143.
55. Bandara E, Liang X, Foytik P, Shetty S, Ranasinghe N, Zoysa KD, Ng WK (2021) Saas—microservices-based scalable smart contract architecture. In: Thampi SM, Wang G, Rawat DB, Ko R, Fan C-I (eds), Security in computing and communications. Singapore, Springer Singapore, pp. 228–243. https://doi.org/10.1007/978-981-16-0422-5_16
56. Bandara E, Ng W, Ranasinghe N, Zoysa K (2019) Aplos: Smart contracts made smart. BlockSys. https://doi.org/10.1007/978-981-15-2777-7_35
57. Wan Z, Hudak P (2000) Functional reactive programming from first principles. In: PLDI '00: Proceedings of the ACM SIGPLAN 2000 conference on programming language design and implementation, vol 35. ACM, pp 242–252. https://doi.org/10.1145/349299.349331
58. Upstill R, Eccles D, Fliege J, Collins A (2013) Machine learning approaches for the discovery of gene–gene interactions in disease data. Brief Bioinform 14(2):251–260. https://doi.org/10.1093/bib/bbs024
59. Rieke N, Hancox J, Li W, Milletarì F, Roth HR, Albarqouni S, Cardoso MJ (2020) The future of digital health with federated learning. Npj Digit Med 3:119. https://doi.org/10.1038/s41746-020-00323-1
60. Chen M, Shlezinger N, Poor HV, Eldar YC, Cui S (2021) Communication-efficient federated learning. PNAS 118(17). https://doi.org/10.1073/pnas.2024789118
61. Lundberg SM, Erion G, Chen H, DeGrave A, Prutkin JM, Nair B, Lee S-I (2020) From local explanations to global understanding with explainable AI for trees. Nat Mach Intell 2(1):56–67. https://doi.org/10.1038/s42256-019-0138-9
62. Lundberg SM, Nair B, Vavilala MS, Horibe M, Eisses MJ, Adams T, Lee S-I (2018) Explainable machine-learning predictions for the prevention of hypoxaemia during surgery. Nature Biomed Eng 2(10):749–760. https://doi.org/10.1038/s41551-018-0304-0
63. Montavon G, Samek W, Müller K-R (2018) Methods for interpreting and understanding deep neural networks. Digit Signal Process 73:1–15. https://doi.org/10.1016/j.dsp.2017.10.011
64. Ribeiro MT, Singh S, Guestrin C (2016) Why should i I trust you?: explaining the predictions of any classifier. [Cs, Stat]. Retrieved from http://arxiv.org/abs/1602.04938

Xueping Liang is an Assistant Professor in the Department of Information Systems and Supply Chain Management at the University of North Carolina at Greensboro. Prior to that, she was an Assistant Professor of Computer Science at Virginia State University. Her research is centered around data provenance mechanisms, cybersecurity, blockchain, privacy protection, and the Internet of Things (IoT). Specifically, she is interested in distributed consensus models in blockchain technology, cyber-resiliency in IoT, and various practical issues in cloud computing security. She received her Ph.D. in Cybersecurity from the University of Chinese Academy of Sciences. She has published more than 30 conference and journal papers and book chapters at reputed venues.

Eranga Bandara worked as a Senior Research Scientist at the Virginia Modeling Analysis and Simulation Center (VMASC) Virginia, USA. His research interests include Distributed Systems, Blockchain, Big Data, Actor-based Systems, and Functional programming. He worked as a Lead Engineer at Pagero AB Sweden. With Pagero AB, he was involved with research and developments in Distributed Systems, Functional Programming, Big Data, Actor-based systems, and DevOps.

Juan Zhao is a Research Assistant Professor in the Department of Biomedical Informatics at Vanderbilt University Medical Center. She achieved a Ph.D. degree in Computer Science at the University of Chinese Academy in Beijing, China. Her current research interests focus on deep learning, machine learning, natural language processing, and blockchain, especially on leveraging such technologies to further the understanding of complex diseases and improve clinical outcomes and treatment. This research combines her educational background in computer science, and machine learning, her programming experience as a senior software engineer, and her training as a postdoctoral fellow in cybersecurity at Tennessee State University and in Biomedical Informatics at Vanderbilt. Dr. Zhao's study was funded by the American Heart Association (AHA) and the National Institutes of Health (NIH). She has published ~20 research papers and achieved four software copyrights. She is also serving as a program committee member in the Association for the Advancement of Artificial Intelligence Conference (AAAI) and IEEE Big Data, the Associate Editor of the Journal of Network Modeling Analysis in Health Informatics and Bioinformatics, and the Special Topic Editor for Frontiers in Big data.

Sachin Shetty is an Associate Director in the Virginia Modeling, Analysis, and Simulation Center at Old Dominion University and an Associate Professor in the Department of Computational Modeling and Simulation Engineering. Sachin Shetty received his Ph.D. in Modeling and Simulation from Old Dominion University in 2007. His research interests lie at the intersection of computer networking, network security, and machine learning. Recently, he has been involved with developing cyber risk/resilience metrics for critical infrastructure and blockchain technologies for distributed system security. His laboratory has been supported by the National Science Foundation, Air Office of Scientific Research, Air Force Research Lab, Office of Naval Research, Department of Homeland Security, and Boeing. He has published over 150 research articles in journals and conference proceedings and four books. He is the recipient of Commonwealth Cyber Initiative Research Fellow, Fulbright Specialist award, EPRI Cybersecurity Research Challenge award, and DHS Scientific Leadership Award and has been inducted into Tennessee State University's million-dollar club.

Considerations for Ensuring Success of Blockchain in Life Sciences Research

Valuing Research Data: Blockchain-Based Management Methods

Wendy M. Charles and Brooke M. Delgado

Abstract Research data sets are not just considered highly valuable for scientific purposes; these data sets could be sold and traded for economic value. Data sets could also be regarded as intangible assets, which do not have physical properties but could provide future economic benefits. With consideration that life sciences organizations possess thousands of siloed data sets, these could be sold to support additional research and could add value when life sciences organizations are appraised. Blockchain-based technologies are increasingly used to manage the control and auditability of both data and asset transactions in ways not possible with traditional databases. This chapter encourages life sciences organizations to view their data silos differently and consider the potential value these can create for the organization. This chapter describes methods to value and monetize health-oriented life sciences research data using common accounting principles. The chapter also describes the assetization of data sets and when data sets could be classified and traded using blockchain as non-fungible tokens. Last, the authors share ethical, legal, and regulatory constraints that should be considered before implementation.

Keywords Blockchain · Data valuation · Intangible assets · Micropayments · Data sales

1 Introduction

Organizations in the life sciences research industries base their research findings and product development on large volumes of data. Data sets are created from aggregating data from feasibility studies, clinical trials, electronic health records, apps, and a variety of mobile devices [1]. While life sciences data could involve any biological

W. M. Charles (✉)
Life Sciences Division, BurstIQ, Inc., Denver, CO, USA
e-mail: wendy.charles@cuanschutz.edu

B. M. Delgado
Finance/Accounting, BurstIQ, Inc., Denver, CO, USA
e-mail: brooke.delgado@gmail.com

or chemical topic [1], this chapter focuses on individuals' health and wellness data. Health-oriented data from individuals provide insights that could be used for future research and/or targeted outreach [2].

Data sets are retained to retrieve data for compliance verification purposes and perform "data mining," which involves extracting and analyzing patterns in large data sets [3]. Often, disparate data sets are stored within proprietary electronic data capture systems or are maintained in organizational data silos where there may not be much data visibility across teams [4]. However, life sciences organizations seek broader sources of data to generate business value and pursue research innovation. For example, Pfizer analyzed volumes of oncology research data, genomic sequencing, and electronic health records to spot trends for targeted drug development opportunities [5]. As a result of these searches, Pfizer identified a small group of lung cancer patients who demonstrated a unique genetic mutation. Pfizer subsequently developed a precision medicine drug, crizotinib (Xalkori), targeted to this patient group [5].

The drive for data in life sciences organizations has been expanded in the United States by the twenty-first Century Cures Act that requires additional uses of existing data as part of the Real-World Evidence Framework [6]. The FDA defines real-world data as data originating from various sources, including electronic health records, disease registries, patient-generated data, and claims and billing activities [7]. Real-world data have been used to generate new research hypotheses, assess trial reliability, identify prognostic indicators, and create research cohorts ([7], p. 7). In August 2021, the FDA approved a new indication for the Astellas drug Prograf (tacrolimus) to prevent organ rejection where the control group was derived, in part, from real-world data [8]. Similarly, the FDA's Center for Device and Radiological Health provided examples of 90 regulatory decisions where national registries or electronic health records served as the primary or secondary sources of device efficacy [9]. An FDA press release noted that "a non-interventional study has the potential to meet FDA's regulatory standards for an adequate and well-controlled clinical study" ([10], p. 2).

Life sciences organizations are also seeking real-world data to save costs on conducting research and reproducing research findings. Real-world data analyses can reveal insights into clinical care and identify opportunities for new scientific research hypotheses [11]. One of the chapter authors (WC) oversaw the regulatory compliance of academic medical centers and noted that the vast majority of research conducted involved retrospective data analyses from the electronic health record system. The author notes that a single researcher can complete these studies within a few months without outside funding. This process of data mining—conducting analyses on existing data—generates so many research ideas and inventions that [3] created data mining models that include the total cost of ownership of direct and indirect expenses of data acquisition.

While many sources of health-oriented data can be sold, including direct sales from patients [1], health data marketplaces [12], and healthcare organization purchases (e.g., [13]), this chapter focuses on how life sciences organizations can value and sell their data. Further, while organizations can select among a few technologies to manage data access and sales, this chapter focuses on the emerging uses of blockchain technologies for this purpose.

This book chapter is organized into the following sections. The first section lays the framework for the current methods that life sciences research organizations use to acquire and exchange health-oriented data. This section also describes the current role of blockchain for life sciences data exchanges. The following section provides definitions and accounting principles that explain how life sciences organizations could value data sets as assets and possible methods for data sales. Next, this chapter describes how blockchain is currently being used to manage and monetize data. Last, this chapter outlines many factors that life sciences research organizations should consider when developing data valuation and sales plans.

1.1   Nature of Health Data

Only a fraction of life sciences organizations constitutes covered entities. A covered entity is "(1) a health plan, (2) a healthcare clearinghouse, or (3) a healthcare provider who transmits any health information in electronic form in connection with a transaction" (45 CFR § 160.103) required to protect and maintain health information under the Health Insurance Portability and Accountability Act (HIPAA). Covered entities are required to protect the privacy and security of health data classified as protected health information (45 CFR § 160.103).

Covered entities may sell protected health information as follows:

Deidentified information. Protected health information can be distributed without restriction when all the 18 identifiers from the Safe Harbor provision are removed or when a person with generally accepted statistical principles determines that the risk is very small that the information could be used by an anticipated recipient to identify individual patients (45 CFR § 164.514(a) and (b)). Recently, 14 U.S. health systems created a new company, Truveta, to aggregate and sell deidentified health information [14].

Identifiable information. Covered entities may not "sell lists of patients or health plan enrollees without obtaining authorization from each person on the list" (45 CFR § 164.508(a)(4); [15]). Specifically, individual authorizations are required for disclosures subject to protected health information planned for sale as defined in 45 CFR § 164.501 and 45 CFR § 164.508(a)(4).

Business Associate Agreement. While not a method of sale, covered entities can receive professional services for their data under a Business Associate Agreement. For example, Ascension Health entered into a Business Associate Agreement with Google—without exchanging funds—for Google to develop mutually beneficial health algorithms [16], and HIPAA [17]. There has been debate, though, as to whether the scope of data shared by Ascension Health met the HIPAA requirement that the information was the minimum necessary to achieve the stated purpose (45 CFR § 514(d)(iii)).

While most life sciences organizations are not covered entities subject to these restrictions, there may be contractual agreements in place with covered entities that provide protected health information for research studies [4]. These contractual agreements may limit what the life sciences organization can do with the health information.

1.2 Health Data Management

Life sciences organizations utilize several different types of technologies to manage data sets, such as research databases [2], clinical trial management systems [18], and servers that manage data archives [4]. However, most of these systems serve effectively as data silos, as there are few mechanisms to integrate, share, or control information—other than with manual intervention [11]. Blockchain technologies are increasingly used to manage life sciences data sets efficiently while still preserving control. Detailed features of blockchain-based systems are offered in the *Data Sales Methods* section.

2 Data as an Asset

Potential profit is hidden in the value that both data and data sets hold [19]. Therefore, data can be viewed as an economic asset due to their inherent economic benefit. Assets can be tangible objects such as real estate, office furniture, or computers [20]. Intangible assets do not have physical properties but still can provide future benefits to the organization and add to the organization's value. Common intangible assets include intellectual property such as trademarks, patents, and copyright [20].

2.1 How to Value Data Assets

2.1.1 Data Value Factors

The value of data is relative to the use, need, and demand for the data. The value can be variable and subjective, leaving the valuation in constant flux and uncertainty depending on who, when, and how data are used. The following factors could impact the value of data.

Complexity. Having a diverse data set can remove biases from results and outcomes. Complex data can add value due to the rarity or difficulty in securing the data.

Size. Data with larger set sizes carry a higher value than data with just one or two pieces of data.

Format. Life sciences data sets represented in the Resource Description Framework are more valuable due to interoperability with variable naming conventions

[21]. The Resource Description Framework offers standards and specifications for data exchange on the Internet [22].

Identifiability. Another aspect of data that creates value is whether data are identifiable. It is more valuable to know individuals' identities to target them for additional opportunities [23]. Deidentifiable data are bought and sold on open marketplaces, but as previously discussed, deidentified data are not as valuable as identifiable data.

Restrictions for reuse. Data may be made available under commercial licensing agreements with or without restrictions for reuse. Data released under restrictive licensing agreements are more valuable because of revenue-generating opportunities [24]. However, some journals, such as Scientific Data, promote the open exchange of data and will not consider publications describing data sets involving restrictions on reuse [25].

Age. Time-dependency is defined by the length of time passed since the data were collected or the time taken to prepare a data set. Data perishability is the devaluation of data as time passes [26]. While it seems intuitive that data value would decline with age, researchers cannot definitively predict the useful life of data because it is unknown which trends will evolve in the future. Both innovation and consumer behavior can impact time-dependency leading to data perishability [26]. Therefore, data can be assumed to have indeterminate lives. Thus, the useful life of data should be reevaluated annually to confirm whether the data are still relevant for future use.

Data versus information. Do data sets offer value in themselves, or do the data need to be aggregated and manipulated into meaningful information before we consider the data sets valuable? Atkinson and McGaughey [27] propose that data are similar to raw materials turned into a finished product. The resulting "finished" information is typically more valuable than the raw data.

Industry. Data value will vary by industry. For example, Google tracks user data (cookies) for targeted advertising, making users' data highly valuable for Google's advertising revenue. Health records are highly desirable and valuable data for health and research organizations [28]. Researchers rely on health records to validate drug and pharmaceutical research hypotheses, which are vital to interpreting drug interactions and efficacy.

2.1.2 Data Valuation Methods

Wang et al. [29] note that health-oriented data have realized only 10–20% of possible value, creating an environment of nearly untapped potential. While there are several possible methods for data valuation, this chapter provides three accounting approaches with examples applicable to life sciences industries.

Cost Approach

Using a cost approach, life sciences organizations could price data values based on the actual or estimated costs to prepare or replace the data set with an exact or similar type of data set [30]. A PricewaterhouseCoopers report about the costs of generating life sciences research data describes the time spent by researchers (salary, time spent, cleaning and processing, and integrating with other data) and storage costs [31].

Historical costs can also be used to estimate replacement costs but require several assumptions: First, this approach assumes that preparing or recreating the data would cost the same as the initial preparation. Next, there is a question as to whether it is less expensive to reproduce an item after the initial strategic development has already taken place [32]. In life sciences, research and development costs, as well as time spent through trial and error, may not need to be repeated to reproduce a data set, making the replacement costs less expensive than the initial costs. Last, costs should consider the role of inflation in the initial data preparation effort.

Because the cost approach can apply to company acquisitions, the Roche acquisition of Flatiron Health provides a meaningful example. Roche paid $1.9B for Flatiron Health to acquire oncology electronic health record systems, health platforms, and analytics services [13]. As a result of acquiring health records of 400,000 cancer patients, the cost approximates $950 per patient record [33].

As an additional consideration, organizations do not list data sets as assets on their balance sheet. Data developed in-house incur costs that are not recognized on the balance sheet. These costs are expensed as they occur according to the Accounting Principles Board Opinion No. 17 [27]. Per Financial Accounting Standards Board (FASB) 141, technology-based intangible assets can be capitalized on the balance sheet if acquired through the purchase of a company [34]. Since most company-generated data assets are not listed on the balance sheet, it can be assumed that the balance sheet understates the company's actual value.

Income Approach

An income-based approach quantifies the potential value of the economic benefit to be generated from the data set [35]. The financial benefit can be determined as "commercial opportunity" and "economic uplift." The commercial opportunity involves the incremental difference between boosted revenue and previous revenue trends for new insights generated by the data [33]. Data may also lead to better strategic planning for trends and future research directions [36]. Therefore, the commercial opportunity would be demonstrated by more profound insights into promising chemical compounds that increase the size of the company's drug pipeline, reduce drug development time, or launch a new indication. With economic uplift, there are benefits to the intended audience for the life sciences products. These may include more accurate or faster diagnoses, improved efficiency, or clearer treatment regimens [33].

As an example of the income approach, clinical research originating from National Health Services (NHS) data are estimated to deliver approximately £4.6B per year in additional economic uplift to NHS patients in the United Kingdom [33]. The data value is estimated to arise from operational savings, enhanced patient outcomes, and more informed spending [33].

It is essential to recognize that the income approach requires organizations to make assumptions about future market share, adoption rate, and long-term market growth [32]. For many types of projections, it may be necessary to analyze the outcomes from several assumptions [32].

Market Approach

The market approach estimates value using data value factors (e.g., age, size, and complexity) and compares data pricing for similar factors in the market. The market approach compares per-record values against industry benchmarks, recent sales on an open market, or prior sales [35]. The estimated value is then multiplied by the number of records.

To create life sciences industry benchmarks for per-record values, EY researchers [33] developed a 2019 report with nine pages of tables of public and private companies' estimated values in different health domains and the estimated value per health record. For example, in the table about publicly traded companies that manage records for episodes of healthcare (p. 25), the stock prices for 12 companies were used to create an estimated value for each company, divided by the number of records, to create a median estimated value of £54 per health record. The authors caution, though, that company valuations and acquisitions include capabilities, such as data and analytic platforms, in addition to the data sets [33].

The supply and demand drivers of the market approach also apply to the black market [37]. Health records are sought for illegal acquisitions to obtain Card Not Present e-commerce transactions, card track data, and billing information [38]. Health records sell for $1 to $1,000 per record on the dark web, depending on completeness [28, 39].

Because market drivers fluctuate, the market approach should be reviewed at least annually against recent sales and industry benchmarks.

3   Data Sales Methods

Life sciences organizations sell data sets using a few different methods.

3.1   *Data Brokers*

Data brokers facilitate data sales between interested parties. Typically, a broker assists companies with finding specific data sets and may also aggregate data [2]. For example, Acxiom (https://www.acxiom.com/) is often described as the largest commercial data broker, primarily focused on marketing [40]. Acxiom [41] claims it has pioneered identity resolution to integrate data from disparate data sets to create insights into individual consumers.

While data brokers can provide a valuable service for organizations seeking niche data, the Federal Trade Commission [42] raised concerns about the transparency of data brokers' methods of obtaining information and efforts to link data sets intended to be deidentified. Therefore, some states, like California (1798.99.80–1798.99.88) and Vermont (9 V.S.A. § 2446), require data brokers to register with the state to ensure the protection of personal information. The use of data brokers has primarily transitioned to data marketplaces.

3.2 Centralized Data Marketplaces

Centralized digital data marketplaces allow data set characteristics to be posted and viewed for sale. With centralized data marketplaces, the platform and data oversight processes are centralized. The marketplace owner provides reports, data services, data storage, access controls, terms of use, and manages monetization among parties [43].

IQVIA (https://www.iqvia.com/) offers a private centralized data marketplace where the organization states that it sells data from over 1B deidentified patients from 45 countries [44]. Truveta (https://www.truveta.com/) is an example of a centralized consortium data marketplace. The startup, founded in 2020, is co-owned by 14 hospital facilities that contribute deidentified data with the vision of saving lives with data [14].

3.3 Decentralized Data Marketplaces

Decentralization can refer to the distribution of data storage, management, and/or monetization. Blockchain-based data marketplaces offer variations of decentralization among private, consortium, and independent marketplaces [43]. Further, participation can be designed with and without intermediaries for consumer-to-consumer (C2C), business-to-consumer (B2C), business-to-business (B2B), business-to-consumer-to-business (B2C2B), and/or business-to-business-to-consumer (B2B2C) transactions [45].

In addition to offering new methods for managing data sharing and integrity, blockchain-based systems can allow transactions with digital currencies.

3.3.1 Features of Blockchain-Based Data Marketplaces

The following are common features of using blockchain for data management and exchange.

Granular consent. Blockchain technologies allow individual patients or research participants to have more control over how their data are used and shared [46] with granular and dynamic consent mechanisms to change and revoke permissions [40].

Governance. Rather than centralized management where there are single points of potential vulnerability [47], blockchain-based data exchanges allow for distributed data stewardship and communication [48].

Transparency. Individuals or organizations benefit from blockchain-based transparency in price formulas [49]. Further, the blockchain-based data exchange facilitates data sharing about data uses and related metadata that can be used to determine trends and future needs [50].

Automation. Because malicious data buyers or sellers may refuse to pay for data [47], smart contracts can automatically control payments and revenue distributions [51]. This capability also ensures efficiencies for data exchanges and resource allocations [52]. Further, smart contracts can ensure that only authorized individuals can contribute data or access specific data sets [46].

Non-standardized data formats. Data exchanges require flexibility for ontology matching, standardization, and the ability to link data sets [50]. Blockchain-based ledgers allow for more data flexibility of storage and designing data mapping schemas [50]. Zhu et al. [51] provide the example that medical and diagnostic information benefits from applying natural language processing to clinical notes stored in varying formats on a blockchain.

Network security. Security is a primary requirement for data marketplaces to ensure honest data exchanges and protect against unauthorized access or transactions. First, there is a need to prevent and detect data reselling without appropriate permissions or licensing [47]. Blockchain data marketplaces are designed to ensure legitimate transactions [53] using key-based asset management that prevents data reselling in a manner similar to preventing cryptocurrency double-spending [49]. Also, blockchain-based exchange protocols mask or sanitize sensitive health data [40, 47] and/or preserve the identity of the data sellers [54].

Data integrity. Marketplaces also aim to ensure that data are not altered during data storage or transmission [53] or mismanaged by a central authority [55]. Blockchain-based data marketplaces offer tamper-resistant and tamper-evident mechanisms to ensure data integrity [53].

Data quality. A challenge with any data purchase is ensuring data quality before purchase. While blockchain cannot prevent the common inaccuracies of health information, blockchain-based mechanisms allow data buyers to examine data for correctness and credibility without releasing the data [55]. Specifically, there are methods of applying algorithms to determine if data display expected (naturally occurring) characteristics and distributions [53] and can also view short segments or control points without revealing the data set [55].

Trust. Ultimately, these characteristics support the need for trust that the data exchange and payment process is fair and consistent with payment terms [56]. Zozus and Bonner [18] remind that blockchains enable data exchanges among individuals or organizations that do not (entirely) trust each other because the transactions are attestable and can be reviewed in audit trails. Blockchain supports the need for trust that the data exchange and payment process is fair and consistent with payment terms [56].

3.3.2 Examples of Blockchain-Based Data Marketplaces

The following descriptions compare and contrast blockchain-based data marketplaces intended to sell information to or from life sciences organizations. However, this list is not comprehensive, and it is possible that some features were misunderstood.

Personal Health Record Management

- Ciitizen (https://www.ciitizen.com), Embleema (https://www.embleema.com/), and PhrOS (https://phros.io/services/health_data_market) are designed as personal health record systems that can also share health information for research.

Monetization to Individuals for Sharing Data

- Datapace.io (https://datapace.io/), Datum (https://datum.org), EncrypGen (https://encrypgen.com/), and Hu-manity.co (https://hu-manity.co/) offer tokens native to their platforms.
- Ciitizen offers a range of monetization options "in the form of direct payment, services, discounts, donations, or other values or to donate this value to an advocacy or research non-profit as directed by the patient" ([57], p. 5).
- Dawex (https://www.dawex.com) creates monetization by allowing buyers to offer bids.
- Embleema provides points that can be exchanged for unspecified "rewards."
- PhrOS promotes health data exchanges using "health points" [58].

Additional Data Functionality

- BurstIQ, Inc. (https://www.burstiq.com/) links data sets together using LifeGraphs with artificial intelligence to create deeper scientific insights [59].
- Datum and Datapace.io focus on linking IoT data sets with other information, including health records [60, 61].
- Nokia   (https://www.nokia.com/networks/services/nokia-data-marketplace/) provides federated learning and federated intelligence capabilities on the platform.

- PhrOS states that the platform can provide bi-directional communication to health-care providers and facilities, including alerts to providers when individuals need immediate attention [62].

Decentralized Network Management

- Datapace.io maintains a decentralized network where individuals or organizations participate in the consensus mechanisms [60].
- Datum describes its platform as a decentralized version of Apple HealthKit for research data collection design [61].
- The Enigma blockchain allows decentralized applications to perform calculations and functions on encrypted data, allowing data to remain private even when using a public blockchain [63].

3.4 Non-Fungible Tokens

Life sciences data sets can also be registered as assets using blockchain technologies as non-fungible tokens (NFTs). Briefly, there are two common types of tokens managed by blockchains. Fungible tokens are similar to currencies in that fungible tokens are interchangeable and can be fractionalized while still retaining value [64]. However, NFTs represent unique items that are not interchangeable or divisible. NFTs are used to represent proof of ownership of digital or physical assets [65] or to establish intellectual property [66]. In an extreme example, Pablo Rodriguez-Fraile recently sold a 10 s video NFT for $6.6 M, even though the video can be watched online at no charge [67]. The seller explained that the NFT's value is not in the viewing but in the provenance and ownership of the asset.

NFTs have been explored for establishing rights to data sets to represent unique original value. Sandner et al. [66] suggest that blockchain-based NFT data management provides critical data set provenance and auditability without allowing access to raw data. Blockchains can be used to exchange NFTs as a new asset class that provides the democratization of data access, allowing life sciences organizations to trade valuable data sets more easily with each other without using data brokers or marketplaces [66]. With a similar rationale, Hapiffah and Sinaga [68] designed the proof of consent for a blockchain-based health record system where each patient's medical record is assigned an NFT. The goal is to provide patients with more evidence and control of their records. However, the regulatory status of NFT-based assets remains unclear. It is also uncertain whether these assets would be subject to sales and capital gains taxes [66].

4 Considerations

The complexities of data valuation and sale are complicated by economic, social, and ethical values. If life sciences organizations are considering data valuation and sale, these values drive considerations of community perceptions, informed consent, data ownership, and privacy [1]. This section aims to raise awareness of topics that should influence the organization's internal and external planning and communication.

4.1 *Ethical Considerations*

As life sciences organizations consider their prospects for valuing data for sale, they are encouraged to ask ethical questions about the appropriateness of selling and exchanging patients' health and life sciences data.

4.1.1 Perceptions of Data Sales

While health-oriented data sales and collaborations with outside companies may be legal, patients/research participants may not support these initiatives and view them as inappropriate. As mentioned above, Google created a health data initiative called "Project Nightingale" that started as a secret collaboration with Ascension, a hospital chain of 2600 healthcare facilities [69]. Without notifying patients or physicians, Ascension shared data from tens of millions of patients, involving patients' names, dates of birth, and complete medical records. The goal of this collaboration was for Google to design algorithms to target individual patients and their care, while Ascension planned to use Google's algorithms to identify additional tests or generate new revenue [69]. While this agreement was conducted under a Business Associate Agreement and was deemed legal, Google was heavily criticized and questioned by members of U.S. Congress—primarily because consumers do not trust how Google manages their data [70]. While the project continued, researchers reported in the journal Nature that the Google controversy would reduce trust in research projects conducted with academic medical centers [71].

4.1.2 Informed Consent

Scholars suggest that the most ethical approach to selling data involves full awareness and approval of all parties affected by the data transaction (e.g., [72, 73]). Indeed, when consumers are provided with detailed information about the planned uses and privacy methods, they are up to 52% more likely to share their information when compared to data opportunities that do not allow individuals to manage their privacy preferences [40]. However, when presented with information about possible data uses, the information is often biased toward the advantages of the data sale or exchange. First, the "informed consent" form provides complex policies, terms, and conditions that an average person cannot understand [40]. Also, when describing seemingly innocuous data purposes such as quality assurance or healthcare innovation, patients could not anticipate the possible unfettered sale and uses of their information [73]. Of additional concern, patients—and consumers—generally do not read through long documents [40]. Additionally, individuals are desensitized to click-through agreements authorizing organizations to use their personal data in exchange for rewards or access to technology [40]. These agreements are written so that the general consumer struggles to interpret the legal ramifications of signing such agreements [40].

While individuals may provide specific data intended for use or sale, Clark et al. [74] point out that there is often hidden data collection. This metadata, or data value-level information, remains linked to data values to provide more information about the data [18]. Unknown to the individual, there may be metadata collected about the individual's Internet usage or even Internet-based patterns of behavior that provide deeper insights and more value to the original information [74].

4.2 Ownership

Life sciences research data are collected, aggregated, structured, and analyzed by many contributors who will each claim a piece of the ownership at different levels along the data value chain [20]. While the legal, ethical, and regulatory factors of data ownership are outside the scope of this chapter, these factors should be considered for a data ecosystem designed for multiple ownership [20]. Given the possibility of fractional ownership, it is difficult to determine who has controlling rights for data assets [65]. Therefore, legal documents and agreements are used to control data.

It is argued that personal data cannot be owned; therefore, the organization that collects it has the right to use it and claim it as an asset [2]. There is also debate as to whether data sets could be considered intellectual property. As a refresher, trade secrets protect the process and design of the research, copyright laws are available for annotations of data; patent laws are available for new inventions that came from the data [75].

While life sciences data sets may be protected under privacy law, they are not necessarily covered by copyright law. In 1991, the U.S. Supreme Court ruled that data in themselves are considered "facts" and cannot be copyrighted [76]. No copyright law protects those facts on behalf of the organization that collected them. However, copyright can be obtained through data being "selected, coordinated, or arranged in such a way that the resulting work as a whole constitutes an original work of authorship" ([76], p. 8). With this consideration, some forms of raw research data cannot be classified as intellectual property. However, research data compiled and arranged to answer a hypothesis may be considered IP [75].

4.3 Data Considerations

For healthcare and life sciences organizations interested in pursuing health data sales and sharing, there are important considerations about managing data. These concerns pertain to data privacy and security protections, while others pertain to data management controls, costs, and quality. This section elaborates on the nature of considerations, while specific recommendations are offered in the Recommendations section.

4.3.1 Privacy and Security

For any data marketplace, there are risks associated with data analysis and storage. There is a delicate balance between the need for data availability for rapid aggregation and queries against safeguarding the data [77]. Both the healthcare organizations that provide health data and the life sciences organizations that receive the data are responsible for assessing benefits and risks. Collins and Lanz [77] encourage these organizations to protect their data as if these are assets.

From a patient perspective, patients report that their primary concern pertains to their data privacy [49]. The demand for privacy is even more pertinent for health-related data where patients view their health information to be private [49]. While consumers report being less concerned about their data being collected from Internet usage [78], individuals require more expectations of privacy in return for sharing their information [40].

Robust technological mechanisms for data security and a thoughtful privacy policy are necessary to reduce the risks of individual reidentification. While some data warehouses and marketplaces state that their deidentification strategies are effective at protecting data (e.g., [79]), the process of deidentification is not infallible. On the contrary, individual patients can be reidentified with only a few key attributes [80] involving little more than basic programming ability [81]. Reidentification often carries under-recognized risks to the patients, such as affecting an individual's eligibility for long-term care insurance [82], stigmatization, discrimination [83], and identity theft [82].

Deidentified data provides some reassurance that data are private, but methods to reidentify data are available. Certain genealogical characteristics can specify an individual's race, gender, and other physical identifying characteristics [84]. There is no guarantee that data can remain deidentified [85]. Therefore, organizations should consider that risks of reidentification exist on a spectrum [84].

When studying DNA, the risks of reidentification are amplified due to the unique characteristics of an individual's genetic code that may render complete deidentification impossible [48]. Of more significant concern, genetic information does not simply reveal information about an individual but also about that individual's biological relatives [74] who did not provide consent for their data to be shared or sold.

This genetic information may describe sensitive health vulnerabilities about these relatives that may also harm the relatives [48].

While blockchain offers additional layers of data security, smart contracts and blockchain access keys have been breached to steal cryptocurrency [48], and blockchains may involve (at least theoretical) modifications. Therefore, blockchains should not be considered absolutely immutable [86]. As a final consideration, encryption may be vulnerable to advances in future computing capabilities [80]. Even if using blockchain, life sciences organizations should still employ appropriate data security measures.

4.3.2 Access and Control

Many informed consent terms do not provide patients with options for choices about future uses or sales of their data, leaving individuals with a single option of take-it-or-leave-it [72]. Worse, adding information about the possible sales or sharing their health information in the notice of privacy practices may be inherently unethical. Klugman [73] notes that telling patients they cannot receive care without allowing their data to be sold or shared is an "abandonment of patients" (approx. p. 1).

For most data repositories, participants have the right to withdraw consent at any time, but to request withdrawal typically requires considerable manual effort [87]. Unfortunately, the process of withdrawing from health repositories used for data sales may also carry penalties. If individuals provide consent to participate in a data repository with the intent to receive payments for the uses of their data, there are questions about the fairness of withdrawal. If an individual allows data sales in exchange for free genetic sequencing, should the company allow data removal before recouping the costs [48]? Blockchain-based DNA marketplace, LunaDNA, specifies that individuals who withdraw consent for future use of their data also lose any ownership shares granted to the individual prior to withdrawal [88]. Therefore, some individuals may feel coerced into allowing future uses of their data for fear of losing these financial benefits.

4.3.3 Data Quality

If data are considered an asset to achieve organizational goals, organizations must focus closely on data reliability and quality. Statement of Financial Accounting Concept 2 denotes reliability as "the quality of information that assures that information is reasonably free from error and bias and faithfully represents what it purports to represent" ([89], p. 6). Additionally, as noted above, accurate and complete information is more valuable [90]. Specifically, while life sciences organizations have typically cleaned data collected and used for research studies, there are considerable costs associated with reviewing and eliminating poor-quality data [90].

4.3.4 Costs

This section provides considerations about the costs associated with data marketplace options. The first two items pertain to the costs for data purchases, while the third section offers decision points for dividing and sharing data with patients and research participants.

Transaction Costs

When planning to purchase data, the purchase costs require advanced planning and decisions beyond the perceived data value. Health data marketplaces offer a wide range of purchase options, such as a one-time purchase, monthly/yearly subscriptions for queries and downloads, and usage-based fees [58]. Because there may be hundreds of data sets without consistent data variable standardization, organizations may pursue custom pricing strategies to obtain data curation services for their specific needs [58].

Operating Costs

The process of selecting and curating life sciences data to drive data sharing and sales can be extremely costly—both for data storage and personnel [91]. The process also requires thoughtful data governance and administration regarding which data should be sold or shared as well as how these decisions align with business values and the nature of informed consent that individuals had provided [91].

When using a blockchain to manage data for exchange and sale, the blockchain cannot run autonomously. The platform requires governance among the participating nodes and funds toward incentive mechanisms for individuals or organizations to share data [92]. When using public blockchains, there are also costs associated with gas or tokens to process data logging and the need for computing resources [92]. Data sellers could offset some of these fees by requiring ongoing subscription fees [93].

4.3.5 Monetization Process

When planning to sell life sciences research data, organizations should consider how they should monetize their data and divide the profit for the various activities that could be performed. Because blockchain technologies allow for monetization for each data access or use, there are additional decisions.

Data Pricing

Query pricing. When determining pricing for data access, life sciences organizations should consider whether they will charge for queries [56]. Queries are often performed to determine data set attributes and feasibility for more extensive studies. As an option, life sciences organizations could provide simple counts and prevent data downloads at no charge.

Research and development pricing. Many data mining activities (or retrospective chart reviews) are conducted without funding [4]. These early research studies may not ever create a profitable product. Alternatively, if a product is created, it may take years to generate profit [90]. Life sciences organizations should consider whether they will allow access to data at no charge in exchange for future payments for a developed product or service.

Horizontal value split. This pricing method divides payment among the parties that contributed to the data collection and processing [56]. The list of contributors could include the individual participants, but it would be more common to include the collaborating research institutions or partners. When data sets are sold, life sciences organizations should consider whether funds will be set aside for partners.

Vertical value split. This pricing strategy allocates payment based on degrees of contribution to a data set [56]. For example, should payments to collaborating research institutions or partners be divided based on the number of subjects enrolled or the value of ancillary services provided, such as data curation and analysis?

Payments to Research Participants

Within life sciences research, it is customary to ask research participants to allow their data and specimens to be used for future research. When providing informed consent for this purpose, patients are asked to provide consent specifying that they will not receive monetization for future uses of their data. As an example, the Colorado Multiple Institutional Review Board's (COMIRB's) biomedical informed consent research template informs prospective research participants that if their data or specimens are used in future research: "There is no plan for you to receive any financial benefit from the creation, use or sale of such a product or idea" ([94], p. 6). Other academic organizations provide similar informed consent template wording because it would be truly challenging to determine how best to compensate patients or individual research subjects for data sales [40] or to determine the value of data contributions used to develop new products [52].

When payments are intended for individuals contributing to deidentified data sets, it may not be possible to compensate the participating individuals unless identity and monetization are managed by a third party [51]. Suppose instead that life sciences organizations plan to manage identities for the purpose of providing payments. These organizations must then also protect individuals' identities and/or financial information [82]. As a final constraint, it is logistically challenging to track participants' contact information over long periods [95]. Regardless of which organization tracks individuals' identities for monetization, it is important to consider that the U.S. Internal Revenue Service [96] classifies research payments as taxable income. Specifically, the organization managing payments must issue IRS Form 1099 when an individual receives $600 or more in a calendar year.

5 Recommendations

To address the complexities of health data sales and sharing for life sciences research, the authors of this chapter offer ethical, data management, and legal recommendations. Potential strategies involving the use of blockchain technologies are added to each section.

5.1 Ethical Recommendations

From an ethical perspective, health data sales and sharing mechanisms intended for life sciences research should be more transparent to patients and the healthcare community. The following are recommendations for both healthcare and life sciences organizations.

5.1.1 Provide Explicit Information for Consent

- Similar to the HIPAA requirement of data sales for marketing (45 CFR § 164.508(a)(4)), research participants should be presented with an additional authorization form to (ideally) opt-in to sharing their data or records for data sales and sharing [80].
- When individuals agree to participate in research that may result in data sales, these organizations should provide complete transparency of the types and nature of sales so that the individual could give genuinely informed consent [97]. The scope of information should also include details about control of the data and the storage duration [83]. The consent form should specify if commercial organizations may purchase their data for targeted marketing [97], for subsequent commercial sale [83], or for secondary research [4].
- The consent form should not offer the false promise that individuals cannot be identified in a data set [85] because individuals can be reidentified with fairly straightforward programming tools [81].
- Provide patients with dynamic consent options for how their data may be used and shared with life sciences companies to allow more patient control [98]. Blockchain-based technologies are ideal for providing dynamic consent because

smart contracts can enable granular sharing options and automate access permissions [99]. Further, the inherent audit trails allow for tracking and oversight [99].

5.1.2 Provide Community Engagement

- Provide more information on the life sciences organizations' websites and in the notice of privacy practices with transparency about the nature of data sales or sharing arrangements. This information could share details of data brokering agreements.
- When providing consent to sell genetic information, the ramifications are broader than the individual, as DNA is shared among other biological relatives who may not know that their shared DNA has been sold [48]. Ahmed and Shabani [48] recommend that individuals discuss these potential risks with family members.

5.2 Data Recommendations

Data recommendations encompass a wide range of topics used to protect and manage data. Due to the possible scope of these recommendations, this chapter focuses on blockchain-based data recommendations but does not negate the appropriate security assessments and risk assessments suitable for any electronic environment that manages sensitive information.

Data privacy and security. Blockchain-based technologies offer new methods to protect the privacy of patient-level information.

- **Granular consent for permissions**. While granular consent was described above as a method of allowing individuals or organizations to provide specific permissions, this mechanism is also a privacy-preserving strategy for limiting the amount of information released and creating automatic expirations [100].
- **Centralized access controls**. Some private blockchain environments offer capabilities where individuals log in to a virtual machine or protected zone where data cannot leave the workspace [101]. For example, blockchain provider BurstIQ provides life sciences organizations with access control options where data recipients cannot download data and may be restricted from seeing individually identifiable information [59].
- **Creation of synthetic data replacements**. Blockchain technologies are being used to replace identifiable information with hashes or synthetic data to mask identifiable information [102].
- **Federated learning systems**. Pharmaceutical companies are protecting identifiable information and intellectual property by using federated learning systems in research collaborations. Rather than share raw data, the organizations use a blockchain to share an algorithm [103]. For example, ten pharmaceutical companies formed a federated learning project called the Machine Learning

Ledger Orchestration for Drug Discovery (referred to as "MELLODDY"), where blockchain-shared machine learning algorithms are trained on each company's molecular compounds to identify potential drugs for development [104].

- **Advanced blockchain-based privacy features**. Blockchain developers have been enhancing privacy-preserving features that are gaining broader adoption but are not yet available in most mainstream blockchain applications. Zero-knowledge proofs permit an individual or organization to attest to specific facts without providing the actual data [52, 105]. For example, a zero-knowledge proof could be used to verify that a person is above a certain age without providing the date of birth. Another promising development, homomorphic encryption, allows for calculations on encrypted data without revealing the original information [106].
- **Allow patient or organizational transparency**. Allow patients to see how their data are used whenever possible by using blockchain technologies for audit trails [56].

5.3 Legal Recommendations

The following legal recommendations are presented to ensure that organizations perform due diligence and meet appropriate regulatory requirements and privacy policies.

5.3.1 Contractual Agreements

- Contractual agreements can provide enforceable limitations about permitted uses of health data that are shared or sold. These agreements should specify that data recipients cannot use data for any purpose not specified in the contractual terms, attempt to link data sets, or attempt to reidentify the individuals represented in the data sets [80]. The general terms can be shared with patients or community groups to educate these groups about the limitations regarding how data can be used [80].
- While this chapter points out that the nature of data ownership is complicated, Birch et al. [2] advocate for determining ownership and decision-making among the collaborating partners. This agreement should also specify the licensing, intellectual property, royalties, and liability. Because data sets can create value, insurance may be needed to protect these assets [107].
- For data strategies involving blockchain consortia, organizations should create detailed governance agreements that describe the blockchain's operation, maintenance, and oversight [52].

5.3.2   Privacy Policies

For covered entities to comply with HIPAA or compliance with other jurisdictional privacy policies, the following is recommended:

- Organizations create detailed and robust privacy policies that outline all planned uses of information, including the likelihood and nature of data sales and removal of individuals' data, where possible [80].
- Organizations should have internal policies for managing requests for removing individuals' data or their permissions for using and sharing their data [40].

5.3.3   Tokens

- Organizations are encouraged to seek legal counsel to determine appropriate ownerships and rights for plans involving the use of blockchain-based tokens that will be traded as assets [65].
- Legal counsel is also recommended when organizations plan to create a blockchain-based exchange where data can be exchanged for utility tokens or security tokens that might be scrutinized by financial regulatory agencies [65].

5.3.4   Thorough Documentation

- Because there are regulatory implications for pricing and selling health information, organizations are encouraged to create detailed documentation about their data pricing and management strategies [32].

6   Future Directions

While blockchain-based data marketplaces and valuation strategies grow in response to the need for more data, organizations are encouraged to be vigilant about future activities with evolving regulatory requirements and advances in research.

6.1   *Regulations*

As noted in the Ethical Considerations section above, the sales and exchange of individuals' deidentified health information without their awareness may be legal under the U.S. HIPAA regulations. Still, HIPAA was designed before the emergence of large-scale health data sales and increasing risks of reidentification [80]. Currently, patients have little to no opportunity to redress the adverse consequences of data reidentification [80]. As U.S. state-level privacy statutes have been introduced to provide similar data protections offered in the General Data Protection Regulation, regulatory initiatives would broaden the nature of health information that could be classified as individually identifiable. These efforts would clarify protections of health information and reduce the risks of inadvertent reidentification.

When data are considered assets with value, Tlacuilo Fuentes [40] points out the need for legal recognition of the trade of personal information. A legal framework should organize such data exchanges by regulating the sale of personal data as a "good" so that this market will have more oversight [40]. The author notes that the sale and exchange of personal information is an international matter, and there should be recognition of the cross-border data trade.

6.2 Future Research

While life sciences organizations express interest in selling or licensing data sets, organizations express uncertainty about whether these sales are appropriate. There are concerns about the optics of such sales to their patient communities—and even the academic community [71]. As with medical data, research participants trust that their research team will protect their data, which could include sensitive health and behavioral information [1]. While human subject research regulations require data remain confidential (21 CFR § 56.111(a)(7); 45 CFR § 46.111(a)(7)) and not released into the public realm (45 CFR § 46.102(e)(4)), these regulations apply only to identifiable information where the identity could be discerned. Aggarwal et al. [108] found that 65.7% of patients/research are willing to share deidentified health data with universities, but only 24.4% are willing to share with commercial organizations that conduct health research (p. 7). Future research would be valuable to provide insights into research participant perceptions of life sciences data sales/exchanges and the circumstances where they would permit data sales and sharing.

As noted with the risks of reidentification, future research is needed about the privacy-preserving methods available using blockchain technologies and best practices. It would also be beneficial to learn if these privacy-preserving practices would increase individuals' willingness to support the use of their health information in life sciences research.

Last, additional investigation is warranted for identifying best practices of NFTs to establish data set provenance and rights. Tokenization is not well understood within life sciences organizations, and there have only been tentative approaches to apply NFTs to data sets thus far. Future research is needed to learn whether NTFs allow for more trust or security with data exchanges and whether these factors influence perceived data valuations [91].

7 Conclusions

Health-oriented data sets are increasingly sold and exchanged as more data are available to meet the need for scientific research [4]. The widespread growth of data marketplaces raises both opportunities and concerns about the best methods to value data sets [32] but also how to create defensible and sustainable economic approaches [77] that respect the best interests and privacy of individuals represented in the data sets [73].

The emerging market for data sales also raises awareness about the need to utilize technologies, such as blockchain, that can provide better balances of transparency and protections [93] while allowing patients/research subjects more control over how their information is used [51]. Blockchain-based data exchanges/marketplaces offer technological capabilities for data management that merit additional exploration for life sciences data. Specifically, the privacy-preserving features and dynamic consent options provide more insights and data controls than traditional data platforms [40]. Blockchain-based data monetization and assetization strategies, such as data NFTs, are still being developed, and future research is needed to determine feasibility and acceptance. Overall, blockchain technologies can accelerate data sales and innovation for the emerging data economy in life sciences research.

7.1 Key Terminology and Definitions

Asset: Any resource owned or controlled by an entity that can be used to produce positive economic value.

Covered entity: "(1) a health plan, (2) a healthcare clearinghouse, or (3) a healthcare provider who transmits any health information in electronic form in connection with a transaction" (45 CFR § 160.103).

Data marketplace: A data marketplace is an architecture or an online transactional location that facilitates data sharing and monetization. A data marketplace can also be a platform where data are shared and analyzed for internal business process optimization.

Digital data: Information created and stored in a computer-mediated environment that can potentially be transmitted as discrete information signals over the Internet and may be subsequently processed and/or stored for a range of known and unforeseen purposes.

Dynamic consent: The ability for individuals to independently change or withdraw informed consent options over time [98].

Health information: Any information, including genetic information, whether oral or recorded in any form or medium, that (1) is created or received by a healthcare provider, health plan, public health authority, employer, life insurer, school, or university, or healthcare clearinghouse and (2) relates to the past, present, or future physical or mental health or condition of an individual; the provision of healthcare to an individual; or the past, present, or future payment for the provision of healthcare to an individual. (45 CFR § 160.103).

Homomorphic encryption: A form of encryption that allows one to perform calculations on encrypted data without decrypting it first. The result of the computation is in an encrypted form. When decrypted, the output is the same as if the operations had been performed on the unencrypted data [106].

Individually identifiable health information: Information that is a subset of health information, including demographic information collected from an individual, and (1) is created or received by a healthcare provider, health plan, employer, or healthcare clearinghouse; (2) relates to the past, present, or future physical or mental health or condition of an individual; the provision of healthcare to an individual; or the past, present, or future payment for the provision of healthcare to an individual; and (i) that identifies the individual; or (ii) with respect to which there is a reasonable basis to believe the information can be used to identify the individual (45 CFR § 160.103).

Intangible assets: Non-monetary assets that are not physical, such as patents, copyright, trademarks, software, and, in some cases, data [2].

Non-fungible tokens (NFTs): Unique assets that cannot be divided and are not interchangeable, such as a photo or physical object [64].

Protected health information: Individually identifiable health information transmitted by electronic media, maintained in electronic media, for transmitted or maintained in any other form or medium (45 CFR § 160.103).

Reidentification: The condition where data thought to be anonymous is linked with other data that allow individuals to be identified.

Smart contract: A segment of code or a small computer program deployed designed to execute automatically when certain conditions are met. Nodes execute the smart contract within the blockchain network; all nodes must derive the same results for the execution, and the execution results are recorded on the blockchain [109].

Zero-knowledge proofs: "A protocol that enables one party, called prover, to prove that some statement is true to another party, called verifier, but without revealing anything but the truth of the statement." ([105], p 204448).

References

1. Demuro P, Petersen C, Turner P (2020) Health "big data" value, benefit, and control: the patient ehealth equity gap. Stud Health Technol Inform 270:1123–1127. https://doi.org/10.3233/SHTI200337
2. Birch K, Chiappetta M, Artyushina A (2020) The problem of innovation in technoscientific capitalism: data rentership and the policy implications of turning personal digital data into a private asset. Policy Stud 41(5):468–487. https://doi.org/10.1080/01442872.2020.1748264
3. Härting R-C, Sprengela A (2019) Cost-benefit considerations for data analytics—an SME-oriented framework enhanced by a management perspective and the process of idea generation. Procedia Comput Sci 159:1537–1546. https://doi.org/10.1016/j.procs.2019.09.324

4. Vezyridis P, Timmons S (2021) E-Infrastructures and the divergent assetization of public health data: expectations, uncertainties, and asymmetries. Soc Stud Sci 51(4):606–627. https://doi.org/10.1177/0306312721989818

5. Chiang A, Million RP (2011) Personalized medicine in oncology: next generation. Nat Rev Drug Discov 10(12):895–896. https://doi.org/10.1038/nrd3603

6. Food and Drug Administration (2018, December) Framework for FDA's real-world evidence program. https://www.fda.gov/media/120060/download. Accessed 12 Sept 2021

7. Food and Drug Administration (2021a, July 16) Real-world evidence. https://www.fda.gov/science-research/science-and-research-special-topics/real-world-evidence. Accessed 12 Sept 2021

8. Food and Drug Administration Center for Drug Evaluation (2021, July 16) FDA approves new use of transplant drug based on real-world evidence. https://www.fda.gov/drugs/news-events-human-drugs/fda-approves-new-use-transplant-drug-based-real-world-evidence. Accessed 7 Oct 2021

9. Food and Drug Administration Center for Devices and Radiological Health (2021, February 26) Examples of real-world evidence (RWE) used in medical device regulatory decisions. https://www.fda.gov/media/146258/download. Accessed 12 Sept 2021

10. Food and Drug Administration (2021b, August 4) Role of RWE in regulatory decision-making. https://www.fda.gov/drugs/news-events-human-drugs/fda-approval-demonstrates-role-real-world-evidence-regulatory-decision-making-drug-effectiveness. Accessed 12 Sept 2021

11. Tang C, Plasek JM, Bates DW (2018) Rethinking data sharing at the dawn of a health data economy: a viewpoint. J Med Internet Res 20(11):e11519. https://doi.org/10.2196/11519

12. Cutler JE (2019, January 29) How can patients make money off their medical data? Bloomberg Law. https://news.bloomberglaw.com/pharma-and-life-sciences/how-can-patients-make-money-off-their-medical-data. Accessed 19 Sept 2021

13. Elvidge S (2018, February 16) Roche buys cancer data company Flatiron Health for $1.9B. Biopharma Dive. https://www.biopharmadive.com/news/roche-buys-cancer-data-company-flatiron-health-for-19b/517285/. Accessed 19 Sept 2021

14. Ross C (2021, February 17) Backed by hospitals, Truveta wades into the business of selling health data. STAT. https://www.statnews.com/2021/02/17/truveta-patient-data-terry-myerson/. Accessed 19 Sept 2021

15. Office for Civil Rights (2009, January 7) Marketing: health information privacy. US Department of Health and Human Services. https://www.hhs.gov/hipaa/for-professionals/privacy/guidance/marketing/index.html. Accessed 19 Sept 2021

16. Fisher M (2019, November 13) Google-Ascension: why is HIPAA probably not being violated? Health IT Consultant. https://hitconsultant.net/2019/11/13/google-ascension-why-is-hipaa-probably-not-being-violated/. Accessed 19 Sept 2021

17. HIPAA Journal (2019, November 12) Google confirms it has legitimate access to millions of Ascension patients' health records. https://www.hipaajournal.com/google-confirms-it-has-legitimate-access-to-millions-of-ascension-patients-health-records/. Accessed 19 Sept 2021

18. Zozus MN, Bonner J (2017) Towards data value-level metadata for clinical studies. In: Lau F, Bartle-Clar J, Bliss G, Borycki E, Courtney K, Kuo A (eds) Building capacity for health informatics in the future, vol 234. IOS Press, Amsterdam, pp 418–423. https://doi.org/10.3233/978-1-61499-742-9-418

19. Li H, Li H, Wen Z, Mo J, Wu J (2017) Distributed heterogeneous storage based on data value. In: Proceedings of 2017 IEEE 2nd information technology, networking, electronic and automation control conference (ITNEC), pp 264–271. https://doi.org/10.1109/itnec.2017.8284985

20. Banterle F (2018) Data ownership in the data economy: a European dilemma (No. 3277330). SSRN. https://doi.org/10.2139/ssrn.3277330

21. Yamamoto Y, Yamaguchi A, Splendiani A (2018) YummyData: providing high-quality open life science data. Database. https://doi.org/10.1093/database/bay022

22. RDF Working Group (2014, February 25) Resource description framework. https://www.w3. org/RDF/. Accessed 27 Oct 2021
23. Robinson SC (2017) What's your anonymity worth? Establishing a marketplace for the valuation and control of individuals' anonymity and personal data. Digit Policy Regul Gov 39:88. https://doi.org/10.1108/DPRG-05-2017-0018
24. Data Repository Guidance (2021) Springer Nature. https://www.nature.com/sdata/policies/ repositories. Accessed 11 Oct 2021
25. Open for Business (2017) Sci Data 4:170058. https://doi.org/10.1038/sdata.2017.58
26. Valavi E, Hestness J, Ardalani N, Iansiti M (2020) Time and the value of data (No. 21-016). Harvard Business School. https://www.hbs.edu/ris/Publication%20Files/WP21-016_277b3482-f84f-4a6c-8dbc-00e6826bf1a2.pdf
27. Atkinson K, McGaughey R (2006) Accounting for data: a shortcoming in accounting for intangible assets. Acad Account Financial Stud J 10(2):85–96. https://www.abacademies. org/articles/aafsjvol1022006.pdf
28. Seh AH, Zarour M, Alenezi M, Sarkar AK, Agrawal A, Kumar R, Khan RA (2020) Healthcare data breaches: insights and implications. Healthcare (Basel) 8(2):133. https://doi.org/10.3390/ healthcare8020133
29. Wang D, Liu W, Liang Y, Wei S (2021) Decision optimization in service supply chain: the impact of demand and supply-driven data value and altruistic behavior. Ann Oper Res. https:// doi.org/10.1007/s10479-021-04018-y
30. Firica O, Manaicu A (2018) How to appraise the data assets of a company. Qual Access Success 19(S3):41–49. https://www.srac.ro/calitatea/en/arhiva/supliment/2018/Q-asContents_Vol.19_S3_October-2018.pdf
31. PwC EU Services (2018) Cost of not having FAIR research data: cost-benefit analysis for FAIR research data (No. KI-02-19-023-EN-N). European Commission. https://doi.org/10. 2777/02999
32. Schwartz R, Platten D, Nadell D (2020) How much is your data worth? Duff & Phelps. https://www.duffandphelps.com/-/media/assets/pdfs/webcasts/how-much-is-your-data-worth.pdf. Accessed 28 Sept 2021
33. Wayman C, Hunerlach N (2019) Realising the value of health data: a framework for the future (No. 003378-19Gbl). Ernst & Young Global Limited. https://assets.ey.com/content/ dam/ey-sites/ey-com/en_gl/topics/life-sciences/life-sciences-pdfs/ey-value-of-health-care-data-v20-final.pdf
34. Financial Accounting Standards Board (2001) Statement of financial accounting standards (No. 141). https://www.fasb.org/jsp/FASB/Document_C/DocumentPage?cid=121822 0124901&acceptedDisclaimer=true. Accessed 27 Oct 2021
35. Hitchner JR (2017) Financial valuation workbook: step-by-step exercises and tests to help you master financial valuation. Wiley. https://play.google.com/store/books/details?id=kAr GDgAAQBAJ
36. Lake P, Crowther P (2013) Data, an organisational asset. In: Undergraduate topics in computer science. Springer, Berlin, pp 3–19. https://doi.org/10.1007/978-1-4471-5601-7_1
37. Stack B (2017, December 6) Here's how much your personal information is selling for on the dark web. Experian. https://www.experian.com/blogs/ask-experian/heres-how-much-your-personal-information-is-selling-for-on-the-dark-web/. Accessed 21 Sept 2021
38. Trustwave Global Security Report (2019) Trustwave. https://www.trustwave.com/en-us/res ources/library/documents/2019-trustwave-global-security-report/. Accessed 21 Sept 2021
39. Chernyshev M, Zeadally S, Baig Z (2018) Healthcare data breaches: implications for digital forensic readiness. J Med Syst 43(7). https://doi.org/10.1007/s10916-018-1123-2
40. Tlacuilo Fuentes I (2020) Legal recognition of the digital trade in personal data. Mex Law Rev 12(2):87–117. https://doi.org/10.22201/iij.24485306e.2020.2.14173

41. Acxiom (2020, June 29) Identity Resolution. https://www.acxiom.com/identity-resolution-solutions/resolution/. Accessed 12 Oct 2021
42. Federal Trade Commission (2014, May) Data brokers: a call for transparency and accountability. https://www.ftc.gov/system/files/documents/reports/data-brokers-call-transparency-accountability-report-federal-trade-commission-may-2014/140527databrokerreport.pdf. Accessed 12 Oct 2021
43. van de Ven M, Abbas AE, Roosenboom-Kwee Z, de Reuver M (2021) Creating a taxonomy of business models for data marketplaces. In: Pucihar A, Kljajić Borštnar M, Bons R, Cripps H, Vidmar D, Perša J (eds) 34th bled eConference digital support from crisis to progressive change conference proceedings. University Maribor Press, pp 313–325. https://doi.org/10.18690/978-961-286-485-9.23
44. IQVIA (2021) Real world data sets. https://www.iqvia.com/solutions/real-world-evidence/real-world-data-and-insights. Accessed 12 Oct 2021
45. Täuscher K, Laudien SM (2018) Understanding platform business models: a mixed methods study of marketplaces. Eur Manag J 36(3):319–329. https://doi.org/10.1016/j.emj.2017.06.005
46. Grabis J, Stankovski V, Zariņš R (2020) Blockchain enabled distributed storage and sharing of personal data assets. In: Proceedings of the 2020 IEEE 36th international conference on data engineering workshops (ICDEW), pp 11–17. https://doi.org/10.1109/ICDEW49219.2020.00-13
47. Dai W, Dai C, Choo K-KR, Cui C, Zou D, Jin H (2020) SDTE: a secure blockchain-based data trading ecosystem. IEEE Trans Inf Forensics Secur 15:725–737. https://doi.org/10.1109/TIFS.2019.2928256
48. Ahmed E, Shabani M (2019) DNA data marketplace: an analysis of the ethical concerns regarding the participation of the individuals. Front Genet 10:1107. https://doi.org/10.3389/fgene.2019.01107
49. Mamoshina P, Ojomoko L, Yanovich Y, Ostrovski A, Botezatu A, Prikhodko P, Izumchenko E, Aliper A, Romantsov K, Zhebrak A, Ogu IO, Zhavoronkov A (2018) Converging blockchain and next-generation artificial intelligence technologies to decentralize and accelerate biomedical research and healthcare. Oncotarget 9(5):5665–5690. https://doi.org/10.18632/oncotarget.22345
50. Hayashi T, Ohsawa Y (2020) TEEDA: an interactive platform for matching data providers and users in the data marketplace. Information 11(4):218. https://doi.org/10.3390/info11040218
51. Zhu L, Dong H, Shen M, Gai K (2019) An incentive mechanism using Shapley value for blockchain-based medical data sharing. In: 2019 IEEE 5th Intl conference on big data security on cloud (BigDataSecurity), IEEE Intl conference on high performance and smart computing (HPSC) and IEEE Intl conference on intelligent data and security (IDS), pp 113–118. https://doi.org/10.1109/bigdatasecurity-hpsc-ids.2019.00030
52. Wang Z, Zheng Z, Jiang W, Tang S (2021) Blockchain-enabled data sharing in supply chains: model, operationalization, and tutorial. Prod Oper Manag 30(7):1965–1985. https://doi.org/10.1111/poms.13356
53. Lawrenz S, Andreas SPR (2019) Blockchain technology as an approach for data marketplaces. In: ICBCT 2019: proceedings of the 2019 international conference on blockchain technology, pp 52–59. https://doi.org/10.1145/3320154.3320165
54. Zhao Y, Yu Y, Li Y, Han G, Du X (2019) Machine learning based privacy-preserving fair data trading in big data market. Inf Sci 478:449–460. https://doi.org/10.1016/j.ins.2018.11.028
55. Nasonov D, Visheratin AA, Boukhanovsky A (2018) Blockchain-based transaction integrity in distributed big data marketplace. In: Shi Y, Fu H, Tian Y, Krzhizhanovskaya VV, Lees MH, Dongarra J, Sloot PMA (eds) Computational science—ICCS 2018. Springer, Berlin, pp 569–577. https://doi.org/10.1007/978-3-319-93698-7_43
56. Laoutaris N (2019) Why online services should pay you for your data? The arguments for a human-centric data economy. IEEE Internet Comput 23(5):29–35. https://doi.org/10.1109/mic.2019.2953764
57. Ciitizen—FAQ (2020, October 28) https://www.ciitizen.com/faq/. Accessed 22 Sept 2021

58. Healthcare Data: Best Databases & Providers (2021) Datarade. https://datarade.ai/data-cat egories/healthcare-data. Accessed 27 Sept 2021
59. BurstIQ Technology (2020, March 30) https://www.burstiq.com/technology/. Accessed 22 Sept 2021
60. Draskovic D, Saleh G (2017, December 28) Datapace: decentralized data marketplace based on blockchain. Datapace.io. https://datapace.io/datapace_whitepaper.pdf. Accessed 19 Sept 2021
61. Haenni R (2017) Datum network: the decentralized data marketplace. Datum. https://datum. org/assets/Datum-WhitePaper.pdf. Accessed 23 Sept 2021
62. Healthcare Blockchain Operating System (2021) Digital Treasury Corporation. https://phros. io/services/health_data_market. Accessed 23 Sept 2021
63. Enigma—Securing The Decentralized Web (2020) https://www.enigma.co/about/. Accessed 23 Sept 2021
64. Ante L (2021) The non-fungible token (NFT) market and its relationship with Bitcoin and Ethereum (No. 3861106). SSRN. https://doi.org/10.2139/ssrn.3861106
65. Stein Smith S (2020) Data as an asset. In: Stein Smith S (ed) Blockchain, artificial intelligence and financial services. Springer, Berlin, pp 213–239. https://doi.org/10.1007/978-3-030-29761-9_17
66. Sandner P, Tóth D, Siadat A, Weber N (2021, July 6) Data tokenization: morphing the most valuable good of our time into a democratized asset. Forbes. https://www.forbes.com/sites/philippsandner/2021/07/06/data-tokenization-morphing-the-most-valuable-good-of-our-time-into-a-democratized-asset/. Accessed 23 Sept 2021
67. Howcroft E, Carvalho R (2021, March 1) How a 10-second video clip sold for $6.6 million. Reuters. https://www.reuters.com/article/us-retail-trading-nfts-insight-idUSKCN2A T1HG. Accessed 12 Oct 2021
68. Hapiffah S, Sinaga A (2020) Analysis of blockchain technology recommendations to be applied to medical record data storage applications in Indonesia. Int J Inf Eng Electron Bus 12(6):13–27. https://doi.org/10.5815/ijieeb.2020.06.02
69. Copeland R (2019, November 11) Google's "Project Nightingale" gathers personal health data on millions of Americans. The Wall Street Journal. https://www.wsj.com/articles/goo gle-s-secret-project-nightingale-gathers-personal-health-data-on-millions-of-americans-115 73496790. Accessed 27 Sept 2021
70. Farr C (2019, November 19) Congressional democrats demand details on Google's use of patient data by Dec. 6. CNBC. https://www.cnbc.com/2019/11/18/google-ascension-health-data-deal-under-scrutiny-by-congressional-dems.html. Accessed 9 Oct 2021
71. Ledford H (2019) Google health-data scandal spooks researchers. Nature. https://doi.org/10. 1038/d41586-019-03574-5
72. Edenberg E, Jones ML (2019) Analyzing the legal roots and moral core of digital consent. New Media Soc 21(8):1804–1823. https://doi.org/10.1177/1461444819831321
73. Klugman C (2018, December 12) Hospitals selling patient records to data brokers: a violation of patient trust and autonomy. Bioethics.net. https://www.bioethics.net/2018/12/hospitals-sel ling-patient-records-to-data-brokers-a-violation-of-patient-trust-and-autonomy/. Accessed 29 Sept 2021
74. Clark K, Duckham M, Guillemin M, Hunter A, McVernon J, O'Keefe C, Pitkin C, Prawer S, Sinnott R, Warr D, Waycott J (2019) Advancing the ethical use of digital data in human research: challenges and strategies to promote ethical practice. Ethics Inf Technol 21(1):59–73. https://doi.org/10.1007/s10676-018-9490-4
75. Data Management: Intellectual Property and Copyright (2021, July 2) Kent State University Libraries. https://libguides.library.kent.edu/data-management/copyright. Accessed 26 Sept 2021
76. Feist Publications, Inc. v. Rural Telephone Service Co., Inc, No. 89-1909 (U.S. Supreme Court, March 27, 1991). https://scholar.google.com/scholar_case?case=119533626969805 6315&q=FEIST+v.+RURAL+1991&hl=en&as_sdt=4006&as_vis=1

77. Collins V, Lanz J (2019) Managing data as an asset. CPA J. https://www.cpajournal.com/2019/06/24/managing-data-as-an-asset/
78. Data Privacy: What the Consumer Really Thinks (2018) Direct Marketing Association. https://dma.org.uk/uploads/misc/5a857c4fdf846-data-privacy---what-the-consumer-really-thinks-final_5a857c4fdf799.pdf. Accessed 20 Sept 2021
79. Erdal BS, Liu J, Ding J, Chen J, Marsh CB, Kamal J, Clymer BD (2018) A database de-identification framework to enable direct queries on medical data for secondary use. Methods Inf Med 51(03):229–241. https://doi.org/10.3414/ME11-01-0048
80. Mandl KD, Perakslis ED (2021) HIPAA and the leak of "deidentified" EHR data. N Engl J Med 384(23):2171–2173. https://doi.org/10.1056/NEJMp2102616
81. Narayan A, Felten EW (2014) No silver bullet: de-identification still doesn't work. Princeton University. https://www.cs.princeton.edu/~arvindn/publications/no-silver-bullet-de-identification.pdf. Accessed 14 April 2021
82. De Sutter E, Zaçe D, Boccia S, Di Pietro ML, Geerts D, Borry P, Huys I (2020) Implementation of electronic informed consent in biomedical research and stakeholders' perspectives: systematic review. J Med Internet Res 22(10):e19129. https://doi.org/10.2196/19129
83. Brall C, Schröder-Bäck P, Maeckelberghe E (2019) Ethical aspects of digital health from a justice point of view. Eur J Public Health 29(Suppl 3):18–22. https://doi.org/10.1093/eurpub/ckz167
84. Time to Discuss Consent in Digital-Data Studies (2019) Nature 572(7767):5. https://doi.org/10.1038/d41586-019-02322-z
85. Chiauzzi E, Wicks P (2019) Digital trespass: ethical and terms-of-use violations by researchers accessing data from an online patient community. J Med Internet Res 21(2):e11985. https://doi.org/10.2196/11985
86. Yaga DJ, Mell PM, Roby N, Scarfone K (2018) Blockchain technology overview (No. 8202). National Institute of Standards and Technology. https://doi.org/10.6028/NIST.IR.8202
87. Wee R, Henaghan M, Winship I (2013) Ethics: dynamic consent in the digital age of biology: online initiatives and regulatory considerations. J Prim Health Care 5(4):341–347. https://doi.org/10.1071/hc13341
88. LunaPBC (2018, November 27) Can I lose shares in LunaDNA? LunaDNA Help Center. https://support.lunadna.com/support/solutions/articles/43000037181-can-i-lose-shares-in-lunadna-. Accessed 9 Oct 2021
89. Financial Accounting Standards Board (2008) Statement of financial accounting concepts No. 2 (No. CON2). https://www.fasb.org/jsp/FASB/Document_C/DocumentPage?cid=1218220132599&acceptedDisclaimer=true. Accessed 26 Sept 2021
90. International Society for Biocuration (2018) Biocuration: distilling data into knowledge. PLoS Biol 16(4):e2002846. https://doi.org/10.1371/journal.pbio.2002846
91. Bendechache M, Limaye N, Brennan R (2020) Towards an automatic data value analysis method for relational databases. In: Filipe J, Smialek M, Brodsky A, Hammoudi S (eds) Proceedings of the 22nd international conference on enterprise information systems. SciTePress, Science and Technology Publications, Lda, pp 833–840. https://doi.org/10.5220/0009575508330840
92. Zha C, Yin H, Yin B (2020) Data ownership confirmation and privacy-free search for blockchain-based medical data sharing. In: Zheng Z, Dai H-N, Fu X, Chen B (eds) Blockchain and trustworthy systems. Springer, Berlin, pp 619–632. https://doi.org/10.1007/978-981-15-9213-3_48
93. Yao L, Jia Y, Zhang H, Long K, Pan M, Yu S (2019) A decentralized private data transaction pricing and quality control method. In: ICC 2019—2019 IEEE international conference on communications (ICC). Shanghai, China. https://doi.org/10.1109/icc.2019.8761577
94. COMIRB (2020, September 17) Biomedical consent template with compound optional procedures and HIPAA authorization. COMIRB Forms, University of Colorado, Anschutz Medical Campus. https://research.cuanschutz.edu/docs/librariesprovider148/comirb_documents/forms/combined-consent-form-and-compound-hipaa-biomedical_9-17-20.doc. Accessed 26 Sept 2021

95. Kleschinsky JH, Bosworth LB, Nelson SE, Walsh EK, Shaffer HJ (2009) Persistence pays off: follow-up methods for difficult-to-track longitudinal samples. J Stud Alcohol Drugs 70(5):751–761. https://doi.org/10.15288/jsad.2009.70.751
96. Internal Revenue Service (2018, November 19) 2019 Instructions for Form 1099-MISC. U.S. Department of the Treasury. https://www.irs.gov/pub/irs-prior/i1099msc--2019.pdf. Accessed 26 Sept 2021
97. Mulder T, Tudorica M (2019) Privacy policies, cross-border health data and the GDPR. Inf Commun Technol Law 28(3):261–274. https://doi.org/10.1080/13600834.2019.1644068
98. Budin-Ljøsne I, Teare HJA, Kaye J, Beck S, Bentzen HB, Caenazzo L, Collett C, D'Abramo F, Felzmann H, Finlay T, Javaid MK, Jones E, Katić V, Simpson A, Mascalzoni D (2017) Dynamic consent: a potential solution to some of the challenges of modern biomedical research. BMC Med Ethics 18(1):4. https://doi.org/10.1186/s12910-016-0162-9
99. Albanese G, Calbimonte J-P, Schumacher M, Calvaresi D (2020) Dynamic consent management for clinical trials via private blockchain technology. J Ambient Intell Humaniz Comput 11(11):4909–4926. https://doi.org/10.1007/s12652-020-01761-1
100. Charles WM (2021) Accelerating life sciences research with blockchain. In: Namasudra S, Deka CG (eds) Applications of blockchain in healthcare. Springer, Berlin, pp 221–252. https://doi.org/10.1007/978-981-15-9547-9_9
101. Wei P, Wang D, Zhao Y, Tyagi SKS, Kumar N (2020) Blockchain data-based cloud data integrity protection mechanism. Future Gener Comput Syst 102:902–911. https://doi.org/10.1016/j.future.2019.09.028
102. Wang T, Wu X, He T (2019) Trustable and automated machine learning running with blockchain and its applications. SAS Institute, Inc. http://arxiv.org/abs/1908.05725
103. Rahman MA, Hossain MS, Islam MS, Alrajeh NA, Muhammad G (2020) Secure and provenance enhanced internet of health things framework: a blockchain managed federated learning approach. IEEE Access 8:205071–205087. https://doi.org/10.1109/ACCESS.2020.3037474
104. Burki T (2019) Pharma blockchains AI for drug development. Lancet 393(10189):2382. https://doi.org/10.1016/S0140-6736(19)31401-1
105. Tomaz AEB, Nascimento JCD, Hafid AS, De Souza JN (2020) Preserving privacy in mobile health systems using non-interactive zero-knowledge proof and blockchain. IEEE Access 8:204441–204458. https://doi.org/10.1109/access.2020.3036811
106. Zhou L, Wang L, Ai T, Sun Y (2018) BeeKeeper 2.0: confidential blockchain-enabled IoT system with fully homomorphic computation. Sensors (Basel) 18(11). https://doi.org/10.3390/s18113785
107. Fernandez RC, Subramaniam P, Franklin MJ (2020) Data market platforms: trading data assets to solve data problems. In: Balazinska M, Zhou X (eds) Proceedings of the VLDB endowment, vol 13. Association for Computing Machinery, pp 1933–1947. https://doi.org/10.14778/3407790.3407800
108. Aggarwal R, Farag S, Martin G, Ashrafian H, Darzi A (2021) Patient perceptions on data sharing and applying artificial intelligence to health care data: cross-sectional survey. J Med Internet Res 23(8):e26162. https://doi.org/10.2196/26162
109. Alharby M, Aldweesh A, van Moorsel A (2018, November) Blockchain-based smart contracts: a systematic mapping study of academic research. In: Proceedings of the 2018 international conference on cloud computing, big data and blockchain (ICCBB). Fuzhou, China. https://doi.org/10.1109/iccbb.2018.8756390

Dr. Wendy Charles has been involved in clinical trials from every perspective for 30 years, with a strong background in operations and regulatory compliance. She currently serves as Chief Scientific Officer for BurstIQ, a healthcare information technology company specializing in blockchain and AI. She is also a lecturer faculty member in the Health Administration program at the University of Colorado, Denver. Dr. Charles augments her blockchain healthcare experience by serving on the EU Blockchain Observatory and Forum Expert Panel, HIMSS Blockchain Task Force,

Government Blockchain Association healthcare group, and IEEE Blockchain working groups. She is also involved as an assistant editor and reviewer for academic journals. Dr. Charles obtained her PhD in Clinical Science with a specialty in Health Information Technology from the University of Colorado, Anschutz Medical Campus. She is certified as an IRB Professional, Clinical Research Professional, and Blockchain Professional.

Brooke Delgado has 14 years of professional finance and accounting experience in various industries such as retail, manufacturing, service, and technology. She currently services as Director of Finance for BurstIQ, a healthcare-oriented blockchain company. Brooke sits on the Accounting Advisory Council at the University of Colorado—Denver, where she participates in guiding the direction of the Accounting Department and community involvement with the accounting program. Brooke obtained her MS Accounting degree from the University of Colorado—Denver and her BSBA Finance degree from the University of Missouri—Columbia. Brooke is a member of the Institute of Management Accountants.

Blockchain Adoption in Life Sciences Organizations: Socio-organizational Barriers and Adoption Strategies

Chang Lu

Abstract The goal of this chapter is to reveal socio-organizational barriers for blockchain adoption in life sciences organizations and delineate the strategies that managers and executives can undertake to facilitate adoption. By drawing on the literature on innovation adoption and organizational culture and studying real-time blockchain adoption, this chapter describes the following socio-organizational barriers for blockchain adoption in life sciences organizations: (1) negative stereotypes of blockchain technology, (2) perceived technological complexity of blockchain, (3) highly institutionalized nature of life sciences organizations, and (4) lack of ecosystem mindset. The chapter also reveals strategies that can facilitate adoption: (1) holistic evaluation of blockchain life sciences use cases, (2) framing blockchain adoption as aligned with the innovative and safety-driven culture in life sciences organizations, (3) unobtrusive implementation, and (4) acting swiftly when the innovative culture gains strength. These insights draw attention to the overlooked socio-organizational aspects of blockchain adoption in life science and offer practical insights to make blockchain adoption a reality.

Keywords Blockchain adoption · Socio-organizational barriers · Adoption strategies · Organizational culture · Resources · Life sciences

1 Introduction

In recent years, blockchain is emerging as a transformative technology to address data challenges in life science. By combining cryptography and decentralized internet, blockchain makes it extremely difficult to change or erase data, producing an exceedingly high degree of data immutability, provenance, and transparency [1, 2]. Researchers have suggested that using blockchain in life sciences could help monitor compliance with regulations, improve patient and trial safety, and enhance the credibility of clinical research [3].

C. Lu (✉)
Blockchain@UBC, University of British Columbia, Vancouver, BC, Canada
e-mail: chang.lu@ubc.ca

© The Author(s), under exclusive license to Springer Nature Singapore Pte Ltd. 2022   175
W. Charles (ed.), *Blockchain in Life Sciences*, Blockchain Technologies,
https://doi.org/10.1007/978-981-19-2976-2_8

Despite the transformative potential of blockchain, the academic and practical discussions about blockchain in life sciences have focused on exploring new use cases and the design and technical parameters of the use cases [4, 5]. Very little attention has been paid to what life sciences organizations experience when adopting blockchain, what socio-organizational barriers they encounter, and what organizational strategies could facilitate adoption.

It is important to address these gaps. As scholars have suggested, the more an emerging technology represents a radical innovation, the more likely it will face strong socio-organizational barriers during adoption [6, 7]. Without a solid understanding of those barriers and the strategies to overcome the barriers and facilitate adoption, we cannot materialize the transformative potential of the technology, regardless of the care given to its technical design [8]. As blockchain represents a decentralized solution to data challenges in life science, radically departing from existing centralized solutions, the socio-organizational barriers and adoption strategies are likely to be crucial to the outcome of mass adoption and must be understood thoroughly.

This chapter reveals the socio-organizational barriers that hinder and the strategies that facilitate blockchain adoption in life sciences organizations. This chapter is based on an empirical study of a real-time blockchain adoption use case in a global life sciences organization. The author followed a standard qualitative case-study approach, collected 28 semi-structured interviews with managers and experts that participated in the adoption process, as well as documents and meeting minutes, and analyzed the data using a grounded theory approach [9].

Blockchain adoption may face the following socio-organizational barriers in life sciences organizations: (1) negative stereotypes of blockchain technology, (2) perceived technological complexity of blockchain, (3) highly institutionalized nature of life sciences organizations, and (4) lack of ecosystem mindset. These barriers compound the uncertainty and ambiguity of blockchain technology, making it difficult for mass adoption to occur. However, drawing from the literature on organizational culture and a real-time blockchain pilot project that gained initial success, the following strategies might increase the likelihood of adoption: (1) holistic evaluation of blockchain life sciences use cases, (2) framing blockchain adoption as aligned with both the innovative and safety-driven culture in life sciences organizations, (3) unobtrusive implementation, and (4) acting swiftly when the innovative culture gains strength. These findings fill the gap that socio-organizational understandings of blockchain adoption are limited and hold the potential to advance the practice of blockchain adoption.

2 Background Literature

Blockchain is referred to as a "distributed ledger with confirmed blocks organized in an append-only sequential chain using cryptographic links" ([1], s. 3.6). This technology represents an enabler for decentralized information systems that are superior in privacy and security while less susceptible to "single point of failure." By using cryptography and algorithm-generated public–private key structure, blockchain can protect user privacy better than traditional password-based systems. Specifically, blockchain makes security attacks and data tampering extremely difficult through hash function and the chained structure of information blocks. Moreover, as a peer-to-peer network where each node holds a copy of all transactions, blockchain avoids the "single point of failure" inherent in centralized data systems. More importantly, these technical properties render it possible to govern information in decentralized mechanisms.

All in all, blockchain is considered to transform information management. Given the importance of information in the digital era, blockchain may bring fundamental changes to business and social systems [2]. As such, scholars and practitioners have explored blockchain use cases in a wide range of sectors, such as finance, agriculture, retail, and life sciences [10].

Use cases in multiple business domains have been proposed in life sciences, including clinical trials, drug supply chains, payments, and consumer health records [3, 4]. So far, the literature has focused on exploring the value propositions of the use cases and prototyping application design. For example, Tseng et al. [11] showed how blockchain could enhance the traceability of drug supply chains and proposed a governance framework based on the Gcoin. Metcalf et al. [5] discussed how blockchain might enable decentralized clinical and genomic data marketplaces, laying out the key value propositions and stakeholders. Recently, the design aspect of blockchain in life sciences has become nuanced, as specific technical measures are proposed to accommodate the characteristics of health data and regulations. For instance, Dubovitskaya et al. [12] suggest storing original clinical data off-chain and only meta transaction data on-chain. Clinical data are often too large to be stored on-chain, and some instances of on-chain storage may not be compliant with the U.S. Health Insurance Portability and Accountability Act (HIPAA). They also propose Member Services as a unique node to certify and grant access to member nodes on the blockchain.

The literature has also begun to explore the adoptability of blockchain in life sciences from the originators of human data. Lu et al. [13] explored consumer attitude towards blockchain-based health records focusing on consumer health records. They demonstrated that consumers are not necessarily ready for the application because of the concerns about losing their private keys and the inability to revoke data access. Khurshid et al. [14] examined clinical data sharing behavior in a simulated setting. They found that people may initially exhibit faulty sharing behaviors and the errors decrease over time, suggesting that users can adapt to blockchain-based clinical data sharing.

Despite the advancements above, the literature lacks an understanding of the socio-organizational barriers for blockchain adoption in life sciences and the strategies that may facilitate adoption. This is a critical omission because the literature on innovation adoption [6–8], as well as on electronic laboratory records [15–20], strongly suggests that blockchain adoption is likely to encounter socio-organizational barriers. The neglect of these barriers could lead to adoption failure.

The literature on innovation adoption has long suggested that new practices or technologies encounter socio-organizational barriers when introduced into a new organization, industry, or country [6–8, 21, 22]. These barriers stem from the inertia of the "old" socio-organizational systems, whether behavioral routines, the entanglement of social networks, entrenched material interests in old systems, or taken-for-granted values and beliefs. The stronger the inertia of the "old," the more barriers new practices or technologies are likely to face [6, 7]. Research has shown that these barriers can significantly delay or change the trajectory of practice or technology adoption. For example, Barley [23] showed that the inertia of the interactions between professionals has strongly affected the adoption of body imaging technology and the technology had to adapt to local clinical practices. Dougherty and Dunne [24] found that scientists with different backgrounds were embedded in different knowledge systems that hindered their collaborative adoption of digital technologies in life sciences laboratories. As such, scholars suggested that due to socio-organizational barriers, new practices or technologies must be made to fit with local structural, cultural, and political arrangements [8].

In the literature on the Electronic Lab Notebooks (ELN), researchers have found that the implementation of ELN in life sciences labs, a technology that matured decades ago, met a variety of socio-organizational barriers. For instance, Kanza et al. [18] found that the low adoption of ELN in life sciences research institutions might be due to the behavioral inertia of lab participants, who would shift back to paper-based notebooks even after being made aware of ELN. Also, Kanza [17] and Zupancic et al. [20] suggest that scientists sometimes hesitate to adopt ELN because they believe that the implementation process would disrupt workflow. These socio-organizational barriers have delayed the adoption of ELN and contributed to the operation inefficiencies experienced by scientists [20].

Blockchain is more technically complex than ELN and represents a paradigm shift in how information is managed in life sciences. Therefore, blockchain is bound to meet more substantial socio-organizational barriers than those encountered by ELN. Without understanding those barriers and coping strategies, blockchain adoption in life sciences may suffer a more challenging path than witnessed by ELN.

3   Research Methods

The aim of this study was to understand the socio-organizational barriers for blockchain adoption and the responding strategies using a qualitative approach. A qualitative approach is well suited to explore emerging or poorly understood phenomena and is commonly used by behavioral researchers to explore the barriers to emerging technology adoption [23, 25, 26]. It can provide rich, detailed descriptions of barriers that early adopters experienced and their actions to tackle them. The qualitative approach also promises to develop more abstract, widely relatable categories of the barriers, informing both researchers and practitioners.

The primary data source involved 28 semi-structured interviews with directors, managers, and executives in the life sciences sector who have been involved in the adoption of blockchain. A purposeful sampling method was adopted [27], targeting experienced life sciences professionals with first-hand experience of blockchain adoption. A pharmaceutical company known to implement blockchain projects was contacted to provide contact information for those who have participated in the blockchain projects; the company named 15 individuals working on blockchain adoption in two locations. Twelve agreed to participate in the interview. In the meantime, interview invitations were sent to 26 individuals identified on LinkedIn after searching for "blockchain" and "life science." Ten decided to participate. In addition, the author attended three multi-day blockchain conferences with sessions discussing blockchain and reached out to the speakers and attendees, inquiring about their professional backgrounds and interest in an interview. Six individuals at those conferences agreed to the interview. All individuals work in middle-senior professional roles, such as manager, Information technology (IT) manager, blockchain consultant, enterprise innovation manager, compliance, and regulatory director. Among others, all individuals have either led or participated in blockchain life sciences projects. Table 1 shows the number of study participants in different professional roles.

Participants were interviewed by following a semi-structured format. Participants were first asked to describe their involvement in the adoption of blockchain in life science. Based on their different levels of involvement, questions were tailored about their socio-organizational barriers to adoption and how they have overcome those barriers. The interviews took between 30 min and one hour. Except for three interviews where notes were taken, all others were recorded and transcribed. Apart from the 28 primary interviews, the author also collected 13 secondary, publicly available interviews were evaluated that broadly discussed the adoption of blockchain in life sciences and health. Blockchain advocates often interview each other and publish the interviews on social media (e.g., YouTube or Vimeo) as part of the effort to prompt adoption. These interviews contain information that helps to understand the nature of blockchain and touch on the barriers for blockchain in life sciences organizations. In addition, the author collected 193 news articles from major blockchain websites (e.g., coinbase.com, blockchain.news.com, cointelegraph.com) that cover

Table 1 Professional roles of primary interview participants	Professional role	Number of participants (n = 28)
	Middle managers in life sciences	11
	Senior Executives in life sciences	2
	Blockchain + life sciences start-up leaders (co-founder or CEO)	9
	Blockchain + life sciences experts or consultants	5

blockchain adoption in life sciences or health. The secondary interviews and news-paper articles constitute a rich, comprehensive pool of background information to support the understanding of the primary interviews.

Data were analyzed using the grounded theory method, following the steps commonly used in the literature [28, 29]. The first step was to extensively read the secondary interviews and documents, by which a general understanding of blockchain technology was gained and its applications in life sciences settings. Next, the author conducted thematic coding of the 28 primary interviews in Nvivo 12.0, summarizing the surface meaning of words, phrases, and sentences concerning the socio-organizational behavioral barriers to adopting blockchain in life sciences orga-nizations. This step generated 812 first-order codes. Axial coding was then conducted to aggregate first-order codes that share underlying meanings into second-order cate-gories. The author then moved back and forth between the first-order codes and second-order categories to continuously adjust the categorization of the first-order codes. When the second-order categories captured the meaning structure of the first-order codes, the second categories were aggregated into the third-order dimensions through multiple iterations.

4 Findings

To understand the barriers of blockchain adoption in life sciences organizations, it is essential to first understand the current state of the blockchain + life sciences ecosystem since the ecosystem is the context in which the barriers are experienced. As such, the findings on the key characteristics of the ecosystem are presented.

4.1 The State of the Blockchain + Life Sciences Ecosystem

The concept of "ecosystem" refers to "a group of interacting firms that depend on each other's activities" ([30], p. 1). According to the literature, a business ecosystem comprises ecosystem boundaries, governance structures, relational networks among participants, and shared practices [7, 31]. Thus, we can understand the nature of a business ecosystem by examining its components.

The qualitative coding suggests that the components of the blockchain + life sciences ecosystem have the following characteristics:

4.1.1 The Ecosystem Boundary is in Constant Flux

Boundaries are conceptual and group distinctions made by individuals and organiza-tions to categorize reality [32]. In the context of a business ecosystem, the system's boundary demarcates legitimate members from those who are not. Boundaries can exist as formal membership rules or informal shared understandings of the resources and identities of new entrants. In the blockchain + life sciences ecosystem, however, such boundaries are in a state of constant flux. New entrants can almost freely claim their participation in the system, regardless of their resource base or organizational identity. As undistinguished entrants crowd into the system, they tend to misinterpret or misuse terminology, ideas, and concepts, increasing the system's opaqueness. As shown by the following field note and interview excerpts:

> The conference is organized in a different format from academic conferences. Many speakers are start-up leaders. I went to the information desk of a start-up whose leader spoke 15 min ago. The information booklet is incomplete. It states the company focuses on blockchain and genomic data marketplace, but there is no business model. (Field note, 20190407)
>
> I'm invited to speak at many conferences. You can say you are a blockchain company and you get to speak. (Interview, blockchain start-up founder)

4.1.2 The Governance Structure is Developing but as yet to Form

The governance structure of a business ecosystem consists of "governance organizations or associations within the field whose sole job is to ensure the routine stability and order of the field" and the rules that these organizations enact ([33], p. 77). The governance organizations formalize rules and regulations, designate power structures, oversee practices, ensure compliance, and penalize misconduct.

In a blockchain + life sciences ecosystem, some organizations are attempting to develop governance structures. For example, there are discussions in the Institute of Electrical and Electronics Engineers about setting industry standards for blockchain in health and life sciences, organizing industry associations (e.g., various consortiums such as MELLODDY), and creating industry-wide task forces. However, all these efforts seem to be at an early stage. No evidence suggests that a mature government structure has been established in the blockchain + life sciences ecosystem.

4.1.3 The Relational Network is Forming but Thin in Content

Relational networks are the backbone of an ecosystem [30, 31]. It consists of relationships among participants, resources, information, and trust exchanged through the relationships—in other words, the content of ties [7, 34]. The higher degree of interconnectedness among system participants, the more stable the system. Besides, with ties characterized by trust, the system is more likely to optimize the utilization of resources [30, 35].

This analysis shows that relational network is forming but continues to evolve in the blockchain + life sciences ecosystem. A few central organizations have emerged, such as ConsenSys, which has footsteps in life sciences, the IBM blockchain team that covers life sciences projects, Microsoft Azure, and BurstIQ. These organizations

actively participate in public speaking, research, business development, and education. As such, these organizations are given higher status by other participants in the
ecosystem. However, the rest of the network seems in flux.

> I know most of the start-ups. But I think most of them have gone out of business except a
> few. (Interview, start-up founder)

Notably, the relational network is thin in content because the resources exchanged
through the ties are limited. Although start-ups, universities, and pharmaceutical
companies have organized seven major consortiums, the consortiums have yet to
deliver substantial outcomes.

4.1.4 Proto-Practices Being Explored

Practices in the context of ecosystems are defined as "behaviors, strategies, ideas,
technologies, or structures that have obtained a social fact quality [that] renders
them as the only conceivable, 'obvious,' or 'natural' way to conduct an organizational activity" ([36], p. 229). Based on this definition, practices have certainly not
taken form in the blockchain + life sciences ecosystem. Nevertheless, proto-practices
are being explored by various organizations. For example, the MELLODDY project
is exploring federated learning (combining blockchain and machine learning) for
molecule data sharing. Even though federated learning is still a developing technology, EncrypGen, and Nebula Genomics are exploring decentralized, blockchain-
enabled genomics data marketplaces. Regardless of the stage of these experimentations, experts understand that the experimentations have not given rise to conclusions
as to what kind of blockchain applications in life sciences are the most feasible, how
to implement them, and what structures could secure implementation.

> I don't think we know the ultimate use case. In fact, I don't know [that] many people
> understand what blockchain can do and what it cannot do. This year, there has been so much
> confusion... (Interview, blockchain expert)

Having described the key characteristics of the blockchain + life sciences
ecosystem, the socio-organizational barriers that early adopters have experienced
will be described.

4.2 Socio-organizational Barriers for Blockchain Adoption in Life Sciences

The interviews and documents show that three types of blockchain use cases can
be distinguished based on primary goals: use cases that aim to improve compliance,
facilitate research data sharing and transactions, and enable individuals to monetize
personal data. These different types of use cases are associated with the increased

necessity to engage multiple stakeholders and build permissioned blockchain solutions. Specifically, the use cases that aim to improve compliance with regulations can be realized on private blockchains inside a particular organization. Those aiming to facilitate data sharing may require establishing "consortiums" among different organizations and creating blockchain solutions that only consortium participants can access. Last, the use cases aiming to enable personal data monetization may need permissioned blockchains that contain delicate smart contracts. Regardless of use cases, early blockchain adoption has commonly experienced the following socio-organizational barriers.

4.2.1 Barrier 1: The Negative Stereotype About Blockchain Technology

Participants shared that the negative stereotype about blockchain technology has become a significant roadblock when they attempt to create "buy-in." Many participants experienced indifference or dismissal from regulators, patients, and executives, due to the widespread perception that blockchain is a "troublesome" technology linked to cryptocurrency scams, opportunistic investing, money laundering, or financial schemes. Besides, since early blockchain promoters often used evangelical language such as "blockchain will change the world," "blockchain will give power to normal people." Since blockchain's first use case—Bitcoin-was created by a mythical figure (Satoshi Nakamoto) and embodies the ideology of decentralization and anarchism, blockchain is often perceived as supporting a "cult" or alike. As one participant depicted: "There is much skepticism about blockchain. When you talk about blockchain, people think about Bitcoin. They think you are a hippy and want to take them into the scheme" (Participant 18). The stereotype that blockchain is associated with schemes or special agendas has made it increasingly difficult for early adopters to promulgate adoption. Study participants reported that when they engage colleagues or stakeholders for support or resources, they are often met with the attitude that "I don't want to hear blockchain anymore." In response, they had to mention the word "blockchain" as little as possible when they attempted to create buy-in. Some participants opted for "distributed trust technology" as a substitute for blockchain, and others emphasized that blockchain advocates must make a strong case that blockchain can add practical value to life sciences, as shown by the following quotes:

> We have done many presentations [about blockchain], but many times, people look at us sideways. It feels like they just don't want to hear about blockchain anymore. So, we have to do all sorts of analysis, cost–benefit analysis, risk analysis, competitive analysis just to make sure they understand where we come from and what is the business value [of blockchain]. (Participant 10)

4.2.2 Barrier 2: Perceived Technological Complexity of Blockchain

The second barrier study participants pointed out is the perceived technological complexity of blockchain, which somewhat contributes to the difficulty overcoming the first barrier. Study participants report that the complexity of blockchain technology has created the dilemma of whether they should explain the technical foundations of blockchain to decision-makers. If they show blockchain's highly developed computational methods and network infrastructure, not only are most of the decision-makers unable to follow and understand how blockchain is technically sound but also the study participants themselves may have technical "blind spots." "Sometimes it gets too technical that I, myself, find it hard to explain" (Participant 2). The display can discredit the blockchain technology and reinforce the impression that the blockchain community is "religious" or backed by some schemes. On the other hand, if they do not explain blockchain in technical terms, decision-makers would not have the opportunity to develop trust for blockchain and the early adopters who promulgate further adoption. This dilemma makes it challenging to educate decision-makers (as well as laypeople) about blockchain, slowing down the purging of the negative stereotypes of blockchain.

4.2.3 Barrier 3: The Highly Institutionalized Nature of Life Sciences Systems

Blockchain is in its infancy, facing substantial technological and economic uncertainties. However, life sciences are highly institutionalized [37, 38], meaning that there are rule systems practices, role identities, and shared understandings of rewarded behaviors. In specific terms, life sciences professionals understand who the authorities are, what regulations or protocols to follow, how to reserve and analyze data, the responsibilities of different roles, and the risks of deviating from those rules. Study participants describe that the highly institutionalized nature of life sciences has made blockchain adoption very challenging. First, as the field of life sciences is heavily regulated by government and professional authorities, using blockchain to alter the structure and process of life sciences (e.g., enable individuals to monetize their genomics data by selling the data on a blockchain-based exchange) runs the risk of deviating from established rules. As participants described, the field of life sciences is highly regulated by authorities. While regulations are intended to be technology-agnostic, life sciences organizations may be hesitant to pursue substantial adoption of blockchain without authority clarification.

> In clinical trials, I think that the authorities are a main actor where you need to convince them and explain why they can trust what you're building. But they are very conservative. The bottom line is patient safety. (Participant 11)

The reliance on organizational decision-makers also creates a "chicken-and-egg" problem; life sciences organizations would not pressure top-down adoption unless they see substantial benefits of blockchain. However, the benefits of blockchain (determined by its distributed nature) can only be realized when there is large-scale adoption. As such, the blockchain projects that study participants worked on are commonly incremental; the change they bring is significantly less than that promised in the blockchain discourse.

Haha, yes, exactly. We have a chicken and egg problem. As a start-up, the only thing we can do is build relationships as much as possible…. We are only taking small steps. (Participant 28)

Second, despite continuous adjustment, the institutionalized practices of life sciences have been largely stable, and the same set of structures and processes have become taken for granted [37, 38]. Study participants describe that this "taken-for-grantedness" of life sciences practices tends to perpetuate itself, supported by shared beliefs such as "this is how we do clinical trials" and "this is how insurance transactions are handled" (Participant 24). As a result, even though advocators of blockchain proposed many blockchain applications, their colleagues and stakeholders often put those proposals aside because they could not conceive of doing life sciences differently with blockchain.

Third, as the life sciences field is highly institutionalized, individuals and organizations have entrenched material and political interests into existing practices. However, blockchain is claimed to transform existing practices by eliminating some manual tasks (e.g., calibrating trial data or cross-validating research output) and re-arranging workflows. Therefore, adopting blockchain would mean a shock to entrenched interests. Study participants report that they worry that blockchain may break down the connection between material, political interests, and existing practices have made some individuals resistant to blockchain.

From a business standpoint too, you have to remember, there's hundreds of thousands of people who have jobs in creating risk profiles for patients in insurance companies, and they [inaudible 00:22:26]. And to tell them that overnight, we're going to completely change the nature of your job, there's going to be a lot of pushback. (Participant 15)

Unfortunately, study participants report that they do not anticipate the highly institutionalized nature of life sciences will change any time soon because the field must be highly institutionalized for pragmatic and ethical reasons. Taking that as a given, blockchain advocators may promulgate adoption through small-scale innovations over a long period.

4.3 Barrier 4: The Lack of an "Ecosystem" Mindset

Another barrier that study participants commonly experienced is the lack of an "ecosystem" mindset among those considering adopting blockchain. An "ecosystem" mindset means that different businesses consider each other as complementors, prioritize the building of system infrastructure, and adopt business strategies that would make the "pie" bigger for all participants to grow [30]. Study participants report that the ecosystem mindset is crucial for blockchain adoption, given that blockchain is a distributed ledger technology. However, participants describe that those at the demand side of blockchain solutions have yet to consider each other as complementors and forge ecosystem-building at the level that blockchain developers would prescribe. Despite the other efforts to build industry-level infrastructure for blockchain adoption, for example, by forging consortiums consisting of research institutions and pharmaceutical companies, participants report that life sciences organizations have not demonstrated the "ecosystem" mindset. In executive meetings, the conversations about using blockchain to facilitate trial data sharing often have yet to yield substantial outcomes; some report that those conversations seemed to be symbolic gestures due to the perpetuating, siloed way of thinking.

> I think interoperability is a technical priority. The Ethereum community knows that there is not going to be a single blockchain… But I think the top priority for many companies is building business value on top of Ethereum [for themselves]. (Participant 6)

On the supply side, established IT firms and blockchain start-ups have yet to do businesses in an ecosystem fashion, continuing to show the "competition" mindset and consider each other as competitors despite some initial conversations about collaboration. Study participants describe that this lack of ecosystem mindset slows down blockchain adoption. Blockchain start-ups who face tremendous, immediate financial pressure cannot explore the most cutting-edge applications by leveraging established IT firms' capabilities. As a result, they go out of business quickly. In the meantime, established IT firms may not develop easily adaptable applications by leveraging the creativity and flexibility of start-ups. Consequently, it becomes challenging to create a supply market for blockchain applications in life sciences. (Note that some participants suggest that academic institutions must be engaged to infuse the ecosystem mindset to blockchain consumers and suppliers.)

> We [as a blockchain-life sciences start-up] can't wait for years because we have a responsibility to our investors to create value for them… We'd like to partner with those big companies, but we haven't seen much interest. I don't think we have reached the point where everyone understands the ecosystem. (Participant 3)

4.4 Adoption Strategies

Despite that the blockchain + life sciences ecosystem is at the early stage of development and that early adopters have experienced the barriers above, interview participants who gained initial success in pilot projects suggested that specific strategies may facilitate adoption. These strategies are described in the following section.

4.4.1 Holistic Evaluation of Blockchain Life Sciences Use Cases

Interview participants suggested that the first step for successful adoption is to perform a holistic evaluation of the proposed use case. Throughout the interview and document data, four factors were described as crucial in determining the adoptability of a blockchain + life sciences use case, and the factors need to be considered holistically.

- **Economic value of the use case**: As many interviewees pointed out, the economic value is the bottom-line factor to consider when adopting blockchain in life sciences, in other words, "what business problems can blockchain solve?" or "how much value can blockchain add to existing business models?" In particular, interviewees cautioned that potential adopters should not adopt blockchain for the sake of blockchain (e.g., adopting blockchain for religious reasons) but for the practical benefits.

- **Readiness for adoption**: Readiness for adoption includes two aspects: technological readiness and behavioral readiness. Technological readiness means the extent to which technological infrastructure is developed for the mass adoption of blockchain, consisting of indicators such as the electrocyclization of data, the availability of blockchain technical knowledge, and the interoperability of blockchains. Behavioral readiness refers to the extent to which those who need to participate in the adoption are willing and conditioned to incorporate blockchain into their current pattern of technology use. Indicators may include perceived benefits of blockchain, perceived ease of adoption, and perceived necessity [5]. As interviewees described, the lack of either technological or behavioral readiness dramatically reduces the likelihood of mass adoption.

  In labs, a lot of work is still done on paper. How can you create a blockchain solution when everything has to be on paper? (Interview, start-up founder)

- **Compliance with existing rules**: The field of life sciences is highly regulated by regulations and organizational rules. There exist taken-for-granted rule systems practices and role identities, as well as shared understandings of rewarded behaviors. Although the multilayer rule system may cause managerial inefficiencies, some interview participants also maintain that the rules are necessary as they help people behave in a way that produces good science.

Adopting blockchain may ultimately lead to the transformation of the rules in life sciences. However, this study suggests that blockchain use cases should comply with existing rules in the foreseeable future. If the use cases are at odds with existing rules, life sciences professionals who must be on board to implement the case may not be inclined to do so as they may fear personal penalty. As mentioned in the previous section, the decision-makers who create and sustain the rules in life sciences are often the same ones who make vital decisions during blockchain adoption. If a blockchain use case violates the rules they created, they are unlikely to endorse it. As such, compliance with existing rules is another factor that determines the adoptability of a blockchain use case.

- **The preservation of data assets, intellectual property, and privacy**: As blockchain is a distributed ledger, the benefits of blockchain rely much on data sharing, and synchronization via blockchain. However, organizations are dispositioned to protect their commercial interests in a business ecosystem, which means protecting the intellectual property crucial for their competitive advantage. Indeed, a vital issue that blockchain experts are concerned about is how to protect data privacy when blockchain is meant to improve data transparency. Smart contracts represent one way to address this issue.

In life sciences, privacy is of paramount importance. Whether protecting clinical research data or genomic data, unrestricted access to the data has severe commercial, ethical, and legal consequences. Thus, blockchain use cases that utilize these data must protect the privacy of each data contributor (node) to the maximum.

> I think there also are some ethical considerations around blockchain that policy could potentially also help with. And, you know, things like anonymity and, you know, the double-edged sword there, right? Anonymity is good to mitigate the risk or preserve privacy. But, you know, anonymity can be abused, right? It's used with ransomware, it's used with extortion, all kinds of things, you know, fraud, and so forth. So, I think policies can help maximize the pros and minimize the cons as well. (Interview, blockchain start-up founder)

One way to consider all these factors holistically is to follow the formula below, where the adoptability of a use case involves multiplying all the factors.

Adoptability = economic value of the use case * readiness for adoption * compliance with existing rules * the preservation of data assets, intellectual property, and privacy

If a use case is rated very low in any of these factors, the adoptability of the use case will be low due to the multiply effect. In contrast, a use case that strikes a balance among these factors will have higher adoptability. By using the proposed formula, decision-makers can gain a deep understanding of the use case and preclude cases with low adoptability.

4.4.2 Framing Blockchain Adoption as Aligned with Both the Innovative and Safety-Driven Culture in Life Sciences Organizations

Interview participants described that it was challenging to legitimize blockchain adoption in life science, in other words, to create buy-in among leaders and key staff. Although some were enthusiastic about blockchain adoption, others focused on traditional role activities and did not demonstrate a strong motivation to understand and incorporate blockchain into their core businesses. Interviewees also reported that the values and beliefs, as part of the culture in life sciences organizations, seem not to support the adoption of blockchain.

> So, I'm very confident in these arguments I'm bringing forward [about the blockchain project]. But because the ideas are too big—maybe too big is the wrong way to describe them—because they're just too hard to comprehend. It's more of a try to get their head around this horizontal type of innovation that's going against their belief system. (Interview, manager)

In response, those who gained initial success in pilot projects carefully framed the adoption as aligned with both the innovative and safety-driven cultures in life sciences. Every organization or sector has a culture, which refers to a system of shared assumptions, values, and beliefs about appropriate and inappropriate behavior

[39, 40]. An organization's or sector's culture is supported by its structure, working processes, and everyday behaviors of the participants. Researchers have consistently found that culture plays a crucial role in successfully adopting new technologies or practices [8]. Specifically, culture can affect what is considered legitimate innovation and whether adoption would receive all levels of endorsement. Researchers also recommend that managers customize innovations to organizational or sector culture to achieve adoption [8, 41].

Although scholars have traditionally considered culture as singular, that is, an organization or field only has one dominant culture, recent work has pointed to the existence of competing cultures within an organization or field [41, 42]. From the perspective of cultural pluralism, the author finds that qualitative data revealed two cultures prevalent in life sciences: innovation-driven and safety-driven. The innovation-driven culture manifests through the ongoing medical experiments and the resources dedicated to innovation, in contrast, the safety-driven culture manifests via the strong focus on compliance to behavioral protocols, hierarchical structures, and the value orientation that treatment should cause no harm.

While two cultures might seem contradictory, some interviewees stated that they had combined them to frame blockchain to legitimize adoption. For instance, they highlighted that the spirit of blockchain adoption is to foster decentralized innovations. Further, the innovations can reduce the risk of non-compliance to data-retention regulations thanks to the near-immutability advantages of blockchain. Also, such innovation can protect life sciences organizations from privacy breaches and reduce the risks of sensitive information being exploited by malicious people. These framings appeal to both the innovation and safety-driven culture in the life sciences field, helping gain traction for blockchain adoption among important stakeholders.

> Well again, it comes down to having a crystal clear business case that could appeal to many different decision-makers meet. Like the "better, faster, cheaper." We're going to save money on this and reduce costs, or you know the risk-averse person. We can reduce non-compliances, and therefore patient risk, and therefore to risk in general. (Interview, manager)

4.4.3 Unobtrusive Implementation

Interview participants also suggest that it is essential to keep the implementation process unobtrusive, meaning that blockchain adoption should be managed so that it does not overtly clash with the existing power structure or the material interests of incumbents. Blockchain project managers need to navigate through the structures and ensure that all the individuals and organizations in the structure—especially those who maintain the structures—are thoroughly engaged. Specifically, health executives, physicians, and researchers that hold high status should be approached early as their endorsement is crucial for the success of blockchain implementation. Also, existing IT departments should have a say about the implementation and should be engaged during the selection process of use cases and the selection of vendors. In addition, compliance officers need to be engaged such that the adoption of blockchain does not challenge any information regulations or science ethics.

4.4.4 Acting Swiftly When the Innovative Culture Gains Strength

Research has suggested that multiple cultures can co-exist in an organization or field. The strength of the cultures (i.e., that extend to the people who enact the culture and the prevalence of the enactment) can wax and wane [42]. In life sciences, safety-driven and innovation-driven cultures often simultaneously guide a particular organization and change in their strength due to resource flows or policy changes. Interview participants suggest that when the innovation-driven culture gains strength, blockchain advocates should take the opportunity and act swiftly to advance adoption.

For example, participants who gained initial success in a corporate blockchain project described that when senior executives who embraced innovations were in charge, they moved quickly to secure financial support from those executives. They also organized themselves quickly around the executives such that the executives were given leadership in the project, which further helped them gain endorsement both within and outside the organization. Participants believed that they successfully identified and exploited the political window of opportunity, which was crucial to the initial success of adoption.

> Yeah, I think when the medical director left, when that happens to you, you lose a key source of support, right? And you have to review the support from the top executive team and that would take a lot of time. After all, there are a lot of background politics going on. (Interview, manager)
>
> So, it was surprising that we made it, in hindsight—especially in the timelines that we're working in. I think there was a whole lot of strategic politics in the background. Our story was very resonating, so we had a lot of leeway to work in the fail fast, fail forward type of mentality and do different things. (Interview, senior manager)

5 Discussion

Understanding the experience of early adopters of blockchain in life sciences is crucial to utilizing the technology's potential to transform the field. This study explored the experiences of the early adopters and identified four primary socio-organizational barriers that they have encountered. Some findings are consistent with previous studies on adopting IT in life sciences; in particular, the barrier "life sciences is highly institutionalized" somewhat resonates with prior research on the barriers for adopting ELN [17, 18, 20]. However, this study highlights three barriers that are unique to the adoption of blockchain: "negative stereotype of blockchain technology," "perceived technological complexity of blockchain," and "the lack of ecosystem mindset." The negative stereotype of blockchain can be attributed to the unique trajectory of how the technology was utilized and became widely known. That blockchain has been negatively stereotyped might be suggestive of its transformational potential. As previous research suggested, genuinely novel, transformational technology often originates from the margins of established institutions [6, 43], the association between the margins and the technology is likely to bring negative stereotypes to the technology.

The "perceived technological complexity of blockchain" barrier is tied to the drastically innovative nature of blockchain. As blockchain represents a decentralized computation system instead of traditional centralized platforms, it requires decision-makers to comprehend the fundamentally technical departure of blockchain. As explained in the previous section, this has created a dilemma for early adopters when they attempt to create "buy-in." The barrier "lack of ecosystem mindset" may also be observed when adopting other technologies that emphasize connectivity and scale of economy, such as the Internet of Things. Even though the barrier "the highly institutionalized nature of life sciences organizations" resonates with previous studies that examine the barriers for ELN adoption, the difficulty that this barrier creates is unique in the case of blockchain. As noted, it creates the "chicken and egg" problem. Namely, decision-makers would only allow mass adoption before they see substantial benefits of blockchain, but such benefits can only be shown after mass adoption has occurred. This problem is peculiar to blockchain adoption because of its distributed nature. The difficulties that the four organization behavioral barriers have created reinforce the perspective that blockchain adoption is not merely a technical issue and focusing on perfecting the technology does not necessitate mass adoption. Furthermore, the barriers identified in this study created difficulties for study participants through "trade-offs" or "dilemmas," highlighting that blockchain adoption involves the delicate balancing of seemingly paradoxical demands and responses, echoing the general discussions about the trade-offs prevalent in blockchain design and implementation [44].

The adoption strategies uncovered in this study relate and contrast the existing literature in the following ways. The first strategy—holistically evaluating blockchain + life sciences use cases—emphasizes that economic value is a crucial parameter for determining a use case's adoptability. This finding echoes the proposition of experts that blockchain adoption in life sciences should not be for "religious" reasons [5]; instead, the adoption must bring practical benefits to various stakeholders. However, the contribution of this finding is that it articulates other factors that need to be accounted for and points out how to do so holistically by introducing the multiplier formula. The multiplier formula contains the economic value of a use case, readiness for adoption, compliance with existing rules, and the preservation of data assets, intellectual property, and privacy. This formula also determines that the overall adoptability would approach zero when any of these parameters approaches zero. It not only accommodates some of the most critical aspects when adopting blockchain in life sciences but also alerts experts and practitioners to consider all those aspects concurrently when evaluating a blockchain + life sciences use case.

The second strategy—framing blockchain adoption as aligned with both the innovative and safety-driven culture in life sciences organizations—challenges how blockchain is framed in academic and industry discourse where blockchain tends to be overly described as "transformational," "innovative," or "disruptive." While blockchain does represent a decentralized way of information storage and transaction [4], and may indeed help legitimize adoption and acquire resources for blockchain ventures in market-driven sectors such as finance or gaming. However, this study found that overly emphasizing blockchain's disruptive property may backfire in life sciences due to the highly institutionalized nature of the field and its risk-averse culture. Rather than simply describing blockchain as disruptive, researchers and practitioners in life sciences may stress blockchain as an innovative tool. Using this tool can strengthen compliance and enhance patient safety without necessitating the dismantling of existing systems. This discursive strategy may facilitate adoption at the present since it reduces the threat perceived by powerful incumbents.

The third strategy, "acting swiftly when the innovative culture gains strength," brings forward the political aspect of blockchain adoption in life sciences, which prior research has not yet addressed. Despite that the literature on innovation adoption has long suggested the crucial role of politics in blocking or facilitating adoption [7, 8], the informatics literature seems not yet to pay plentiful attention to organizational politics when studying the adoption of ELN, not to mention blockchain. However, when blockchain adoption occurs, as described by this study, it would need a political window of opportunity where the advocators embrace the innovation-driven culture—as opposed to the safety-driven culture—attain powerful positions and exercise their power to support the adoption. Such political windows of opportunity may disappear as the advocators lose power, and therefore it is crucial to take swift actions to make adoption happen when the windows are present. This insight opens the research avenue for studying the appropriate speed of adoption, specifically for blockchain as well as other nascent technologies in life sciences.

5.1   Limitations and Future Directions

Despite that this study is one of the first to investigate the socio-organizational barriers for blockchain adoption in life sciences and delineate adoption strategies, it is not without limitations. Due to privacy concerns, it was not possible to sit in executive meetings to directly observe the decision-making process regarding blockchain adoption and document how these barriers manifest in micro-level social interactions. Also, due to the limited period of this study, the author was unable to base the findings on blockchain + life sciences projects that have achieved substantial success. Hence, the adoption strategies presented in this study might be limited in applicability.

Future research may observe in real-time the blockchain + life sciences projects that are widely adopted, achieving a high return on investment and penetration rate, and compare the adoption strategies with those found by this study. Moreover, although consumers are important stakeholders in many blockchain life sciences applications (e.g., blockchain-based genomics marketplaces), this study did not include consumers in the pool of interview participants. Future research should investigate what adoption barriers might stem from the consumer side and how adoption strategies may engage consumers. In addition, as blockchain becomes more widely accepted and understood, some of the socio-organizational barriers that adopters experience may change over time. For instance, the barrier—negative stereotypes of blockchain and perceived technical complexities—may weaken. Future research may investigate what new barriers might emerge and how they may differ from those discovered in this study.

6 Conclusion

Blockchain can transform how information is stored and shared and thus holds strong potential to improve the efficiency of life sciences research and the quality of scientific discoveries. However, early adopters have experienced significant socio-organizational barriers for adoption, which calls for balanced attention to the engineering problems and socio-organizational processes that blockchain adoption may incur. The adoption strategies uncovered in this study may serve as steppingstones for academics and practitioners to re-think blockchain adoption in life sciences.

References

1. International Organization for Standardization (2021) Blockchain and distributed ledger technologies—reference architecture (ISO Standard No. 23257:2021 [under development]). https://www.iso.org/standard/75093.html
2. Werbach K (2018) The blockchain and the new architecture of trust. MIT, Cambridge, MA. https://doi.org/10.7551/mitpress/11449.001.0001
3. Kuo T, Kim H, Ohno-Machado L (2017) Blockchain distributed ledger technologies for biomedical and health care applications. J Am Med Inform Assoc 24(6):1211–1220. https://doi.org/10.1093/jamia/ocx068
4. Kuo T, Zavaleta RH, Ohno-Machado L (2019) Comparison of blockchain platforms: a systematic review and Life science examples. J Am Med Inform Assoc 26(5):462–478. https://doi.org/10.1093/jamia/ocy185
5. Metcalf D, Bass J, Hooper M, Cahana A, Dhillon V (2019) Blockchain in life science: innovations that empower patients, connect professionals and improve care. Taylor & Francis Group, Boca Raton, FL. https://www.routledge.com/Blockchain-in-Healthcare-Innovations-that-Empower-Patients-Connect-Professionals/Dhillon-Bass-Hooper-Metcalf-Cahana/p/book/9781032093888
6. Leblebici H, Salancik GR, Copay A, King T (1991) Institutional change and the transformation of interorganizational fields: an organizational history of the U.S. radio broadcasting industry. Adm Sci Q 36(3):333–363. https://doi.org/10.2307/2393200
7. Scott WR (2014) Institutions and organizations: ideas, interests, and identities, 4th edn. Sage, Los Angeles. https://us.sagepub.com/en-us/nam/institutions-and-organizations/book237665
8. Ansari SM, Fiss P, Zajac EJ (2010) Made to fit: how practices vary as they diffuse. Acad Manag Rev 35:67–92. https://doi.org/10.5465/amr.35.1.zok67
9. Corbin JM, Strauss A (1990) Grounded theory research: procedures, canons, and evaluative criteria. Qual Sociol 13:3–21. https://doi.org/10.1007/BF00988593
10. Tapscott D, Tapscott A (2016) Blockchain revolution: how the technology behind Bitcoin is changing money, business, and the world. Penguin, New York. https://doi.org/10.5555/3051781

11. Tseng JH, Liao YC, Chong B, Liao SW (2018) Governance on the drug supply chain via Gcoin blockchain. Int J Environ Res Public Health 15. MDPI AG: 1055. https://doi.org/10.3390/ije rph15061055
12. Dubovitskaya A, Baig F, Xu Z, Shukla R, Zambani PS, Swaminathan A, Jahangir MM, Chowdhry K, Lachhani R, Idnani N, Schumacher M, Aberer K, Stoller SD, Ryu S, Wang F (2021) ACTION-EHR: patient-centric blockchain-based electronic health record data management for cancer care. J Med Internet Res 22(8):e13598. https://doi.org/10.2196/13598
13. Lu C, Batista D, Hamouda H, Lemieux V (2020) Consumers' intentions to adopt blockchain-based personal health records and data sharing: focus group study. JMIR Form Res 4(11):e21995. https://doi.org/10.2196/21995
14. Khurshid A, Holan C, Cowley C, Alexander J, Harrell D, Usman M, Desai I, Bautista JR, Meyer E (2021) Designing and testing a blockchain application for patient identity management in healthcare. JAMIA Open 4(3):ooaa073. https://doi.org/10.1093/jamiaopen/ooaa073
15. Ghannam R (2020) Do you call that a lab notebook? IEEE Potentials 39(5):21–24. https://doi.org/10.1109/MPOT.2020.2968798
16. Guerrero S, Dujardin G, Cabrera-Andrade A, Paz-Y-Miño C, Indacochea A, Inglés-Ferrándiz M, Nadimpalli HP, Collu N, Dublanche Y, De Mingo I, Camargo D (2016) Analysis and implementation of an electronic laboratory notebook in a biomedical research institute. PLoS ONE 11(8):e0160428. https://doi.org/10.1371/journal.pone.0160428
17. Kanza S (2018) What influence would a cloud based semantic laboratory notebook have on the digitisation and management of scientific research? Doctoral thesis, University of Southampton. https://eprints.soton.ac.uk/421045/1/Final_Thesis.pdf
18. Kanza S, Willoughby C, Gibbins N, Whitby R, Frey JG, Erjavec J, Zupančič K, Hren M, Kovač K (2017) Electronic lab notebooks: can they replace paper? J Cheminformatics 9(1):31. https://doi.org/10.1186/s13321-017-0221-3
19. Taylor KT (2006) The status of electronic laboratory notebooks for chemistry and biology. Curr Opin Drug Discov Dev 9(3):348–353. http://www.atriumresearch.com/library/Taylor_Electronic_laboratory_notebooks.pdf
20. Zupancic K, Pavlek T, Erjavec (2021) Digital transformation of the laboratory: a practical guide to the connected lab. Wiley-VCH, Weinheim. https://doi.org/10.1002/9783527825042
21. Ansari S, Garud R, Kumaraswamy A (2016) The disrupter's dilemma: TiVo and the U.S. television ecosystem. Strateg Manag J 37(9):1829–1853. https://doi.org/10.1002/smj.2442
22. Sine WD, Lee BH (2009) Tilting at windmills? The environmental movement and the emergence of the U.S. wind energy sector. Adm Sci Q 54(1):123–155. https://doi.org/10.2189/asqu.2009.54.1.123
23. Barley S (1986) Technology as an occasion for structuring: evidence from observations of CT scanners and the social order of radiology departments. Adm Sci Q 31(1):78–108. https://doi.org/10.2307/2392767
24. Dougherty D, Dunne DD (2012) Digital science and knowledge boundaries in complex innovation. Org Sci 23(5):1467–1484. https://doi.org/10.1287/orsc.1110.0700
25. Compagni A, Mele V, Ravasi D (2015) How early implementations influence later adoptions of innovation: social positioning and skill reproduction in the diffusion of robotic surgery. Acad Manag J 58(1):242–278. https://doi.org/10.5465/amj.2011.1184
26. Stake RE (2010) Qualitative research: studying how things work. Guilford Press, New York, NY. https://www.routledge.com/Qualitative-Research-Studying-How-Things-Work/Stake-Usinger-Erickson-Merriam-Lincoln/p/book/9781606235454
27. Patton MQ (2014) Qualitative research and evaluation and methods, 4th edn. Sage, Saint Paul, MN. https://us.sagepub.com/en-us/nam/qualitative-research-evaluation-methods/book232962
28. Gioia DA, Corley KG, Hamilton AL (2013) Seeking qualitative rigor in inductive research: notes on the Gioia methodology. Organ Res Methods 16:115–131. https://doi.org/10.1177/1094428112452151
29. Glaser BG, Strauss AL (1967) The discovery of grounded theory: strategies for qualitative research. Aldine Transaction, New Brunswick & London. https://www.google.com/books/edition/The_Discovery_of_Grounded_Theory/rtiNK68Xt08C

30. Jacobides MG, Cennamo C, Gawer A (2018) Towards a theory of ecosystems. Strateg Manag J 39:2255–2276. https://doi.org/10.1002/smj.2904

31. Zietsma C, Groenewegen P, Logue DM, Hinings CR (2017) Field or fields? Building the scaffolding for cumulation of research on institutional fields. Acad Manag Ann 11(1):391–450. https://doi.org/10.5465/annals.2014.0052

32. Lamont M, Molnár V (2002) The study of boundaries across the social sciences. Annu Rev Sociol 28:167–195. https://doi.org/10.1146/annurev.soc.28.110601.141107

33. Fligstein N, McAdam D (2012) A theory of fields. Oxford University Press, New York. https://doi.org/10.1093/acprof:oso/9780199859948.001.0001

34. Uzzi B (1996) The sources and consequences of embeddedness for the economic performance of organizations: the network effect. Am Sociol Rev 61(4):674–698. https://doi.org/10.2307/2096399

35. Chen M-J, Miller D (2015) Reconceptualizing competitive dynamics: a multidimensional framework. Strateg Manag J 36:758–775. https://doi.org/10.1002/smj.2245

36. Gondo MB, Amis JM (2013) Variations in practice adoption: the roles of conscious reflection and discourse. Acad Manag Rev 38(2):229–247. https://doi.org/10.5465/amr.2010.0312

37. Colyvas JA, Powell WW (2006) Roads to institutionalization: the remaking of boundaries between public and private science. Res Organ Behav 27:305–353. https://doi.org/10.1016/S0191-3085(06)27008-4

38. Smith-Doerr L (2005) Institutionalizing the network form: how life scientists legitimate work in the biotechnology industry. Sociol Forum 20:271–299. https://doi.org/10.1007/s11206-005-4101-7

39. Chatman JA, Cha SE (2003) Leading by leveraging culture. Calif Manag Rev 45(4):20–34. https://doi.org/10.2307/41166186

40. Kerr J, Slocum JW (2005) Managing corporate culture through reward systems. Acad Manag Exec 19(4):130–137. https://doi.org/10.5465/ame.2005.19417915

41. Reay T, Hinings CR (2005) The recomposition of an organizational field: health care in Alberta. Organ Stud 26(3):349–382. https://doi.org/10.1177/0170840605050508722006

42. Besharov M, Smith W (2014) Multiple institutional logics in organizations: explaining their varied nature and implications. Acad Manag Rev 39:364–381. https://doi.org/10.5465/amr.2011.0431

43. Cattani G, Ferriani S, Lanza A (2017) Deconstructing the outsider puzzle: the legitimation journey of novelty. Org Sci 28:965–992. https://doi.org/10.1287/orsc.2017.1161

44. Lemieux V, Feng C (2021) Building decentralized trust: multidisciplinary perspectives on the design of blockchains and distributed ledgers. Springer Nature Switzerland AG. Cham, Switzerland. https://doi.org/10.1007/978-3-030-54414-0

Dr. Chang Lu is a postdoc research fellow at Blockchain@UBC, the University of British Columbia. His theoretical research focuses on technology adoption, organization and institutional change, and the interplay between culture and power. Currently, he studies these topics in the context of adopting blockchain technology in life sciences, examining the adoption processes at both organizational and institutional levels. He has published several articles in leading management journals and taught senior undergraduate and MBA students Organizational Strategy and Organizational Behavior. He supervises master's and MBA students for their research projects and is currently creating education materials for executives about blockchain in life sciences. He earned his PhD in Strategic Management and Organization, School of Business from the University of Alberta. Prior to his academic career, he worked as an HR professional in China and Europe.

Blockchain Governance Strategies

Denise McCurdy

Abstract Collaboration among ecosystem members is essential for well-functioning blockchains, especially for the life sciences industry. With many start-ups eager to capitalize on the burgeoning informatics industry, non-cooperative business practices holding siloed data hostage prevents blockchain technology from reaching its full potential. To help break down these barriers, thoughtful and informed collaboration among life sciences partners using best-practice governance models is needed. Like much of the underlying technology, these governance models are still emerging. This chapter explores the special considerations needed to manage successful blockchain deployments for life sciences ecosystems. Recognizing good governance's essential role in guiding collaborative behavior, the chapter concludes with governance models specific to life sciences ecosystems.

Keywords Blockchain · Collaboration · Ecosystems · Governance · Life sciences

1 Introduction

Blockchain ecosystems are particularly communal. Due to the nature of the technology, successful blockchain deployments require partners to coordinate their activities more closely than almost any other prior technology. The high degree of coordination and the governance needed to manage it is especially acute for the life sciences industry. Ecosystem sharing of participant data—considered the basic building block—among life sciences partners can be technically challenging and is subject to lengthy legal and regulatory process reviews. Robust regulatory requirements for some in the life sciences ecosystem, but not others, and a highly fragmented and siloed industry add to the challenges of sharing data [1], p. 677).

However, breaking down the silos requires thoughtful and well-designed governance mechanisms among partners. These governance mechanisms–the practical

D. McCurdy (✉)
Center for Engaged Business Research, Georgia State University. Grove Gate Consulting, Inc., Atlanta, GA, USA
e-mail: dmccurdy@grovegc.com

© The Author(s), under exclusive license to Springer Nature Singapore Pte Ltd. 2022   197
W. Charles (ed.), *Blockchain in Life Sciences*, Blockchain Technologies,
https://doi.org/10.1007/978-981-19-2976-2_9

agreements struck by ecosystem partners–are challenging for ecosystems traditionally disinclined and discouraged to share. Governance models to assist ecosystem
partners are vital aids on the path to fuller data sharing. These models are in short
supply. This lack of visibility of good governance models is a critical shortfall, as
research suggests that poor governance choices are the most significant reasons for
ecosystem failure [2].

Other research with industry leaders in blockchain concurs that "the true benefits
of blockchain are realized when multiple stakeholders collaborate…this challenge
of collaboration among stakeholders will be compounded in sectors that are highly
fragmented, owing to the need to develop consensus among multiple stakeholders
with unique needs and requirements" [1, p. 677]. Additionally, in a 2015 report, the
World Economic Forum states that "The most impactful [distributed ledger tech-
nology] applications will require deep collaboration between incumbents, innovators, and regulators, adding complexity and delaying implementation" [3, p. 18]. Of
course, more topically, COVID-19 is a compelling reminder of the need for disparate
partners in ecosystems to work together: "It is hard to overstate the role drug makers,
working in tandem with the federal government, played in bringing the pandemic to
heel in the U.S." [4], p. 2).

It is fair to say that we are in the very early days of in-production blockchain
for life sciences. We have much to learn. What we *have* discovered is the importance of managing, or governing, these relationships for life sciences partners in a
blockchain ecosystem. This chapter attempts to remedy the gaps in our knowledge
and explores ecosystem governance for blockchains in life sciences. We begin with
clear definitions of governance and blockchain governance. We then discuss the
types of decisions required of ecosystem partners. We explore governance strategies
and typical roles of life sciences partners, emphasizing the special considerations
required for these ecosystems. We conclude with recommendations for ecosystem
partners in life sciences, using evidence-based research as our guiding principle, as
we explore the main objective of this chapter: *What is an appropriate blockchain
governance model for life sciences ecosystems?*

2   Defining Governance

Considered fashionable, popular, and notoriously slippery to define [5], governance,
in its simplest form, is a set of agreements by members to achieve the desired state
for a particular condition or state of affairs. There are countless other definitions
of governance; it is a deep reservoir well outside the scope of this chapter. There
is an equally large number of governance *forms*. Governance can be collaborative,
democratic, networked, private, multilevel, and adaptive, to name but a few [5].
Governance can be further described with the use of a qualifier, expanding both the
definition and form. For example, the United Nations prefixes their definition of
governance with "good" to include the unvoiced and disenfranchised. This equitable

flavor of governance is central for ecosystem governance, and therefore this chapter includes the concept of minority voices, for example, resource-constrained start-ups.

> It is participatory, consensus-oriented, accountable, transparent, responsive, effective, and efficient, equitable and inclusive, and follows the rule of law. It assures that corruption is minimized, the views of minorities are taken into account, and that the voices of the most vulnerable in society are heard in decision-making. It is also responsive to the present and future needs of society [6, p. 1].

It is essential to distinguish governance from *government*. For this chapter, governance is the operating model by partners to do business for their ecosystem. In this view, governance co-exists with government as governments can—and do—impose their sovereign imprimatur on governance agreements with obligatory regulations [7].

3   A Deeper Dive: Blockchain Governance

Governance for blockchain has its antecedents in Information Technology (IT) studies. In previous research, IT governance has been defined as a "framework for decision rights and accountabilities to encourage desirable behavior in the use of IT" [8, p. 6]. According to Weill, governance is essential to IT and IT investments because "good IT governance leads to superior returns on IT investments [8, p. 1]." Weill's definition invokes three significant dimensions of IT governance that are pertinent to ecosystems: decision rights, accountability, and incentives [9, 8]. These key findings from IT literature extend to blockchain ecosystems and are crucial to driving collaborative behavior.

Definitions for *blockchain* governance include the type of decisions that partners must make before deployments, such as voting rights, accountability, and incentives [9]. In this definition, governance for blockchain is more focused on "what" types of decisions with which to agree. Blockchain governance can also refer to "how" such decisions are arrived at or "the methods by which blockchain communities and key stakeholders arrive at collective action" [10, 11]. The more general definition is preferred, as it captures the broad scope of governance activities required in blockchain: "the means of achieving the direction, control, and coordination of stakeholders within the context of a given blockchain project to which they jointly contribute" [12, 11].

3.1   On-Chain Governance

Blockchain governance can be considered on-chain or off-chain. With on-chain governance, many of the decisions are codified onto the blockchain network itself [9, 13]. The rules are typically stored in smart contracts, including the ability to modify rules. Developers propose changes through software code updates depending upon the consensus protocols employed. Rules within the underlying code determine the updates, also called "the rule of code" [14, 15].

3.2 Off-Chain Governance

Off-chain governance is a more social construct. It includes collective decisions and rules that ecosystem partners make as they deploy blockchain [15, 16]. Decisions can be codified into the blockchain of choice, but these decisions are struck off-chain through more traditional contractual arrangements. Blockchain experts believe that "governance system design is one of the highest leverage activities known…initial design is important, but over a long enough timeline, the mechanisms for change are most important" [17], Why Blockchain Matters section). These mechanisms for change, including the initial design and subsequent actions initiated by ecosystem members, are a critical component of good ecosystem governance.

4 Types of Ecosystem Governance Decisions

Regardless of the definition used, blockchain requires many practical decisions by members of the ecosystem. These decisions can be codified "on-chain" or "off-chain," following the more familiar processes among business partners. Typical decisions required of blockchain ecosystems include the following:

- **Technical**—What blockchain network type is best for the desired use case? How are legacy concerns managed? What are the performance requirements and are there potential latency issues with the blockchain network of choice? What are the change management processes and procedures? If something technically goes awry, what are the fallback plans?
- **Voting rights**—Which members of the ecosystem get to vote? How are votes recorded? How many votes are needed to reach quorum? What makes a majority?
- **Dispute resolution**—Do ecosystems agree on the use of arbitration, online dispute resolution, invoice reconciliation systems of record, or judicial/legal systems? Are these dispute resolutions equitable for all members?
- **Agreement on standards**—Are the standards established, or are they still evolving? What standards are mandatory for all partners? What standards are compulsory for a subset of partners? What are the costs for adopting specific standards, and will these costs be borne by all partners?
- **Commercial**—What are the start-up costs? Who bears the cost? What elements of existing systems can be re-used? What are the capital and operational expenses for a fully functional blockchain deployment? What are the capital and operational

expenses for minimal viable ecosystems and/or proofs-of-concept? Are incentives aligned across the ecosystem for all partners?

- **Legal**—How do partners ensure compliance with regulators (industry-specific)? What happens if regulatory bodies have conflicting requirements? Are smart contracts legally binding or triggers to an external agreement? How is intellectual property managed? How are competitor concerns addressed?
- **Security**—How do members ensure compliance with agreed security protocols? What are the ramifications if partners do not adhere to security protocols? What constitutes a security breach?
- **Human Resources**—How compatible are the cultures of the firms? Is there a skills shortage? Are all organizations in an ecosystem open to change? How disruptive is blockchain to existing operational and supplier relationships?
- **Outcomes**—Are there agreed technical, performance, and outcome metrics? Are these metrics or key performance indicators (KPIs) equitable for all partners? Will new ecosystem partners adopt the metrics as part of the onboarding process?

These decisions can be further categorized as business or operational concerns, though the term "blockchain governance" often covers the raft of decisions. It is important to note that process owners with deep functional expertise are needed to scope their firm's requirements adequately. These process owners are needed for each organization of an ecosystem.

5   Common Blockchain Governance Strategies

Blockchain governance, like governance in general, ranges from highly centralized to fully democratic. The roles, decisions, and indeed blockchain network type can vary dramatically. In the upcoming sections, the most common governance strategies are explored, with the cautionary note that blockchain technology, and the associated decisions to support it, are still evolving [18].

5.1   Founder Led/Benevolent Dictator

At one end of the span-of-control spectrum is the Founder or Benevolent Dictator. In this form of governance, the original creator/lead developer creates the blockchain platform, the software code and has authority over the initial decisions [19]. The Benevolent Dictator could be a person, such as Satoshi Nakamoto, a pseudonym for the person or group who developed the fundamentals of Bitcoin [20], or Vitalik Buterin [21], the creator of Ethereum. Benevolent Dictators can also be enterprise firms who wish to launch a particular form of blockchain and associated use cases for commercial use. One example of a commercial Benevolent Dictator is Maersk's initial blockchain initiative prior to TradeLens [19]. Another example of a Benevolent

Dictator is Chronicled, the blockchain provider of the MediLedger Network, which offers a product verification use case to prevent counterfeit pharmaceuticals [22]. The Benevolent Dictator (also known as a lead firm) approach can move faster than the more decentralized decision-making models, as they can act as a catalyst, architect, and guide [23], p. 44). Still, they run the risk of mistrust by future participants if the incentives appear inequitable in any way.

5.2   Core Development Team

The Core Development Team is typically a group of the most active software programmers who initiate a particular blockchain network and commit to timely code releases. Though individual programmers can suggest changes and contribute code via open-source mechanisms, the Core Development Team has the final say in the code. Examples of core development teams include Hyperledger Project, Bitcoin, and Ethereum [24, 25, 26].

5.3   Federations or Alliances

Alliances are made up of like-minded partners who agree to equitably share the decisions required to deploy a blockchain for a decided use case. Initial partners discuss design considerations, including the use case, the type of desired blockchain network to support the use case, required standards, equitable incentives, dispute resolution mechanisms, and a number of decisions, as noted above [27]. Examples of life sciences alliances include MediLedger, PharmaLedger, Machine Learning Ledger Orchestration for Drug Discovery (MELLODDY), and the Triall Foundation. Alliances tend to be more democratic, as decisions are collaboratively reached among partners. However, democratization comes at a price, as it does in governments. Alliance arrangements can be more time-consuming, as consensus can be messy and slow [28].

6   Ecosystem Roles

There are different strategies with which to govern a blockchain ecosystem, and there are also other roles within an ecosystem. In the life sciences industry, as in other industries, an ecosystem partner can take the role of a founder or initiator, a member, an observer, and an operator [19]. A brief description is provided below (see Fig. 1):

- A founder such as MediLedger, for example, typically establishes the blockchain and makes critical decisions regarding member voting rights, onboarding, capital requirements, and other foundational choices. Founders almost always operate nodes.
- Members join a particular blockchain ecosystem for the use case; for example, a track and trace ecosystem for the pharmaceutical industry will likely attract those with existing relationships and vested interests in that use case, such as a pharmacy or a start-up. Members often have nodes, but it generally is not a requirement. Members can be start-ups that provide wearable fitness products, pharmaceutical firms, regulatory agencies, for example (see Fig. 9.1).
- Observers typically have read-only access to the blockchain ledger. With access to either the entire dataset or a portion, observers tend to be regulators, auditors, or other roles from the public service sector.
- Operators run the blockchain network, adhering to previously agreed service level agreements and contractual commitments. IBM, for example, can be an operator and part of the life sciences solution. Another example is Chronicled, which runs the blockchain network for MediLedger. Operators almost always have node privileges. Operators are often network service providers or blockchain-as-a-service providers.
- Suppliers provide technology support to the blockchain ecosystem, such as auditors or tax providers.

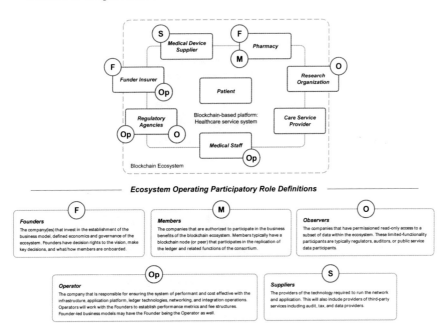

Fig. 1 Typical life sciences blockchain ecosystem (Adapted from [29, 19, 30] and Microsoft)

7 Typical Ecosystem for Life Sciences Blockchain

A typical life sciences ecosystem includes patients, pharmacies, funders, regulatory bodies, care providers, and others (refer to Fig. 1). The challenges to ensure that all partners are in lockstep to deploy an interconnected technology platform are apparent even to the non-technical eye. For example, a small start-up with a desirable health wearable device may not have the Health Insurance Portability and Accountability Act (HIPAA) requirements of the clinician from whom they obtain data. (Note: HIPAA would only apply if the start-up receives data from the clinician, not if the start-up is sending data to the clinician. This example reinforces the complexity of the regulatory challenges borne by some in the life sciences industry). Still, as part of the ecosystem, the start-up must understand all partners' up and downstream requirements. As such, ecosystems in life sciences require careful mapping of the risks and rewards for each partner. As mentioned above, examples of life sciences blockchain consortiums include the following:

- MediLedger
- MELLODDY
- PharmaLedger
- Triall Foundation

As mentioned earlier, MediLedger is a blockchain-enabled platform for the pharmaceutical and life sciences sectors. MediLedger helps firms comply with the Drug Supply Chain Security Act of 2013 (DSCSA), which requires verification of returned pharmaceuticals to reduce counterfeits [31]. MediLedger stores proofs that transactions have been verified instead of the transactions themselves. MediLedger uses a Benevolent Dictator and "consensus through collaboration" governance model [32, 19].

MELLODDY, based in the European Union (EU), is a privacy-preserving consortium for pharmaceutical companies that shares the successful compounds for drug development without exposing underlying proprietary information. MELLODDY's governance structure is made up of an executive committee, work packages (i.e., work groups), and a general assembly comprised of representatives from all partners [33]

PharmaLedger is sponsored by the Innovative Medicines Initiative and the European Federation of Pharmaceutical Industries and Associates. PharmaLedger includes global pharmaceutical companies and public and private entities from academia, research organizations, patient representative organizations, and others [34].

Triall is a blockchain-based software provider for life sciences partners to better manage clinical trial data. Triall aims to streamline new vaccine development, as "clinical trials involve increasing amounts of electronic systems and data" [35]. Triall's governance structure is made up of a non-profit foundation that is responsible for the ecosystem. The foundation's board of directors, initially consisting of seven members, manages the foundation to meet applicable regulatory requirements.

7.1   Special Considerations for Life Sciences

In many respects, the governance decisions required of life sciences blockchain ecosystems resemble ecosystems of other industries. However, the combination of human research protection regulations and an array of players eager for market share contribute to specific concerns for life science ecosystems. Some of these concerns are technical, while other challenges are within the governance realm. For example, data security policies may need to extend beyond the borders of a blockchain ecosystem; security breaches in one firm may affect all other partners, introducing unacceptable risk. Blockchain resources are scarce, and the participating firms in the ecosystem need adequate blockchain training and human resource support. Interoperability is vital, as each partner in a life sciences ecosystem will have some internal (and often, quite dated) legacy systems. Compliance with regulatory bodies is essential for life sciences ecosystems. These governance considerations are discussed below:

7.1.1   Compliance with the Regulators

Life sciences ecosystems can be especially challenging due to regulatory requirements not found in other ecosystems. These decisions include compliance with state, federal, and global regulators, restrictions on node placement, the need to maintain patient privacy from blockchain design inception, compliance verification, and, of course, how to manage and report breaches. Understanding the costs of non-compliance is critical, as blockchain (and smart contracts) are often not scoped adequately [36]. The EU requires privacy protection (General Data Protection Regulation, or GDPR) and the ability to track and trace pharmaceuticals (Falsified Medicines Directive, or FMD) (European Medicines Verification [37]. The United States (US) has anti-money laundering, Know Your Customer, and intellectual property protection, covered in other chapters. In pharmaceutical manufacturing, the US signed into law a sweeping DSCSA that requires the pharmaceutical industry to improve supply chain traceability [38]. Another notable concern, though less discussed, is the provenance of the original data. This is the age-old "garbage in, garbage out" conundrum. Blockchain can resolve many issues regarding data tampering due to the immutable nature of the technology. It cannot, however, solve for insufficient data as it enters the blockchain.

The regulatory requirements discussed here are not an exhaustive list. Instead, they highlight the oversight required by ecosystem partners to ensure compliance for their firms and the regulatory requirements for all partners in a blockchain ecosystem.

7.1.2 Data Management and Security

Data are at the center of any blockchain ecosystem. This is especially true for life sciences ecosystems, as sensitive data are subject to breaches on a routine basis [39, 40, 41]. Data must be simultaneously accessible to authorized parties and secure [39]. Though many believe that blockchain's distributed design is more resistant to malicious attacks, no technology is entirely safe. Therefore, the ability to trust data contained on a blockchain is critical. This trust "lays the foundation of the willingness of users to participate in and execute transactions" [42, p. 7].

Data governance decisions include vetting the partners who need data, sharing data, ensuring data security, reviewing access to data, and liability for data leaks. A data breach from one partner will likely affect the entire ecosystem. Safeguards and remedies are essential to prevent a complete ecosystem failure.

7.1.3 Interoperability

Interoperability requires ecosystem partners to agree to exchange information across blockchain network types, such as from a public to a permissioned ledger or across different permissioned ledgers. Included in the interoperability equation are the various wearable devices and trackers that are increasingly available. This holistic view requires a transformation from the existing silo and fragmented approach to fully integrated platforms [36]. The ability to integrate processes, systems, and cross-jurisdiction regulatory requirements will be especially critical in the life sciences industry.

Adding to the confusion is the abundance of organizations developing standards; blockchain standards are being developed by standards development organizations, industry organizations, regulators, and customized standards baked into legacy systems. In advance of standards development, we cannot expect partners to retrofit costly legacy systems based on evolving industry-accepted standards. Though blockchain promises to overcome these challenges by allowing for uniform data exchanges, ecosystem partners need to be aware of—and sensitive to—difficulties in standards uplift.

Other interoperability considerations include sovereign regulatory requirements and where to store data (as some countries restrict offshoring). Ideally, protections for sensitive data involve a balance of accessibility and privacy; accessible to any member of the research team at any time it is required while preserving the confidentiality of research participants. Interoperability can be accomplished with agreements by ecosystem members, as contents of a block of data can be shared with the appropriate members, allowing the original member to maintain data ownership or control.

7.1.4   Implementation Costs and Legacy System Concerns

Research suggests that blockchain could save the industry up to $100 billion per year in costs related to IT, operations, support functions, personnel, and data breaches by 2025 [43, 44]. Within life sciences alone, blockchain technology could provide a $3 billion opportunity by 2025 [45]. Equally high implementation costs moderate these appealing numbers. This does not include the transformational uplift to blockchain technology. This can give pause to even the most stalwart CIO. The requirement for multiple IT systems upgrades, both within a life sciences organization and across an entire ecosystem, is daunting. Of course, outdated IT systems and associated processes are not unique to life sciences.

7.1.5   Organizational Barriers

An organization's readiness to change requires members to collectively resolve to change and believe in their capability to do so [46]. Life sciences organizations are notoriously reluctant to embrace change, including technology changes. Several reasons contribute to this reticence. Shortages of skilled technical resources and other technical concerns are factors [47]. A loss of productivity during transitions and the risk of data breaches are also barriers [48, 49].

7.1.6   Incentives to Join

Blockchain's promise to reduce costs and increase efficiencies may increase life sciences ecosystem members to accept the previously discussed risks. However, the allure of blockchain is not enough; partners will need incentives to join and remain in a blockchain ecosystem. Incentives to join an ecosystem may be driven by regulators, such as the recent mandates to track pharmaceutical and medical supplies with the US DSCSA and the EU FMD [50]. Incentives may also be more commercially focused; the pharmaceutical industry is well aware of blockchain's ability to track drugs, potentially reducing $200 billion of annual losses due to counterfeit drugs [44].

7.1.7   Onboarding

Onboarding participants to a (permissioned) blockchain ecosystem involves membership support, technology alignment, and operationally enabling each partner to the network (Depository Trust & Clearing [51]. Potential partners will typically request entrée to a blockchain ecosystem that fulfills their unique business value propositions. In life sciences, these new partners are typically universities, academic medical centers, contract research organizations, pharmaceutical companies, device companies, and laboratories. Understanding the criteria for each partner requires careful construction, as each partner is likely to have unique considerations. Onboarding new members, and the legal, regulatory, and member-signing documents, is not likely to be a cookie-cutter exercise. After members are legally onboarded and have signed the requisite paperwork, the technical onboarding takes place. In this onboarding phase, partners test their network connectivity, determine their node participation, install the appropriate node software, synchronize databases, and read/write to the blockchain network (depending on their role, as discussed earlier.) Finally, the participants move to operations and, depending upon their role, ensure that the agreed security, privacy (especially critical to life sciences), and resiliency of the blockchain network is performed as agreed.

8   Recommendations

This chapter acknowledges the challenges of deploying blockchain for life sciences and discusses the importance of ecosystems, and governing those ecosystems, leading to this question: *What is an appropriate blockchain governance model for life sciences?* The author advocates a hybrid governance model, combining the benevolent dictator approach using collaborative governance principles [52, 22, 27]. This ecosystem arrangement provides the necessary structure to initiate a blockchain ecosystem while considering the challenges specific to life sciences. After all, ecosystems are made up of real people who make decisions, and these decisions depend on what participants are made aware of (and collectively manage) [53].

Specifically, life sciences blockchain ecosystems can be improved by embracing these collaborative activities:

- Start with a shortlist of partners with which to work, and not the technology. Does each firm have a positive track record with the other partners? Do the ecosystem partners have substantial change and project management capabilities? Are they committed to a long-term engagement? Are all levels of management supportive of the proposed ecosystem?
- Equitably design incentives for initial ecosystem partners. Plan for future members and include attractive on-ramps for new partners. Take care to identify ecosystem incentives as well as individual partner goals. Explore prior research supporting incentives for adopting and using technology [49].
- Focus on a specific use case representing an opportunity, a challenge, or a response to regulatory requirements. Incorporate metrics, KPIs, and other specific outcomes that align with the agreed use case.
- Align agreed on incentives to the outcomes and vet roadblocks early in the process.
- Identify the existing standards used by all partners and the "must-have" regulatory requirements to meet government expectations.
- Agree on the minimum information needed to share and design a lightweight, minimum viable ecosystem (MVE). Budget for increasingly complex components added to the MVE.
- Involve legal and business process teams in the early stages of the design.

- Join industry standards alliances specific to life sciences and standard development organizations, such as the International Organization for Standardization and the Institute of Electrical and Electronics Engineers.

9   Future Directions

Given the evolving nature of blockchain technology, flexibility is the most crucial consideration. Digital transformations will include a basket of technologies, such as artificial intelligence, machine learning, cloud computing, quantum computing, 6G mobile applications, and other emerging tech. These technologies will not be used in isolation but instead will likely work as part of a more comprehensive solution for the life sciences. For example, mobile apps for wearable devices collecting health data will likely evolve to 6G as the underlying technology improves. The method by which mobile apps interface and record events on a particular type of blockchain ledger is also likely to change. This co-evolving of technologies that must liaise successfully to meet the needs of a holistic solution (technically and from a legal/regulatory perspective) only increases the need for foundational governance.

Keep informed of evolving standards. Investigate rulings of regulatory bodies, including those not found in supplier ecosystems today (because your blockchain ecosystem may include non-traditional partners). Stay informed of new legal opinions and continuously assess compliance with state, federal, and global regulators.

10   Conclusions

Blockchain in the life sciences sector promises to protect patient data, safeguard supply chains, improve the collection of clinical trial data, and even incent healthier patient habits [54]. Yet, these technological promises are unlikely to gain traction and adoption among life sciences partners without fundamental agreements based on sound governance models. These governance models, required to guide collaborative activities, are in short supply. They are even more critically required given the recent events of the COVID-19 pandemic that accelerated the need for successful collaboration between governments, research institutions, and the private sector [25].

This chapter describes the typical governance structures and roles in permissioned blockchain ecosystems for the life sciences industry. Emphasis is placed on specific collaborative activities for life sciences partners. The chapter concludes with a recommendation for a hybrid governance model comprised of a Benevolent Dictator approach bolstered by cooperative principles. This combination–the Benevolent Dictator with deep collaboration–provides the necessary structure and foundational support to launch successful blockchain ecosystems. A blockchain ecosystem that can connect everyone must have broad acceptance before it does connect everyone

[55]. After all, getting governance right is the most critical component for sustainable blockchain deployments.

References

1. Sharma L, Olson J, Guha A, McDougal L (2021) How blockchain will transform the healthcare ecosystem. Bus Horiz 64(5):677–682. https://doi.org/10.1016/j.bushor.2021.02.019
2. Pidun U, Reeves M, Schüssler M (2020, June 22) Why do most business ecosystems fail? Boston Consulting Group. https://www.bcg.com/en-us/publications/2020/why-do-most-business-eco systems-fail
3. McWaters RJ, Bruno G, Galaski R, Chatterjee S (2016) The future of financial infrastructure: an ambitious look at how blockchain can reshape financial services. World Economic Forum. https://www3.weforum.org/docs/WEF_The_future_of_financial_infrastructure.pdf
4. Grant C (2021, June 25) A faster future for drug development. The Wall Street Journal. https:// www.wsj.com/articles/covid-vaccines-will-deliver-long-term-health-boost-for-drugmakers-11624636800
5. Ansell C, Torfing J (eds) (2016) Handbook on theories of governance. Edward Elgar Publishing. https://doi.org/10.4337/9781782548508
6. Sheng YK (2009) What is good governance? United Nations ESCAP. https://www.unescap. org/resources/what-good-governance
7. Levi-Faur D (2012) The Oxford handbook of governance. Oxford University Press. https://doi. org/10.1093/oxfordhb/9780199560530.001.0001
8. Weill P, Woodham R (2002) Don't just lead, govern: implementing effective IT governance. *SSRN* (No. 317319).https://doi.org/10.2139/ssrn.317319
9. Beck R, Müller-Bloch C, King JL (2018) Governance in the blockchain economy: a framework and research agenda. J Assoc Inf Syst 19(10):1020–1034. https://doi.org/10.17705/1jais.00518
10. Carter N (2018, June 28) An overview of governance in blockchains. YouTube. https://www. youtube.com/watch?v=D1NeTN_AR18
11. Van Pelt R, Jansen S, Baars D, Overbeek S (2020) Defining Blockchain Governance: A Framework for Analysis and Comparison. Inf Syst Manag 38(1):21–41. https://doi.org/10.1080/105 80530.2020.1720046
12. Markus ML (2007) The governance of free/open source software projects: Monolithic, multidimensional, or configurational? J Manag Gov 11(2):151–163. https://doi.org/10.1007/s10997-007-9021-x
13. Ray PP, Dash D, Salah K, Kumar N (2021) Blockchain for IoT-based healthcare: background, consensus, platforms, and use cases. IEEE Syst J 15(1):85–94. https://doi.org/10.1109/jsyst. 2020.2963840
14. De Filippi P, McMullen G (2018) Governance of blockchain systems: governance of and by distributed infrastructure. Blockchain Research Institute and COALA. https://hal.archives-ouv ertes.fr/hal-02046787/document
15. Reijers W, Wuisman I, Mannan M, De Filippi P, Wray C, Rae-Looi V, Cubillos Vélez A, Orgad L (2018) Now the code runs itself: On-chain and off-chain governance of blockchain technologies. Topoi. https://doi.org/10.1007/s11245-018-9626-5
16. Revisiting the on-chain governance vs. off-chain governance discussion. (2018, May 22). Pool of Stake. Medium. https://medium.com/%40poolofstake/revisiting-the-onchain-govern ance-vs-off-chain-governance-discussion-f68d8c5c606
17. Ehrsam F (2017, November 27) Blockchain governance: Programming our future. Medium. https://medium.com/%40FEhrsam/blockchain-governanceprogramming-our-future-c3bfe30f2d74
18. Massessi D (2019, January 25). Blockchain governance in a nutshell. Coinmunks. https://med ium.com/coinmonks/blockchain-governance-in-a-nutshell-67903c0d2ea8

19. Lacity M, Steelman Z, Cronan P (2019) Blockchain governance models: insights for enterprises. (BCoE 2019–02). University of Arkansas: Blockchain Center of Excellence. https://cpb-us-e1.wpmucdn.com/wordpressua.uark.edu/dist/5/444/files/2019/11/BCCoEWhitePaper022019OPEN.pdf

20. Nakamoto S (2008) Bitcoin: a peer-to-peer electronic cash system. https://nakamotoinstitute.org/bitcoin/

21. Buterin V (2013) Ethereum white paper. GitHub Repos, 22–23. http://kryptosvet.eu/wp-content/uploads/2021/05/ethereum-whitepaper-kryptosvet.eu_.pdf

22. Mattke J, Maier C, Hund A, Weitzel T (2019) How an enterprise blockchain application in the U.S. pharmaceuticals supply chain is saving lives. MIS Q Executive 18(4):245–261. https://doi.org/10.17705/2msqe.00019

23. Williamson PJ, De Meyer A (2012) Ecosystem advantage: How to successfully harness the power of partners. Calif Manage Rev 55(1):24–46. https://doi.org/10.1525/cmr.2012.55.1.24

24. Chen Y (2018) Blockchain tokens and the potential democratization of entrepreneurship and innovation. Bus Horiz 61(4):567–575. https://doi.org/10.1016/j.bushor.2018.03.006

25. Levy V (2021) 2021 Global life sciences outlook. Deloitte Insights. https://www2.deloitte.com/global/en/pages/life-sciences-and-healthcare/articles/global-life-sciences-sector-outlook.html

26. Parkin J (2019) The senatorial governance of Bitcoin: Making (de)centralized money. Econ Soc 48(4):463–487. https://doi.org/10.1080/03085147.2019.1678262

27. McCurdy D (2020) The role of collaborative governance in blockchain-enabled supply chains: a proposed framework. [DBA, Georgia State University]. https://scholarworks.gsu.edu/bus_admin_diss/131/

28. Iansiti M, Lakhani KR (2017) The truth about blockchain. Harv Bus Rev, Jan-Feb. https://hbr.org/2017/01/the-truth-about-blockchain

29. Chang SE, Chen Y (2020) Blockchain in health care innovation: Literature review and case study from a business ecosystem perspective. J Med Internet Res 22(8):e19480. https://doi.org/10.2196/19480

30. Mitra R (2020, April 24) What is blockchain governance: Ultimate beginner's guide. Blockgeeks. https://blockgeeks.com/guides/what-is-blockchain-governance-ultimate-beginners-guide/

31. Food and Drug Administration (FDA) (2018, October 30) *Drug Supply Chain Security Act.* U.S. Department of Health and Human Services. https://www.fda.gov/Drugs/DrugSafety/DrugIntegrityandSupplyChainSecurity/DrugSupplyChainSecurityAct/default.htm

32. Jensen T, Hedman J, Henningsson S (2019) How TradeLens delivers business value with blockchain technology. MIS Q Executive 18(4):221–243. https://doi.org/10.17705/2msqe.00018

33. MELLODDY (2021) Project Organisation. https://www.melloddy.eu/project-organization

34. PharmaLedger (2021) About us—the project. https://pharmaledger.eu/about-us/the-project/

35. Triall (2021) *Clinical trials are going digital.* https://www.triall.io/#resources

36. World Economic Forum (2020) *Redesigning trust: Blockchain deployment toolkit.* https://widgets.weforum.org/blockchain-toolkit/index.html

37. European Medicines Verification Organisation (2021) Introduction to the European Medicines Verification System (EMVS). https://emvo-medicines.eu

38. Food and Drug Administration (FDA) (2021) Drug Supply Chain Security Act (DSCSA). U.S. Department of Health and Human Services. https://www.fda.gov/drugs/drug-supply-chain-integrity/drug-supply-chain-security-act-dscsa

39. Dagher GG, Mohler J, Milojkovic M, Marella PB (2018) Ancile: privacy-preserving framework for access control and interoperability of electronic health records using blockchain technology. Sustain Cities Soc 39:283–297. https://doi.org/10.1016/j.scs.2018.02.014

40. Durneva P, Cousins K, Chen M (2020) The current state of research, challenges, and future research directions of blockchain technology in patient care: Systematic review. J Medical Internet Res 22(7):e18619. https://doi.org/10.2196/18619

41. HIPAA Journal (2018, January 8) Healthcare Data Breach Statistics. https://www.hipaajournal.com/healthcare-data-breach-statistics/

42. Riasanow T, Burckhardt F, Soto Setzke D, Böhm M, Krcmar H (2018) The generic blockchain ecosystem and its strategic implications. In: AMCIS 2018 proceedings: strategic and competitive use of information technology (SCUIT). Twenty-fourth Americas conference on information systems, New Orleans. https://aisel.aisnet.org/amcis2018/StrategicIT/Presentat ions/13
43. BIS Research (2018) Global blockchain in healthcare market focus on industry analysis and opportunity matrix—analysis and forecast, 2018–2025. https://bisresearch.com/industry-rep ort/global-blockchain-in-healthcare-market-2025.html
44. Donovan F (2019, August 5) Healthcare blockchain could save industry $100B annually by 2025. HIT Infrastructure. https://hitinfrastructure.com/news/healthcare-blockchain-could-save-industry-100b-annually-by-2025
45. Guenther C, Pierson M, Modi N (2018) In blockchain we trust: transforming the life sciences supply chain. Accenture. https://www.accenture.com/us-en/insight-blockchain-innovations-life-sciences
46. Weiner BJ (2009) A theory of organizational readiness for change. Implement Sci 4(1):67. https://doi.org/10.1186/1748-5908-4-67
47. Kruse CS, Kristof C, Jones B, Mitchell E, Martinez A (2016) Barriers to electronic health record adoption: A systematic literature review. J Med Syst 40(12):252. https://doi.org/10.1007/s10 916-016-0628-9
48. Hermes S, Riasanow T, Clemons EK, Böhm M, Krcmar H (2020) The digital transformation of the healthcare industry: Exploring the rise of emerging platform ecosystems and their influence on the role of patients. Bus Res 13(3):1033–1069. https://doi.org/10.1007/s40685-020-00125-x
49. McClellan SR, Casalino LP, Shortell SM, Rittenhouse DR (2013) When does adoption of health information technology by physician practices lead to use by physicians within the practice? J Am Med Inform Assoc 20(e1):e26–e32. https://doi.org/10.1136/amiajnl-2012-001271
50. Pisa M, McCurdy D (2019) Improving global health supply chains through traceability. CGD Policy Paper 139. www.cgdev.org
51. Depository Trust & Clearing Corporation (2019) Distributed ledger technology governance for private permissioned networks. https://www.dtcc.com/dtcc-connection/articles/2019/july/30/governing-dlt-networks
52. Ansell C, Gash A (2007) Collaborative governance in theory and practice. J Public Adm Res Theory 18(4):543–571. https://doi.org/10.1093/jopart/mum032
53. HIPAA Journal (2021, January 19) 2020 Healthcare data breach report: 25% Increase in breaches in 2020. https://www.hipaajournal.com/2020-healthcare-data-breach-report-us/
54. Baillieu J, Emmanuel J (2020) Exploring and deploying blockchain solutions in the life sciences and healthcare sectors. Eur Pharm Rev 25(4:59–62. https://www.europeanpharmac euticalreview.com/article/126761/exploring-and-deploying-blockchain-solutions-in-the-life-sciences-and-healthcare-sectors/
55. Smith T, Nelson J, Amaya C, Thyagarajan A, Synenki M, Weinstein J, Montalto M, Pandey P, Kalakota R, Konersmann T, Roy A, Fox J, Housman D (2018) Blockchain to blockchains in life sciences and health care. Deloitte Insights. https://www2.deloitte.com/content/dam/Del oitte/us/Documents/life-sciences-health-care/us-lshc-tech-trends2-blockchain.pdf

Dr. Denise McCurdy is a veteran of the Information and Communications industry with over 30 years of experience working with Fortune 100 firms. With a professional and personal interest in emerging technologies, Dr. McCurdy returned to academics to investigate blockchain, completing her doctorate in business from Georgia State University. Dr. McCurdy's research includes blockchain governance using collaborative modeling, technology in policing, blockchain, and artificial intelligence for the life sciences industry.

Life Sciences Intellectual Property Through the Blockchain Lens

Michael Henson

Abstract Intellectual property (IP) refers to a broad range of intangible assets derived through creations of the mind. Intangible assets are not physical and, as such, are distinguishable from other assets like real estate and personal property. Intangible assets can be owned by individuals or entities and encompass everything from inventions, to expressive works, to designations for goods and services. With the recent emergence of blockchain, life sciences organizations are realizing that blockchain can be an attractive tool to manage assets while protecting and enforcing IP rights. This chapter provides an overview of the different facets of IP protection for blockchain in life sciences and explains how life sciences organizations can utilize blockchain technologies to help procure, maintain, and enforce their IP.

Keywords Intellectual property · Blockchain · Patents · Trademarks · Copyrights · Life sciences · Open source

1 Introduction

Protecting the creative efforts of inventors, scholars, scientists, and the like is integral to the evolution of life sciences research and continued innovation. While there can be a variety of intangible assets a life sciences organization owns, the most common ones are patents, copyrights, trademarks, and trade secrets. For many life sciences companies, their IP is more valuable than their tangible assets, and extensive resources are often allocated to protect and enforce IP rights to remain competitive. Blockchain-related technologies provide an opportunity for life sciences organizations to explore new techniques for procuring, managing, and enforcing their IP in ways that can help differentiate themselves from competitors. Numerous efforts are underway to utilize blockchain to enhance business practices, and many life sciences organizations have already begun pursuing various forms of blockchain-related IP to stake their claims.

M. Henson (✉)
Partner, Intellectual Property Group, Perkins Coie LLP, CO Denver, USA
e-mail: MHenson@PerkinsCoie.com

2   The Emergence of Blockchain in Life Sciences

The expansion of blockchain technologies in recent years has piqued the interest of many, including those in the life sciences industry. Patent filing activity is a good indicator of this, and the below tables are illustrative. In relative terms, blockchain technology is fairly new, and patent issuance data in the space is less than a decade old. Studies in recent years indicate that blockchain patent applications filed in the United States have grown appreciably, if not exponentially [1]. For example, in the early years of blockchain, the number of patent applications filed from 2011 to 2016 increased from 6 to 540, while issued patents alone increased from 3 to 62 over the same period [2]. Research indicates that worldwide issued patents in the blockchain space currently exceed 10,000 [3].

Derwent Innovation, a leading patent research and analysis platform, has analyzed trends in blockchain applications. As shown in Fig. 1, Derwent estimated the compound annual growth rate between 2013 and 2018 to be 285.6% [4]. Figure 2 provides a breakdown of blockchain innovations by country, indicating that China and the United States have led the way [4].

Similar trends have occurred in the life sciences industry, specifically. Life sciences is a highly competitive space. As with many other technology sectors, there is an impetus to file patent applications early and often to avoid losing the exclusivity of patent protection to rivals racing to the patent office. Often, adequate protection of innovations entails filing numerous patent applications, resulting in large portfolios as companies attempt to protect various aspects of the underlying

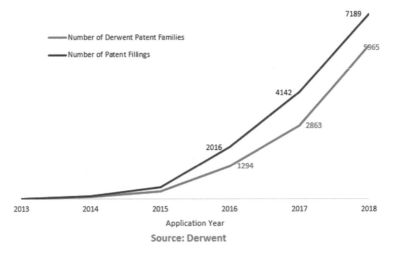

Fig. 1 Derwent patent families and filings 2013–2018. Data from Derwent Innovation, provided by Clarivate. Derwent Innovation and Clarivate are trademarks of their respective owners and used herein with permission

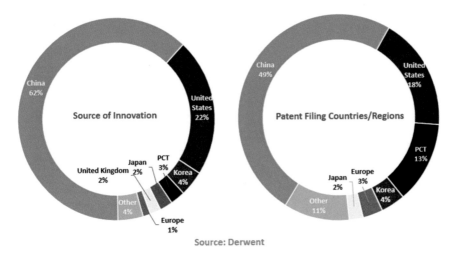

Source: Derwent

Fig. 2 Breakdown of blockchain innovations by country 2013–2018. Data from Derwent Innovation, provided by Clarivate. Derwent Innovation and Clarivate are trademarks of their respective owners and used herein with permission

technology. To the extent such innovations incorporate blockchain-related technologies to improve processes, it will create additional patent filing opportunities for life sciences organizations in the years to come.

The charts below summarize life sciences blockchain patent filing activities worldwide, in which the term "blockchain" is specifically mentioned in at least one of the patent claims. Figure 3 [3] shows an appreciable increase since 2012 in life sciences patent assets. Analyzing patent activity since about 2012 is a suitable starting point because it corresponds to when the U.S. Supreme Court issued its groundbreaking decision [5], which has come to be known as the *Alice* test. This decision changed the landscape of how computer-implemented inventions, such as those which now leverage blockchain technologies, are analyzed for patentability.

The values shown in Fig. 3 were obtained by conducting searches on September 13, 2021 in the AcclaimIP database for patent assets incorporating the term "blockchain" in the specification and claims and having a priority date on or after January 1, 2012term. Life sciences blockchain assets were then identified by narrowing the dataset to those having one or more of the following CPC or IPC classes on the face of the patent: A23, A61, C02, C05, C07, C12, and C40. The resulting dataset was additionally reviewed for assignees in life sciences industries or with selected life science terms in their name (e.g., medicine, medical, genetic, pharma, and hospital).

Most patent applications are not publicly accessible until 18 months after they are filed, while others may not be publicly accessible until issuance. Accordingly, since there is an 18-month lag in the results, we will not have a complete picture for all of 2020 (let alone 2021) until mid-2022, but 2020 is already on pace to surpass results from 2019.

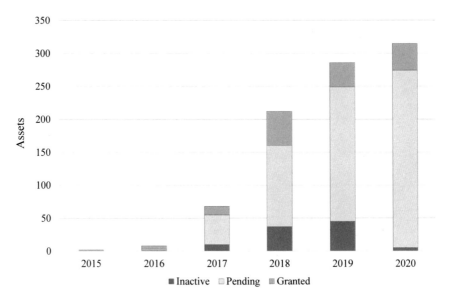

Fig. 3 Life sciences blockchain assets by filing years 2012–2020

Figure 4 [3] shows the relative activity for those same patent assets emanating from the United States compared to the rest of the world. In recent years, the U.S. has

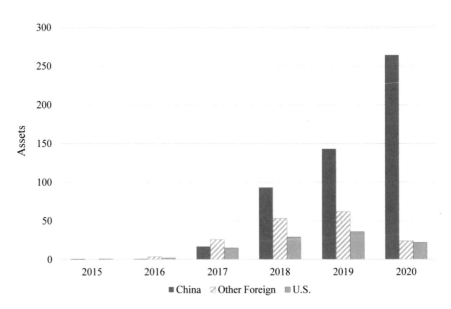

Fig. 4 Active life sciences blockchain assets by filing years 2012–2020 by geographic region

begun to lag further behind the rest of the world, and Fig. 4 reveals this is primarily due to the high volume of cases coming out of China.

Finally, Table 1 [3] shows the top patent filers of blockchain-related technologies in the life sciences sector.

Table 1 Organizations with the most blockchain patent filings by geographic region

Assignee	China	Other Foreign	U.S.	Total
PING A LIFE INSURANCE CO OF CHINA	96			96
SHAANXI MEDICINE CHAIN BLOCK CHAIN GROUP CO	53			53
PING AN MEDICAL & HEALTHCARE MAN CO	49			49
MERCK PATENT GMBH	3	20	8	31
[INVENTOR / NO DATA]	15	7	5	27
SHAANXI MEDICINE CHAIN BLOCKCHAIN GROUP CO	22			22
NCHAIN HOLDINGS LTD	4	12	3	19
ABMAX BIOTECHNOLOGY CO	17			17
SHAANXI MEDICINE CHAIN GROUP CO	13			13
SIEMENS AG	4	5	3	12
AI BIOMATERIAL HEALTHTECH LTD	1	8	2	11
CRIRM ADVERTISEMENT		10		10
VOICE LIFE INC	2	6	1	9
BEIJING ANBAOKANG BIOLOGICAL MEDICINE TECH CO	8			8
MASSACHUSETTS MUTUAL LIFE INSURANCE			8	8
MACROGENICS INC	3	4		7
CHINA LIFE INSURANCE CO SHANGHAI DATA CENTER	6			6
FUBON LIFE INSURANCE CO		6		6
HANGZHOU LIANZHONG MEDICAL TECH CO	6			6
NAN SHAN LIFE INSURANCE CO	2	4		6
TAIPEI MEDICAL UNIV	2	2	2	6
EBAO INTERNET MEDICAL INFORMATION TECH BEIJING CO	6			6
PING AN HEALTH INSURANCE CO	6			6
RESMED CORP	1	3	1	5
ZHONG AI HEALTH TECH GUANGDONG CO	5			5

3 The Intersection of Blockchain and Life Sciences IP Rights in the United States

The protection of IP rights in the United States stems first and foremost from the U.S. Constitution, which grants Congress the enumerated power "To promote the progress of science and useful arts, by securing for limited times to authors and inventors the exclusive right to their respective writings and discoveries" [6]. Clause 8 constitutes the source of Congressional power to enact legislation respecting patents and copyrights and is, therefore, often referred to as the "Patent and Copyright Clause." Procuring, managing, and enforcing IP rights are vital to life sciences organizations. Blockchain provides an opportunistic platform to improve the efficacy of these efforts, in part by providing nearly immutable records of the life cycles of IP rights. While procuring IP rights varies by the type of IP involved and by jurisdiction, the central tenets of IP rights protection in the U.S. are introduced along with examples of how blockchain technologies are being used and considered to further the efforts of life sciences organizations.

3.1 Patents

A patent is a grant made by a governmental agency to an inventor that confers upon him/her the exclusive right to practice an invention for a limited period of time. A United States patent grants an inventor the sole right to make, use, sell, offer to sell, or import the patented invention during the term of the patent, typically twenty (20) years from the date on which the patent application is filed [7]. It is important to appreciate that a patent does not actually grant the patent holder the right to do anything, but rather the right to prevent others from doing something, namely, infringing the patent. In this respect, a patent can be described as a negative, intangible personal property right. To the extent a commercially successful blockchain-related technology of a life sciences organization can be patented, the patent(s) can corner the market and exclude competition. With these rights in perspective, what are the prerequisites for obtaining a patent, and what type of blockchain-related applications have been patented in the life sciences space?

In the United States, there are three types of patents one can be awarded: utility patents, design patents, and plant patents. Utility patents are the most common and can be obtained for any new and useful process, machine, manufacture, composition of matter, or any new and useful improvement [8]. Design patents protect new, original, and ornamental designs for articles of manufacture. Plant patents allow for the protection of asexually reproduced, distinct, and new varieties of plants.

Regardless of which type of patent is pursued, there are three requirements in the United States for obtaining a patent, and similar requirements exist in other countries. Assuming an innovation is not deemed to be primarily directed to a law of nature, a natural phenomenon, or an abstract idea, it will be patentable provided it is useful, new, and non-obvious. If an innovation fails to satisfy any one of these tests, it will not be patentable. The U.S. Supreme Court and lower courts have struggled for years to articulate what constitutes an "abstract idea" under our patent system. What actually constitutes an abstract idea is beyond the scope of this chapter but often arises in computer-implemented inventions, including blockchain-related life science innovations.

Usefulness, or utility, is generally a low threshold and simply means that an innovation has some purpose or performs its intended function. However, in the life sciences arena, this can sometimes prove challenging. For example, in silico techniques (i.e., those generated through computerized experimentation) can lead to the discovery of information on genes and proteins before their practical or technical application is known. Related to this is the role blockchain and artificial intelligence (AI) can play in advancing medical devices that use algorithms to enable medical treatment by monitoring and diagnosing patients via, for example, a handheld device or telemedicine. These create evolving legal issues regarding utility and inventorship, and more specifically, the role that computer-implemented innovations play in potentially supplanting human ingenuity.

To determine if a blockchain-related innovation is new, the inquiry focuses on whether it previously existed in its exact form. Though simple on its face, the concept of novelty can present challenging and technical legal issues. Even if a blockchain innovation is useful and novel, it still must satisfy the third test of patentability, referred to as non-obviousness. That is, innovation must not be obvious in view of the prior art. The test of non-obviousness (referred to as "inventive step" in many other countries) is subjective and demands that the subject matter of a patent contribute more to the state of the art than a mere technical advancement over it. Rather, the contribution must not be something that would be obvious to a person skilled in the art to which the invention pertains.

3.1.1 Blockchain-Related Patents in the Life Sciences Arena

The practically immutable nature of blockchain makes it an attractive component for many life sciences sectors. For example, healthcare professionals are constantly trying to comply with Health Insurance Portability and Accountability Act (HIPAA) requirements and privacy issues in general. Blockchain applications can be incorporated into privacy data storage, dissemination, and permissioning systems to help further these objectives. Additionally, where traceability of drugs, their sources, constituents, etc. is of interest, blockchain can help provide a tamper-resistant and tamper-evident audit trail that is more robust than existing approaches. These represent only a few applications where blockchain can intersect with life sciences.

Blockchain-related patents in the life sciences space touch on a variety of applications, some of which are represented below.

Patient Management

U.S. Patent No. 10,896,749 [9] relates to a drug monitoring system that comprises a data receiver for a patient's pharmacokinetic (PK) profile. An interactive user interface displays a time-varying therapeutic plasma protein level of the patient based on an administered dose of a drug and the PK profile of the patient. A QR code having patient information is generated and then encrypted using AES-256 encryption and stored on a blockchain using public-key cryptography standards.

U.S. Patent No. 10,541,807 [10] facilitates healthcare system security and interoperability. This patent describes a permissioned private blockchain platform maintaining health-related parameters for a plurality of patients. Machine learning (ML) techniques are employed to assess real-world information comprised of multi-dimensional blocks in the blockchain to determine effective pharmaceutical dosages, target drugs based on demographics, drug interaction information, or efficacies associated with various modes of drug administration.

U.S. Patent No. 11,069,448 [11] relates to expedient collaborative decision-making in medicine by addressing disagreement among experts for a medication regime by optimizing doctors' consensus resolution.

U.S. Patent No. 11,0496,012 [12] relates to managing therapy data for a diabetic therapy system within a distributed ledger. The diabetic therapy system validates data obtained from multiple computing nodes and then administers an insulin dosage to a patient based on the validated data.

U.S. Patent No. 10,991,463 [13] provides a method for making AI-based medical treatment plan recommendations. The method involves storing a patient's healthcare data in a blockchain database and updating the healthcare data with health records and service notes from medical care visits. Utilizing the patient's health information, the system then determines the success of therapies, calculates the probability of disease progression, ranks therapies based on a probability of successful treatment, and automatically transfers payment in a native cryptocurrency to a healthcare provider according to the terms of a smart contract.

Managing Patient Data

U.S. Patent No. 11,055,419 [14] provides a decentralized data authentication system that integrates blockchain technologies, independent verification software, a cloud-based decentralized certificate authority, and a centralized redundant database system. Together, these components form data portability and longevity systems designed to integrate lifetime health records accessible by the patient, provider, and payer using public/private keys. This system also provides a data connectivity cloud that encrypts and stores the signature data in a Merkle tree created from the record data.

U.S. Patent No. 10,923,216 [15] provides a health status platform that receives test results of a biological sample collected from a patient indicating the presence of an infectious disease. The system implements a blockchain architecture and incorporates an end-to-end encryption system that receives encrypted venue access requests for validation and provides an encrypted certificate of origin to a venue access manager.

U.S. Patent No. 11,003,791 [16] provides a computer-implemented method of managing health information such as genomic data comprising DNA sequence information or RNA sequence information. The data analysis applications permit a data miner to access and analyze the health information in an encrypted format while maintaining the privacy of the data contributor miners. The data contributors may include pharmaceutical companies, medical laboratories, or hospitals that use various methods to perform research on aggregated contributor data. Health information is securely exchanged among the various stakeholders, including data owners or contributors, data requestors or miners, and medical providers such as hospitals, clinics, and research laboratories.

Healthcare Data Validation

U.S. Patent No. 10,340,038 [17] provides validation systems for healthcare data transactions. When a health-related transaction is conducted, the healthcare parameters (e.g., clinical evidence and outcomes) are sent to validation nodes that generate a new block for the transaction via proof-of-work consensus.

U.S. Patent No. 11,048,788 [18] provides a system for publishing authenticated digital content, content authentication, and validation via multi-factor digital tokens. The system receives an electronic file including a digital image and biometric information associated with a person, such as omic sequence data including a DNA sequence, an RNA sequence, or an amino acid sequence.

Drug Discovery

U.S. Patent No. 10,937,068 [19] discloses a system for determining an assessment value of a document related to current research work in drug discovery. Information indicative of entities and semantic inter-relationships related to the technical field of the current research work is accessed to determine a status factor indicative of the novelty of the document with respect to publicly available knowledge. The assessment value is determined in a cryptocurrency for enabling future transactions of the document, such as a sale, using a blockchain.

Dental Health Management

U.S. Patent No. 10,930,377 [20] discloses a blockchain method for managing dental records. The method entails receiving signals from a dental device associated with a dental activity being performed and detecting dental feature indicators with associated confidence levels by analyzing the dental signals. Dental feature indicators are aggregated to compute a multi-dimensional feature vector. Transactions are then created and sent to validating peers on a blockchain network that includes a record of a patient's dental-related features and events throughout the patient's life.

Tracking Healthcare Data and Other Information

U.S. Patent No. 11,017,892 [21] discloses a method for tracking an ingestible medication device for monitoring compliance with a medication regimen, dosage information, or pharmaceutical prescriptions. External devices include patient wristbands, caregiver handheld devices, and a healthcare provider computer system that records information to a blockchain. The method retrieves a medication record for the human subject from the healthcare provider system corresponding to a first medication-tracking blockchain stored on a plurality of network nodes of the healthcare provider system.

U.S. Patent No. 11,017,883 [22] provides blockchain-based systems and methods for tracking donated genetic material such as an egg, an embryo, sperm, blood, tissue, stem cells, a genome, DNA, RNA, nucleic acid, or an organ. Blockchain-based records associated with the donated genetic material are generated to create an audit trail.

U.S. Patent No. 10,943,302 [23] relates to determining an insurance risk score or an insurance cost for an individual. The system involves the steps of (i) receiving an individual's health parameter data from sensors within wearable devices, (ii) validating the received health parameter data based on a predefined validation rule, (iii) recording the health parameter data and a unique ID associated with the individual in a blockchain, and (iv) retrieving from the blockchain a plurality of instances of health parameter data associated with the unique ID to generate a risk score associated with the individual.

U.S. Patent No. 10,929,901 [24] discloses a method of tracking the provenance of fur, leather skin, reptile skin, or ostrich skin. The process involves imaging a live animal, removing a DNA sample from the live animal, electronically storing information associated with the DNA sample in a computer-based system, and storing sale information in a distributed ledger or blockchain to associate it with an identifier number and the DNA sample.

U.S. Patent No. 10,943,680 [25] describes safe, efficient, and fraud-proof continuous retrieval of health data captured by a health tracker such as a wearable device. The method comprises receiving a request to update a record associated with a user blockchain that comprises identification information associated with the health tracker. Upon receipt of health data from a health tracker server, the validity of the user's latest blockchain is verified, and a new block instance is generated corresponding to the received data.

U.S. Patent No. 10,942,956 [26] relates to detecting medical fraud using blockchains for medical prescription verification and dispensing. The method receives prescription data from a device associated with a prescribing entity, generating a validation code for the prescription, adding the validation code to the prescription data, and appending them to a blockchain that can then be queried.

3.1.2 Using Blockchain to Facilitate Procurement and Maintenance of Patents

Assessing Inventorship

Many aspects of a life sciences organization's overall IP strategy rely heavily on accurate recordkeeping. It can be important, for example, to properly document the evolution of patentable innovations and the respective contributions of those involved. The "audit trail" afforded by a blockchain makes it a natural fit to effectuate this purpose. Such audit trails can also help resolve disputes over ownership, authorship, and dates of use and facilitate due diligence efforts during merger and acquisition transactions. To these ends, creative blockchain constructs that leverage permissioned and permissionless architectures could provide publicly available repositories for information while also addressing confidentiality concerns [27].

When applying for a blockchain-related life sciences patent, it is necessary to identify those individuals who made inventive contributions. Often, this entails more than one person's efforts within an organization or even the collective efforts of numerous individuals from multiple organizations. This is particularly true for individual entities, joint ventures, research and development corporations, and the like. Where there exist inadequate or disparate recordkeeping systems, the ability to accurately document each person's contribution can prove quite challenging. A tamper-resistant blockchain can greatly facilitate documenting the innovation process by recording the dates, times, levels of each person's involvement, when certain milestones were reached, and the overall evolution of a technical solution (as well as those that fell by the wayside). This type of information could be useful in various other contexts, for example, to establish priority of invention, demonstrate prior art, compensate employees based on relative contributions, facilitate licensing arrangements, and IP ownership allocation, to name a few [28].

Assisting Patent Examiners

Analyzing the patentability of blockchain-related life sciences inventions can be an intense process prone to human errors. Examiners at the United States Patent and Trademark Office (USPTO) have various databases at their disposal to search for prior art in an attempt to ascertain if a claimed invention is useful, new, and non-obvious. These databases are not currently secured through blockchain. While some maintain this is unnecessary since existing registration systems are already managed at the governmental level [29], others argue that doing so could have obvious advantages by providing universally accessible and pragmatically immutable repositories of information for examiners at patent offices throughout the world [30]. For example, the dates of prior art references can be critical when assessing patentability for life sciences innovations. One can envision an infrastructure in which governmental agencies automatically record and timestamp references on a blockchain as they become publicly available to provide indisputable records of provenance. AI methods and smart contracts could then be used in conjunction with the blockchain(s) to streamline prior art analysis. IPwe is one company that is focused on such capabilities, and its patent platform provides a secure, open blockchain repository of worldwide patents, along with tools for identifying, researching, evaluating, and transacting patents [31].

While utility patents are the most commonly pursued types of patents, design patent protection is also quite prevalent. However, it can be particularly challenging for examiners to determine whether the design of a useful article of manufacture is patentability distinguishable over the prior art since a design patent application essentially comprises a series of representations of the article from different perspectives. The examiner is tasked with determining whether such a depicted article is novel and non-obvious in view of pre-existing articles. This challenge is exacerbated by the volume and variety of prior art in which a similar article might be encountered, for instance, other design or utility patents, trade journals, and marketing materials. Minimally, if other patent assets were stored as Computer Aided Designs (CADs) on a blockchain, an AI engine could crawl the prior art records and perhaps make an initial assessment for an examiner whether the pending design is patentable, or at the very least, identify for the examiner what prior art items might be relevant to the analysis [32].

While it is not anticipated that such blockchains will replace the need for examiners altogether, as the human element will remain important to any patentability analysis, blockchain technologies can reduce turnaround time and aid examiners in becoming more efficient. From the applicant's standpoint, this will expedite the examination process and address the numerous backlog issues prevalent in many popular patent offices throughout the world, such as the USPTO and the European Patent Office (EPO), not to mention the potential cost savings advantages from both a registration and maintenance point of view.

3.2 Trademarks and Trade Dress

Trademarks are used to indicate the sources of goods or services and to distinguish them from others. Akin to a trademark is trade dress, which refers to the visual appearance of products or their packaging designed to help consumers recognize the source of the products. It is through trademarks that consumers can connect a product or service to a company. As such, trademarks help promote goodwill and reflect brand recognition. A variety of things can function as a trademark. Generally, a trademark can be any word, name, symbol, device, sound, color, or even a smell. Trademarks often comprise words, designs, or logos and can essentially be anything provided they are used in commerce, are distinctive, are non-functional, and act as source identifiers.

Trademarks represent very valuable intangible assets for life sciences companies and powerful marketing tools. As with other industries, companies in the life sciences space rely heavily on trademarks to protect their branding and goodwill. Some spend enormous amounts in legal fees to procure, maintain, and enforce their marks. Often, pharmaceutical companies in particular take advantage of trademark laws to effectively extend their dominance beyond patent expirations. Such name recognition can be critical and ultimately may become a hurdle that competitors simply cannot overcome despite their ability to produce the same drug under a generic brand.

3.2.1 Using Blockchain to Facilitate Procurement and Maintenance of Trademarks

Blockchain technologies could also prove beneficial to life sciences organizations where trademarks are concerned. In determining the viability of a proposed mark, it is important to consider whether it might interfere with an existing mark. In the U.S., this is known as the "likelihood of confusion" test and takes into account the visual, audible, and connotational similarities between marks, as well as the relatedness between their goods and services.

Different databases are used throughout the world to search trademarks, and this can lead to increased costs and potential inconsistencies in assessing marks for trademark clearances, the likelihood of confusion analysis, due diligence, and the like. However, similar to patents, CAD and blockchain/AI resources could be employed to assess the viability of trademarks, either at the local (i.e., state), national, or international level. Blockchain could bridge the gap between these disparate systems and provide a universal platform for analysis [32]. The blockchain could also store any changes made to existing registrations, such as amendments to the recitation of goods or services, dates of first use and first use in commerce, status changes, or name changes. This would eliminate the need for one to search individual records of marks at the respective trademark offices and could also streamline the process of providing evidence in legal proceedings involving trademarks [27].

Indeed, life sciences organizations performing due diligence before applying for patents or trademarks could benefit from blockchain IP repositories to independently assess the prospect of protection prior to filing. Thus, one can imagine a deployment environment that allows an organization to conduct inquiries into an open, permissionless blockchain. Alternatively, there is potential for a hybrid model that allows an organization to query an open blockchain for such information while maintaining requisite privacy levels for individual transactions.

Additionally, many jurisdictions require trademarks to be maintained at certain intervals following registration. Depending on the jurisdiction, evidence of use is needed to renew, maintain, or demonstrate incontestability of marks [27]. Blockchain technology, if endorsed by administrative bodies or courts, could simplify the process of providing evidence of a variety of trademark-related information, such as dates of first use and first use in interstate commerce, ownership, chain of title, goodwill, and acquired distinctiveness or secondary meaning associated with marks in the marketplace [27]. If recorded on a blockchain and coupled with smart contracts, information regarding the state of the relevant market could be utilized by organizations to assess the likelihood of confusion as part of an infringement analyses. Given the sensitivity of such an analysis and the privileges attached, organizations could integrate appropriate permissioning to protect confidentiality.

3.3 Trade Secrets

An area of protection that is sometimes overlooked is trade secrets. Trade secrets for life sciences organizations can encompass a broad range of information to give a company a competitive advantage. A company's secrets may include business methods, techniques, devices, know-how, client lists, chemical processes and formulations, compilations, marketing strategies, etc. It is common, for example, that in arriving at a patentable blockchain-related invention, a life sciences organization might discover certain technical information or know-how that does not ultimately become part of the patent application. Such information may be well suited for trade secret protection to the extent it provides a competitive advantage.

While formal registration with governmental bodies is not a prerequisite to protecting trade secrets, certain measures must be taken to ensure they remain protected. Generally, to maintain trade secret status, information must actually be secret, have commercial value, be disseminated on a limited need-to-know basis, and reasonable efforts must be made to ensure the information remains secret.

Most states in the U.S. have adopted the Uniform Trade Secrets Act (UTSA), which protects against "theft, bribery, misrepresentation, breach or inducement of breach of a duty to maintain secrecy, or espionage through electronic or other means" [33, p. 4]. The U.S. Defend Trade Secrets Act [34] governs federal jurisdiction over trade secrets, while the Trade Secrets Directive governs in the EU [35].

3.4 Copyrights

Under U.S. law, a creator owns the copyright to creative work. Creative works can include, for example, music, art, literary works, sculptures, computer programs, as well as compilations of works and derivative works. Copyright protection manifests upon the creation of work provided it is fixed in a tangible medium of expression, regardless of the medium. A copyright holder enjoys various rights with respect to the work, including the exclusive rights to reproduce the work, prepare derivative works, publicly perform, and publicly display the work. These rights can be held or licensed by the copyright holder in whole or in part, but critically, unless the rights are expressly assigned or licensed away, they remain with the copyright holder. In the United States, the Copyright Act forms the basis of copyright protection [36].

Blockchain can play an integral role in memorializing the evolution of copyrights. Upon creating a work, users could store their work, or a hash referencing the work, on a blockchain, thereby providing a timestamp establishing possession and ownership whether or not it is ever formally registered with the Copyright Office. Relevant records about copyrights can be found in administrative governmental bodies, individual organizations, or third parties maintaining information repositories. These disparate systems are not interoperable and can be expensive to maintain and secure. Some maintain that creating a blockchain system from scratch to manage IP rights in copyrights would be less expensive than trying to synchronize relevant records from multiple sources in an attempt to create a universal, interoperable one [32]. The system would resemble traditional permissionless blockchains, allowing system users to be nodes on the platform, thereby contributing to the system's security without attendant costs. These same users would have visibility into the entire chain, and the costs of identifying rights holders would be significantly reduced. Users could also self-manage their own IP rights and deploy smart contracts to control all aspects of transactions relating to their rights, thereby reducing operating fees and eliminating the need for third-party management.

Recently, there has been a growing interest in the use of non-fungible tokens (NFTs) to represent rights in digital goods. An NFT is a type of cryptographic token that represents ownership of digitally scarce (or unique) goods, such as collectibles, and can be implemented on any blockchain that supports smart contract programming. Like their fungible counterparts, NFTs certify ownership of assets and are transferable. However, unlike fungible tokens, NFTs are not interchangeable since they represent items with unique qualities. These characteristics of NFTs make them an appealing option for managing the ownership and transfer of digital assets, which might prove beneficial in some applications to life sciences organizations.

3.4.1 Open-Source Software

Computer programming code is a type of expressive work to which copyright protection can attach. Most end users never see the underlying source code behind an application. In order to use proprietary software, users must accept certain terms and conditions, typically by agreeing to terms of a license agreement displayed the first time the software is executed on one's machine. These licensing terms set forth the parameters governing the permitted uses/non-uses of the installed software.

Open-source software (OSS) generically refers to software distributed along with its underlying source code. OSS is, thus, publicly accessible for all to see. This open approach allows authorized programmers to inspect, modify, enhance, and distribute the code as they choose, subject to certain original rights that may attach to the code. In this regard, OSS is distinct from "proprietary" or "closed source" software in that the originator is the only entity that can modify or maintain exclusive control over the code. As with proprietary software, users of OSS often must agree to license terms and conditions to use the software, but the specifics of these terms and conditions can be quite different. In fact, there is a common misconception that access to OSS grants a user the ability to use the software without restriction, but this is not the case in most situations. Open-source licenses, and there are many, stipulate how users can use, modify, and distribute software. For the most part, the public can use open-source software for any purpose they choose. The full text of the most common open-source licenses may be found at http://www.choosealicense.com.

There are many blockchain open-source projects in existence today. Well-known projects include Hyperledger, Enterprise Ethereum, Corda, Quorum, and OpenChain. Among OSS licenses, MIT, Apache 2.0, and GPLv3 are perhaps the most popular for blockchain-related applications, but numerous others are available. Permissive licenses are the closest one can get to absolute free use of the software. MIT, Apache 2.0, and BSD are popular permissive licenses. The MIT license, for example, has very few restrictions other than notice requirements on copies and derivative works. Closer attention, however, must be paid to restrictive licenses. Restrictive licenses, in general, are actually quite popular despite their more stringent provisions. With restrictive licenses, it is important to understand the implications relating to derivative works since these licenses attach the same or similar terms as the original license to derivative works, in effect allowing the licensor to control the downstream distribution of the software. This approach is known as "copyleft." Popular copyleft licenses are GPLv2, GPLv3, LGPL, and AGPLv3. Certain restrictive licenses such as the GPL also require the source code for derivative works to be released back to the licensor.

Since open-source licenses are generally characterized as either permissive or restrictive (i.e., copyleft), life sciences companies utilizing blockchain technology should carefully consider their open-source options. For example, suppose a life sciences blockchain project leverages existing blockchain source code such as Ethereum or Hyperledger Fabric. In that case, it is imperative to understand any and all licenses that govern the use of such platforms and to appreciate that different licensing schemes may apply. Moreover, some blockchain platforms have adopted either permissive or restrictive licenses for their network code. In contrast, others have adopted a mixture with different criteria depending on the software component involved, such as the core codebase, APIs, and middleware. Ethereum is one such example [37].

Blockchain companies that utilize or develop software generally have policies in place for the development and licensing of software. Some companies require their employees' projects to have permissive licenses to use the project in its closed source applications. Alternatively, a company may want the exclusive right to use a project in its closed source software and, thus, require a copyleft license. Depending on corporate objectives, certain licenses may be preferred over others. Considerations might include whether the project will use dependencies, whether it will be used by a company that may have specific licensing policies for open-source projects, and whether the project will benefit from the contributions of others who may not want their contributions subject to restrictions. Consider, for example, whether a blockchain project might use dependencies. Programmers will appreciate that many projects have dependencies, i.e., libraries linked either statically or dynamically during runtime. Each library will have its own license. If the licenses are permissive, a user can generally use whichever license is desired. However, some of these dependent licenses may be "copyleft" such that the same terms are provided to downstream users.

While the distributed nature of blockchains makes it understandable that many blockchain platforms are subject to open-source licenses, certain business propositions in the life sciences arena may dictate a preference for proprietary licenses. Organizations will need to assess the pros and cons of each on a case-by-case basis. For example, some of the most potentially impactful legal implications can arise when proprietary software is combined with open-source software. Specifically, and as noted above, some open-source licenses require that any derivate works be covered by the terms of the original license when the software is distributed. If not taken into account, this can significantly impact the value of a life sciences company's software or the company itself. One possible way to address such ramifications is to avoid triggering a "distribution" provision by deploying proprietary code via cloud-based services or SaaS. On the other hand, a "distribution" can occur under some open-source licenses merely by accessing covered software via a network. As such, in the blockchain space, unintended consequences may arise when proprietary code is combined with open-source software and deployed on nodes. It is, therefore, prudent for life sciences organizations involved in the development or use of blockchain applications to fully understand the licenses that might attach to code developed or incorporated at various levels of the software stack, particularly in light of the various role permissions the participated nodes, oracles, or IoT devices may have.

Patents and Open Source—Can They Coexist?

The interaction between patents and open source is often misunderstood, and this misconception can be particularly true for blockchain-related inventions, whether in the life sciences space or others. As noted above, one of the most potentially detrimental legal risks with open-source software is its impact on proprietary software. That is, certain open-source licenses mandate that if other software includes, is derived from, or is combined with open-source software, then that "other" software must be licensed under the same terms as the open-source license when it is distributed. This is known in the open-source community as "tainting" of proprietary software. The effect of this is that licensees can copy, modify, and redistribute the software free of charge and require access to the source code to permit such rights to be exercised. Thus, the source code the developer assumed was propriety is not.

While open-source licenses are principally copyright licenses, which confer certain rights to licensees, patent issues that may arise with such licenses cannot be ignored. A common misconception is that open-source software cannot be patented. In actuality, it can be patented provided it meets the requirements for patentability discussed above, irrespective of how it is licensed. However, various open-source licenses have provisions that expressly require the licensee to grant patent licenses to others. This can be a serious ramification if overlooked. Another patent-related ramification can arise by using certain open-source licenses which attempt to deter licensees from asserting patent infringement claims that relate to the use of an open source.

Moreover, depending on the license, certain penalties may apply if one tries to assert a patent infringement claim. These consequences may include, among other things, revocation of the patent license granted to the licensee or the inability of the licensee to use the open-source software in the future. Yet another legal consideration is whether the use of certain open-source licenses can trigger implied licenses.

Various circumstances can trigger patent-related licensing provisions, so it is prudent to carefully scrutinize the particular licenses attached to open-source software incorporated into other proprietary software. Absent a careful review, life sciences organizations may run the risk that patents they now own or may acquire in the future will be subject to licensing requirements, thereby potentially eviscerating a competitive advantage. Furthermore, the scope of such licenses may encompass not only what the life sciences organization contributed but also downstream modifications by others. Accordingly, it is important for life sciences organizations to fully understand their use cases to avoid any unintended consequences.

4 Transferring IP Rights Through Blockchains

There are many situations in which a life sciences organization may need to transfer complete or partial ownership of IP to others. The need for transfer often occurs in connection with mergers and acquisitions, bankruptcy proceedings, insolvency proceedings, and various other situations. Depending on the nature of a transaction, an entity might assign or license IP on an exclusive or non-exclusive basis. Jurisdictional considerations also come into play as IP rights can be transferred on a territorial basis. Regardless of the circumstances, life sciences companies can leverage blockchain technologies to facilitate the transfer of IP rights. Demonstrating IP ownership is a prerequisite to transferring title to it, whether by assignment or license. Blockchain can provide a tamper-proof digital registry to establish provenance and ownership of IP. One blockchain initiative that addresses this issue is the Open Music Initiative ("OMI") by the Berklee College of Music in Boston, Massachusetts, which seeks to properly identify IP rights holders using blockchains as a repository for such rights. OMI plans to build its own Application Programming Interface (API) to allow other products and services to use the repository, greatly enhancing its use both in IP transactions and enforcement activities [38].

Licensing transactions can involve numerous obligations by both the licensor and the licensee. For example, a licensee may have recurring royalty payment obligations to the licensor that are tied to the volume of product sales covered by a patent, the number of goods or services associated with a trademark, or the number of copyrighted works distributed. A licensee may also be required to mark patented products or ensure that trademarks and software meet certain specifications as to quality and reliability and be used by the licensee in accordance with agreed-upon terms. Both parties will likely need to maintain suitable business records for auditing purposes. Certain events might also trigger provisions within these transactions, such as reverting an exclusive license to a non-exclusive license if the licensee cannot meet minimum sales requirements or if a breach has occurred based on actions or inactions by one of the parties. While these factors and others are certainly not unique to life sciences organizations, they are undoubtedly relevant and applicable [28].

Blockchains can be designed to not only record this pertinent information throughout the life of the license agreement, but the license terms themselves. Certain license terms and provisions could also be coded into smart contracts and securely stored and deployed on the blockchain to automate many of the parties' respective obligations, such as reporting and payment requirements, late penalties, and notice requirements, etc. [39].

5   Managing IP Rights Through Blockchain

As noted above, blockchain can be used in various ways to help manage IP rights by recording the life cycle of rights and aspects of legal transactions involving IP, such as licensing arrangements. As such, if properly recorded and maintained, blockchain could provide a robust, traceable repository of information that could be relevant and probative in a variety of disputes pertaining to IP rights. In this regard, blockchain could be used as an enforcement tool to establish infringement or misappropriation of an IP right (patent, copyright, trademark, or trade secret). Similarly, blockchain could serve as a defensive tool to aid in establishing a lack of ownership, invalidity, or non-infringement of an IP right. There are numerous examples of projects underway that relate to the use of blockchain for digital rights management, addressing such areas as remuneration to artists, content registration, content tracking, storage, dissemination, transfer, and enforcement, to name a few [40, 41].

For example, where trade secrets are concerned, blockchain can be a useful tool to manage the protection of them [28]. The encrypted nature of blockchain data, made possible through hashing, can provide the requisite security of information. Further, a permissioned blockchain environment naturally lends itself to ensuring limited dissemination of information on an as-needed basis, for example, by restricting who can view data on the ledger or write data to it. The blockchain could also record the evolution of information through timestamping and provide an audit trail of access to it, essentially recording the life cycle of the trade secret. This historical accounting could be used to establish the reasonableness of measures a life sciences organization has in place to protect the trade secret and verify the identity of anyone who executed an NDA and was given access to the trade secret [42]. Historical records could also inform investigations into the extent that a trade secret is misappropriated or help resolve disputes about who was the first to create a trade secret. At the same time, care should be taken to ensure that sensitive information is not inadvertently revealed, and serious consideration might be given to storing only the hash of information on the blockchain, not the underlying information itself. In this manner, the rightful owner of the information could prove such status and whether the information had been tampered with simply by encrypting the information stored off chain and reproducing its hash.

6 Blockchain in Adversarial Proceedings Involving IP Rights

The credibility of evidence is fundamental to any legal proceeding, and laying a proper foundation is a prerequisite to admitting evidence in a tribunal. Laying a foundation establishes the authenticity of information by demonstrating personal knowledge, chain of title, and relevance to conclude that the information being proffered is what its proponent claims.

These prerequisites for the admission of evidence can be satisfied by certain characteristics of blockchain technology, which makes it a suitable tool for this purpose. Because blockchain transactions are timestamped and linked through hashing to create an immutable record of transactions, it can be used to verify the dates and parameters of transactions. Where data are only accessible on a permissioned blockchain to those with certain access privileges, blockchain can be used to verify, for example, who had access to documents, signatories to documents, and chain of title. While adoption by legal regimes of blockchains for use in providing evidence in legal proceedings has been slow to come, at least one Chinese court allowed information on a blockchain to be used to establish the credibility of evidence in a copyright infringement proceeding [43].

Because blockchain technology is relatively new, there have been few litigation proceedings involving life sciences blockchain-related patents to date. However, since blockchain use cases touch on virtually all types of industries, a natural assumption is that industry players will accumulate stockpiles of blockchain-related IP and eventually converge in lawsuits akin to patent wars which have historically followed the introduction of other disruptive technologies such as the Internet, semiconductors, and smartphones. Therefore, it is anticipated that blockchain will play a more impactful role in establishing evidence in legal proceedings as reliable mechanisms of proof [28] and perhaps helping to mitigate future litigation [41].

6.1 Anticounterfeiting

While the procurement, management, and maintenance of IP rights are crucial to life sciences organizations, the resources devoted to these aspects would not be well spent if organizations are not disciplined in policing such rights. The global nature of commerce and technological advancements have contributed to the pervasive nature of counterfeiting and other IP circumvention methods. Supply chains and manufacturing are perhaps the most popular uses of blockchain today due to the blockchain's ability to record an incorruptible chain of information relating to the provenance and life cycle of goods and services. By recording information about the evolution and movement of a product's details throughout its manufacturing process and supply chain, almost every aspect of a product can be recorded, viewed, interrogated, and objectively verified. Efforts are currently underway between the private sector and U.S. customs officials to utilize blockchain to intercept illicit goods entering the United States through its ports of entry [44].

Blockchain is an ideal tool for detecting counterfeits, particularly in the pharmaceutical industry where regulations mandate tracing and tracking of goods through commerce. Similarly, trademark owners could also leverage these benefits to monitor parallel imports to help identify knock-offs. Storing the provenance of goods on a blockchain enables interested parties to instantly verify the authenticity in real time and thwart the efforts of those attempting to circumvent IP rights [32].

7 Future Directions

Widespread acceptance of blockchain in the life sciences arena is yet to occur, but there is a growing trend to protect such innovations in the United States and worldwide. There is a distinct possibility that administrative governmental agencies such as the USPTO, the U.S. Copyright Office, and the EPO will continue exploring the prospect of implementing blockchain technologies in conjunction with their existing systems to further improve, streamline, and secure their operations. Moreover, while it is too early to predict to what extent, if any, blockchain records will be deemed reliable from an evidentiary standpoint in adversarial proceedings involving IP, it is certainly foreseeable. No doubt, there will be numerous obstacles to overcome along the way, such as costs of implementation, adoption, efficiency, and security, to name a few. However, numerous states have already passed blockchain-related legislation beneficial to industry participants, and organizations such as the IEEE have begun initiatives to characterize the space. In short, there is growing momentum to utilize blockchain constructs to help streamline many applications. Life sciences organizations would be well-advised to consider the extent to which blockchain can benefit their business endeavors.

8 Conclusions

Whether implemented as standalone architectures or as complements to existing platforms, the features of blockchains make them attractive for many aspects of IP rights in both the public and private sectors. A well-architected blockchain platform can record a tamper-evident and tamper-resistant, timestamped record of events about various IP rights important to life sciences organizations. Depending on the particular right involved, blockchains can be used to record the evolution, ownership, maintenance, management, and commercialization of IP, as well as provide toolkits for life sciences organizations to use in analyzing and enforcing IP rights. When coupled with smart contracts, Internet of Things (IoT) devices, and AI/ML engines, these capabilities can be enhanced to provide extremely robust data storage, tracking, and analysis capabilities and significantly increase the value of an organization's intangible assets.

References

1. Weinick JM, Cheng RA (2018, September 12) An outlook on the blockchain patent arms race. Law.com. https://www.law.com/njlawjournal/2018/09/12/ip1-0917-an-outlook-on-the-blockc hain-patent-arms-race/
2. Rosario N (2018, March 15) An update on the blockchain patent landscape. Law360. https://www.law360.com/articles/1022783/an-update-on-the-blockchain-patent-landscape
3. AcclaimIP (2021) Patent search & analysis software. Anaqua, Inc. https://www.acclaimip.com/
4. Xie A, Wang T (2019, July 23) Revealed: the countries leading the race for blockchain patents. IAM; Law Business Research. https://www.iam-media.com/patents/revealed-countr ies-leading-race-blockchain-patents
5. Alice Corp. v. CLS Bank International, 573 U.S. 208 (2014). https://www.supremecourt.gov/opinions/13pdf/13-298_7lh8.pdf
6. U.S. Constitution (1789) Article 1, Sect. 8, Clause 18. https://constitution.congress.gov/bro wse/article-1/section-8/
7. 35 U.S. Code § 271(a)
8. 35 U.S. Code § 101
9. Spotts G, Pichler R, Nelson M (2021) Drug monitoring tool (U.S. Patent No. 10,896,749). U.S. Patent and Trademark Office. http://patft1.uspto.gov/netacgi/nph-Parser?patentnumber=10896749
10. Morimura J, Lee J, Kondru R, Doyle T, Shen L (2020) System and method for healthcare security and interoperability (Patent No. 10,541,807). U.S. Patent and Trademark Office. http://patft1.uspto.gov/netacgi/nph-Parser?patentnumber=10541807
11. Bent OE, Nsutezo SSF, Nzeyimana A, Pore M, Tryon K, Walcott A (2021) Multi agent consensus resolution and re-planning (U.S. Patent No. 11,069,448). U.S. Patent and Trademark Office. http://patft1.uspto.gov/netacgi/nph-Parser?patentnumber=11069448
12. Lee KS, Hinton CA, Constantin AE (2021) Therapy data management system (U.S. Patent No. 11,049,601). U.S. Patent and Trademark Office. http://patft1.uspto.gov/netacgi/nph-Parser?pat entnumber=11049601
13. Kutzko JD, Wright WCA (2021) Computer-implemented system and methods for predicting the health and therapeutic behavior of individuals using artificial intelligence, smart contracts and blockchain (U.S. Patent No. 10,991,463). U.S. Patent and Trademark Office. http://patft1.uspto.gov/netacgi/nph-Parser?patentnumber=10991463
14. Williams C, Fiscella JA, Galiano AM (2021) Decentralized data authentication system for creation of integrated lifetime health records. (U.S. Patent No. 11,055,419). U.S. Patent and Trademark Office. http://patft1.uspto.gov/netacgi/nph-Parser?patentnumber=11055419
15. White NE, Powell GE, Marshall VL, Lane MT, Clerici JC (2021) Health status system, platform, and method. (U.S. Patent No. 10,923,216). U.S. Patent and Trademark Office. http://patft1.uspto.gov/netacgi/nph-Parser?patentnumber=10923216
16. Wang S, Wang X, Tang H, Wang W, Farahanchi A, Zheng H (2021) System for decentral-ized ownership and secure sharing of personalized health data (U.S. Patent No. 11,003,791). U.S. Patent and Trademark Office. http://patft1.uspto.gov/netacgi/nph-Parser?patentnumber=11003791
41. Witchey NJ (2019) Healthcare transaction validation via blockchain, systems and methods (U.S. Patent No. 10,340,038). U.S. Patent and Trademark Office. http://patft1.uspto.gov/net acgi/nph-Parser?patentnumber=10340038
18. Witchey L, Sanborn JZ, Soon-Shiong P, Witchey NJ (2021) Content authentication and vali-dation via multi-factor digital tokens, systems, and methods (U.S. Patent No. 11,048,788).

U.S. Patent and Trademark Office. http://patft1.uspto.gov/netacgi/nph-Parser?patentnumber=11048788

19. Bhardwaj G, Keskar A, Gahlot T (2021) Assessment of documents related to drug discovery. (U.S. Patent No. 10,937,068). U.S. Patent and Trademark Office. http://patft1.uspto.gov/netacgi/nph-Parser?patentnumber=10937068

20. Pickover CA, Weldemariam K (2021) Dental health tracking via blockchain (U.S. Patent No. 10,930,377). U.S. Patent and Trademark Office. http://patft1.uspto.gov/netacgi/nph-Parser?patentnumber=10930377

21. Knas M, John J (2021) System and method for ingestible drug delivery (U.S. Patent No. 11,017,892). U.S. Patent and Trademark Office. http://patft1.uspto.gov/netacgi/nph-Parser?patentnumber=11017892

22. Escala W (2021) Blockchain-based systems and methods for tracking donated genetic material transactions (U.S. Patent No. 11,017,883). U.S. Patent and Trademark Office. http://patft1.uspto.gov/netacgi/nph-Parser?patentnumber=11017883

23. Haile M, Canova FF, Gao DZ (2021) Systems, methods, and computer program products for risk and insurance determination (U.S. Patent No. 10,943,302). U.S. Patent and Trademark Office. http://patft1.uspto.gov/netacgi/nph-Parser?patentnumber=10943302

24. Rubman S, Graham D (2021) Secure platform and data repository for fur or skin commodities (U.S. Patent No. 10,929,901). U.S. Patent and Trademark Office. http://patft1.uspto.gov/netacgi/nph-Parser?patentnumber=10929901

25. Knas M, John J, Ferry R, Gibadlo K (2021) Intelligent health-based blockchain (U.S. Patent No. 10,943,680). U.S. Patent and Trademark Office. http://patft1.uspto.gov/netacgi/nph-Parser?patentnumber=10943680

26. Bastide PR, Dunne J, Harpur L, Loredo RE (2021) Detecting medical fraud and medical misuse using a shared virtual ledger (U.S. Patent No. 10,942,956). U.S. Patent and Trademark Office. http://patft1.uspto.gov/netacgi/nph-Parser?patentnumber=10942956

27. Clark B, Burstall (2019) Crypto-pie in the sky? How blockchain technology is impacting intellectual property law. Stanford J Blockchain Law Policy. https://stanford-jblp.pubpub.org/pub/blockchain-and-ip-law/release/1

28. Hauck R (2021) Blockchain, smart contracts and intellectual property. Using distributed ledger technology to protect, license and enforce intellectual property rights. Leg Issues Digit Age 1(1):17–41. https://doi.org/10.17323/2713-2749.2021.1.17.41

29. Ruzakova OA, Grin ES (2017) Application of blockchain technologies in systematizing the results of intellectual activity. Perm Univ Her Jur Sci 4(38):508–520. https://doi.org/10.17072/1995-4190-2017-38-508-520

30. Ishmaev G (2017) Blockchain technology as an institution of property. Metaphilosophy 48(5):666–686. https://doi.org/10.1111/meta.12277

31. IPwe (2018, December) The world's first global patent market. https://aboutipwe.com/wp-content/uploads/2018/12/IPwe-Corporate-Presentation.pdf

32. Gürkaynak G, Yılmaz İ, Yeşilaltay B, Bengi B (2018) Intellectual property law and practice in the blockchain realm. Comput Law Secur Rev, 34(4), 847–862. https://doi.org/10.1016/j.clsr.2018.05.027

33. Uniform Trade Secrets Act with 1985 Amendments (1985) Uniform Law Commission, National Conference of Commissioners on Uniform State Laws. https://www.uniformlaws.org/HigherLogic/System/DownloadDocumentFile.ashx?DocumentFileKey=e19b2528-e0b1-0054-23c4-8069701a4b62

34. Defend Trade Secrets Act (2016) Public Law, 114–153 (18 U.S.C. § 1836). https://www.congress.gov/114/plaws/publ153/PLAW-114publ153.pdf

35. Directive (EU) 2016/943 of the European Parliament and of the Council of 8 June 2016 on the protection of undisclosed know-how and business information (trade secrets) against their unlawful acquisition, use and disclosure. (2016). Off J Eur Union, Legis. 157. https://eur-lex.europa.eu/eli/dir/2016/943/oj

36. 17 U.S.C. § 101 *et seq* (1976)

37. Licensing (2021) Ethereum Wiki. https://eth.wiki/en/archive/licensing

38. Zhao S, O'Mahony D (2018, December 10) BMCProtector: a blockchain and smart contract based application for music copyright protection. In: ICBTA 2018: proceedings of the 2018 international conference on blockchain technology and application. https://doi.org/10.1145/3301403.3301404

39. Open source software tools for smart legal contracts (2020) Accord Project; The Linux Foundation. https://accordproject.org/

40. Barata SL, Cunha PR, Vieira-Pires RS (2020) I rest my case! The possibilities and limitations of blockchain-based IP protection. In: Siarheyeva A, Barry C, Lang M, Linger H, Schneider C (eds) Advances in information systems development. Springer International Publishing, pp 57–73. https://doi.org/10.1007/978-3-030-49644-9_4

41. Yanisky-Ravid S, Kim E (2020) Patenting blockchain: Mitigating the patent infringement war. Albany Law Rev 83(2):603–630. http://www.albanylawreview.org/issues/pages/article-information.aspx?volume=83&issue=2&page=0603

42. Clark B, Polydor S (2018, May 1) How blockchain can protect trade secrets. Intellectual Property Magazine. https://www.intellectualpropertymagazine.com/patent/how-blockchain-can-protect-trade-secrets-129763.htm

43. Hangzhou Internet Court Province of Zhejiang People's Republic of China Judgment (Case No.: 055078 Zhe 0192 No. 81). (2018). Dennemeyer Group. https://go.dennemeyer.com/hubfs/blog/pdf/Blockchain%2020180726/20180726_BlogPost_Chinese%20Court%20is%20first%20to%20accept%20Blockchain_Judgment_EN_Translation.pdf

44. Henson M (2020, June 1) Blocking the fake supply chain. Intellect Prop Mag, June, 36–37. https://www.intellectualpropertymagazine.com/trademark/blocking-the-fake-supply-chain-141605.htm

45. Clark B, Wilkinson-Duffy R (2018, July 18) Time to get smart? CITMA Rev 443:8–12. https://www.citma.org.uk/resources/citma-review-magazine/citma-review-2018/citma-review

46. Sheraton H, Clark B (2017) Blockchain and IP: crystal ball-gazing or real opportunity? Pract Law. https://uk.practicallaw.thomsonreuters.com/w-010-1622

47. Tian C (2017, July 27) The rate of blockchain patent applications has nearly doubled in 2017. CoinDesk. https://www.coindesk.com

Regulatory Compliance Considerations for Blockchain in Life Sciences Research

Wendy M. Charles

Abstract Life sciences organizations are increasingly considering or utilizing blockchain as part of electronic systems used in life sciences research. However, these organizations may not be familiar with how life sciences regulations are applied to various uses of blockchain. There is additional confusion about whether some of the features inherent in blockchain, such as audit trails, may meet regulatory requirements for electronic records and signatures. This chapter explores how various blockchain features could meet U.S. Food and Drug Administration (FDA) regulatory requirements for electronic records and signatures, with cautions about necessary documentation expectations.

Keywords Blockchain · Electronic system · 21 CFR § 11 · Audit trails · Regulatory compliance · Validation

1 Introduction

Electronic technologies manage data for nearly every type of research in life sciences research. Technologies are used for every step of data collection or entry—through transmission, storage, analysis, and (possibly) submission to regulatory authorities. Regulatory agencies struggle to keep up with emerging technologies, such as blockchain, and face limited staffing and capacity to update education and documentation [1]. Within this context of evolving technologies, the regulations are written to be technology agnostic so that—to the extent possible—regulatory requirements would apply to any new technology developed to achieve medical purposes or manage regulated data. Because the integrity of data created and used by these systems could be compromised, any system must be reviewed and thoroughly validated for its intended use [2].

The uses of blockchain in life sciences research have been discussed throughout this book to describe how blockchain can achieve multiple intended uses. The reader

W. M. Charles (✉)
BurstIQ, Life Sciences Division, Denver, CO, USA
e-mail: wendy.charles@cuanschutz.edu

is advised, though, that some blockchains are used merely to connect other commercial electronic record systems and that the blockchain might not process regulated data [2]. The following sections interpret current regulations' applicability and how blockchain technologies may or may not meet regulatory requirements. While nearly every country specifies requirements for systems that maintain electronic records and signatures, this chapter focuses on the specific regulatory criteria related to U.S. Food and Drug Administration (FDA)-regulated research criteria.

1.1 Regulatory Agency Uses of Blockchain

Some life sciences research organizations may be hesitant to deploy blockchain-based systems out of concern that the FDA may be unfamiliar or resistant to emerging uses of blockchain technologies. This section provides examples of the FDA's involvement in blockchain projects.

The FDA Office of Hematology and Oncology Products and the U.S. Department of Health and Human Services Innovation, Design, Entrepreneurship, and Action Laboratory created the Information Exchange and Data transformation project (INFORMED) [3]. Among its objectives, INFORMED provides a blockchain-based infrastructure to securely test new ideas [4]. Data aggregated for this project are used for predictive analytics and to improve data curation and standardization [5].

In advance of the 2023 deadline for the U.S. Drug Supply Chain Security Act (DSCSA), the FDA requested proposals for pilot studies to improve the efficiency of tracking and verifying prescription medications. Dr. Scott Gottlieb, the FDA Commissioner from 2017–2019, encouraged the use of blockchain technologies.

> We're invested in exploring new ways to improve traceability, in some cases using the same technologies that can enhance drug supply chain security, like the use of blockchain. To advance these efforts, the FDA recently recruited Frank Yiannas, an expert on the use of traceability technologies in global food supply chains. He'll be working closely with me on ways for the FDA to facilitate the expansion of such methods, such as blockchain technology, to further strengthen the U.S. food supply [6, p. 1].

In 2019, the FDA joined DSCSA consortia projects with pharmaceutical companies to learn if blockchain technologies could meet the specified requirements [7]. For one project, blockchain was used as the backend of a drug track and trace system to document the step-by-step transfer of drug products between parties in the drug supply chain with the goal of near-real-time track-and-trace [8].

Also in 2019, the FDA released the Technology Modernization Action Plan (TMAP) as a roadmap for enabling new manufacturing and information technologies [9]. The summary document mentions blockchain four times, listing blockchain among "state-of-the-art solutions" (p. 1), and that "distributed ledger solutions like blockchain will be critical to support FDA's track-and-trace priorities" (p. 2).

While some life sciences organizations may be waiting for early adopters to gain experience with blockchain in FDA submissions, these collaborative blockchain projects and the Agency's mention of blockchain in the TMAP suggest that the Agency is becoming familiar with blockchain and recognizes this technology's potential benefits.

1.2 Regulatory Applicability

This section describes how regulatory criteria may apply to different aspects of electronic systems: data that constitute "electronic records" and "electronic signatures." Technologies that generate electronic records or signatures are categorized as "electronic systems." While there are many sections of regulations about human research protections and requirements that apply to individually identifiable data (e.g., the Health Insurance Portability and Accountability Act) or personally identifiable information (e.g., the General Data Protection Regulation and the California Consumer Protection Act), this chapter instead focuses on regulations for expectations of electronic systems used by life sciences organizations. Due to space limitations, this chapter does not address international regulations but instead focuses on regulations and guidance documents for electronic systems set forth by the FDA.

For drugs, biologics, and medical devices to be marketed in the United States, the Food and Drug Amendments Act [10] includes 14 categories of primary regulations to ensure adequate product development, testing, protection of human subjects, and data integrity [11, 12]. Depending on the product, the primary regulations—referred to as "predicate rules"—specify expectations for the nature of data collected and retained to ensure Agency evaluation of product safety and efficacy [13]. Within the U.S. Code of Federal Regulations (CFR), FDA regulation 21 CFR § 11—often referred to as "Part 11"—is a companion regulation to the predicate rules. Part 11 describes technical and procedural controls required if an organization plans to use electronic records and electronic signatures [14]. Part 11 was intended to allow for the widescale use of electronic technologies while ensuring the confidentiality and authenticity of electronic records [15].

1.2.1 Electronic Record

The FDA defines an "electronic record" as "any combination of text, graphics, data, audio, pictorial, or other information represented in digital form that is created, modified, maintained, archived, retrieved, or distributed by a computer system" (21 CFR § 11.3(b)(6)). It is important to note that these electronic records are maintained under predicate rules to perform regulated activities [13], must meet the same elements of data quality expected of paper records (i.e., ALCOA: "attributable, legible, contemporaneous, original, and accurate") [16], and must be available for the FDA to verify data quality and integrity (21 CFR §§ 312, 511.1(b), and 812).

1.2.2 Electronic Signature

An "electronic signature" is "a computer data compilation of any symbol or series of symbols executed, adopted, or authorized by an individual to be the legally binding equivalent of the individual's handwritten signature" (21 CFR § 11.3(b)(7)). Electronic signatures were given full legal authority to execute contracts or sign electronic records in the United States through the Electronic Signatures in Global and National Commerce Act [17]. Subsequently, electronic signatures were also considered legally binding by the FDA.

1.2.3 Electronic System

"Electronic systems" can refer to "computer hardware, software, peripheral devices, networks, cloud infrastructure, personnel, and associated documents (e.g., user manuals and standard operating procedures (SOPs))" [18, p. 5]. Electronic systems could include: commercial off-the-shelf electronic systems, customized systems owned or managed by sponsors, or services outsourced by the sponsor [15]. While electronic systems are used to manage and automate nearly every process in a clinical investigation, this chapter focuses on the nature of systems of most significant concern for product quality and public safety: systems that manage drug or device manufacturing, quality assurance, drug distribution, and research participant safety [10].

1.2.4 Summary

As of Fall 2021, the author could not locate any FDA guidance documents about blockchain, nor could she find any FDA Warning Letters on blockchain as a component of an electronic system maintaining electronic records or electronic signatures. Considering that these regulations are intended to apply to any electronic system for records "created, modified, maintained, archived, retrieved, or transmitted, under any records requirements outlined in Agency regulations" (21 CFR § 11.1(b)), the inclusion of blockchain is interpreted within the context of existing regulations for electronic records and signatures. When blockchains are part of the network or cloud infrastructure used to process electronic records and signatures subject to regulation, all requirements of Part 11 apply [2].

2   Regulatory Review and Documentation

This section describes FDA's requirements for electronic systems and, where pertinent, how blockchain-based systems compare with features of commercially available non-blockchain-based software systems. Where applicable, this section also advises about questions that sponsors, contract research organizations (CROs), or clinical investigators may be asked during an FDA inspection (21 CFR § 11.1(e)) and how to ensure that all parties are prepared to answer these questions about uses of blockchain.

2.1 System Design and Documentation

When life sciences organizations utilize electronic systems to create, modify, maintain, archive, retrieve, or transmit clinical data (21 CFR § 11.3(b)(6)), thorough documentation should specify the software and hardware used [10]. A list of systems often includes electronic data capture (EDC) systems, electronic consent, electronic clinical outcome assessments [19], or electronic health record (EHR) systems [20].

As described in the *Regulatory Applicability* section above, blockchain technology would be included among the electronic systems if an electronic record was created, modified, maintained, archived, retrieved, or distributed (21 CFR § 11.3(b)(6)) by the blockchain. Because blockchain technologies comprise only a component of an integrated electronic system, life sciences organizations may be unclear about the role of the blockchain in processing electronic records. Currently, some blockchain-based systems are not designed to process electronic records but instead automate inventory, payments, or other efficiencies that are not subject to regulatory requirements [21]. However, blockchain-based systems are increasingly designed and utilized to manage regulated records, including managing informed consent [22], collecting and transferring source data [23], and storing data [24].

2.1.1 Data Flow and Architecture Diagrams

With the recognition of the complexities of system integrations, the FDA has specified that sponsors and clinical investigators should create documentation of specific uses of hardware and software in the physical environment [10]. In addition to the hardware and software descriptions, these entities should also document the physical and logical parameters for data flow and visibility among authorized parties [20].

In particular, the FDA recommends that organizations diagram the flow throughout components, access control points, and processing [25]. The roles of specific components are often most visible in diagrams involving swim lanes or journey maps that document the flow of data through a process. These diagrams are commonly drawn to document design features and systems integrations during software development but should also be updated, as necessary, to assess and mitigate risks to data integrity [26]. Figures 1 and 2 display simplified swim lane diagrams that show the flow of information through systems and the layers of blockchain. Figure 1 displays the role of blockchain to create efficiencies for automating payments and site inventory. Study data are stored and processed in the EDC system. The blockchain does not create, modify, maintain, archive, retrieve, or transmit data for any purpose subject

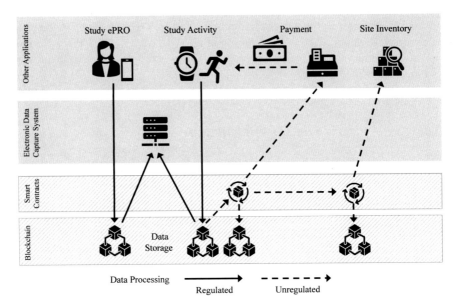

Fig. 1 Simplified swim lane diagram showing the roles of blockchain layers that do not involve the processing of regulated data

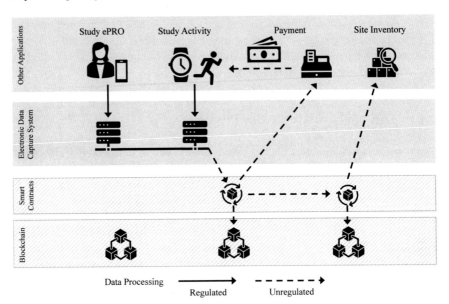

Fig. 2 Simplified swim lane diagram showing the flow of regulated data through blockchain layers

to regulation. In contrast, Fig. 2 shows how blockchain would serve a critical role in processing regulated data and should meet all applicable FDA regulations for electronic records.

Data flow diagrams are also necessary for the distributed nature of blockchain-based storage nodes located in multiple geographic locations. In an FDA Q&A guidance document about electronic records and electronic signatures, the FDA noted that there are no limitations on geographic storage locations if using appropriate controls. "However, it is critical for sponsors and other regulated entities to understand the data flow and know the location of the cloud computing service's hardware in order to conduct a meaningful risk assessment regarding data access, integrity, and security" [15].

2.1.2 Protocol

In addition to creating internal diagrams of data flow, sponsors should include written descriptions of computerized systems used during a clinical investigation. It is also expected that these descriptions would be included in the sponsor's investigational plan and data management plan with a description of the security measures [27]. The study-specific protocol should describe how electronic systems are used to achieve the specific provisions of the protocol, including methods of collecting and recording data [16]. Specifically, the protocol should describe each step where "a computerized system is used to create, modify, maintain, archive, retrieve, or transmit source data" [16, p. 3].

The need to list the role of a blockchain in the protocol or other study-related plan depends on the role of the blockchain, whether blockchain technologies are performing a unique role in the flow of data, and whether the protocol relies on these technologies to achieve study-specific purposes.

2.1.3 Standard Operating Procedures

Life sciences organizations are expected to create SOPs that describe the methods by which an organization maintains, tests, and provides technical support for an electronic system [15]. The FDA Guidance Document, Computerized Systems Used in Clinical Investigations [16] provides a list of recommended SOPs, including system set-up and installation, system operating manual, system operating instructions, security measures, change control, roles and responsibilities, and system maintenance (p. 10).

When blockchain technologies are designed to process regulated data, the SOPs should reflect the nature of blockchain creation and ongoing use. Unlike commercial off-the-shelf software offered as a complete package, many blockchain services are offered as a platform-as-a-service or in-house design. These customized blockchains will likely require ongoing professional services to design smart contracts, perform maintenance/upgrades, and integrate other electronic systems. Therefore, blockchain programmers should make available their manuals, operating instructions, and SOPs for their security, maintenance measures, and training.

Because there are requirements for SOPs in nearly every FDA guidance document, SOPs will also be described throughout this chapter.

2.1.4 Recommendations for Audit Preparation

During FDA inspections, sponsors/CROs, and clinical investigators are asked to specify which electronic systems were used to "create, modify, maintain, archive, retrieve, or transmit" electronic records (21 CFR § 11.3(b)(6); [13, 19]). FDA Consumer Safety Officers have been advised to start asking whether these organizations are utilizing blockchain technologies.

Life sciences organizations that may receive an FDA inspection should prepare lists of software, hardware, locations of hardware, diagrams of system integrations, and data flow [13, 19]. Further, SOPs describing the uses of the system, including maintenance and management controls, should be available to the inspector upon request (21 CFR § 11.10(k)(2); [13, 19]). This advanced preparation will likely require initial negotiation and ongoing communication with the blockchain vendors or programmers when including blockchain technologies.

2.2 System Protection Features

When using a blockchain-based technology as a component of an electronic system processing regulated data, the same security and data protection requirements apply to the blockchain. This section describes the regulatory expectations for system-level protections, including access controls, storage backups, and retention requirements that apply to the blockchain when this technology processes regulated records.

2.2.1 Data and System Protections

Current electronic systems used for life sciences research involve open systems with permissible access from outside the organization's internal network (21 CFR § 11.30); therefore, proactive measures are required to prevent access by unauthorized parties. For example, if data are transmitted wirelessly to an EDC system or if a system is accessed remotely, the transmission should be encrypted during transit to prevent signal interception (21 CFR § 11.30; [15]). There are also expectations to prevent and detect malicious computer software, such as viruses, spyware, ransomware, and worms that could damage the functioning of electronic systems and impact study data [16, 15]. As examples of prevention and detection mechanisms, there should be firewalls and regular software scans for viruses and other harmful software [13, 19]. If there are error messages or system failures, the SOPs should describe methods for implementing corrective actions and documentation of outcomes [19]. SOPs should also describe operational processes involving these preventative and detection mechanisms. All SOPs should be available for inspection by the FDA (21 CFR §§ 11.1(e) and 11.10(b); [16].

Blockchain-based systems are generally viewed as very secure, but life sciences organizations should not develop a false sense of security. The author frequently asserts that blockchain is software, not magic. Blockchain technologies are designed by humans, programmed by humans, administered by humans, and operated by humans. Therefore, blockchain design and programming are vulnerable to human errors such as software bugs [28]. Also, when a blockchain connects to other layers or components of a computerized system, there may be vulnerabilities with application programming interfaces or incompatibilities [29].

While some proponents of blockchain technologies argue for a decentralized approach to node and system management—that is, no central oversight—FDA expectations for general protections, such as encryption, firewalls, intrusion detections, and scanning for malicious software, require a centralized approach to ensure overall system integrity. Di Francesco Maesa et al. [30] add that using a blockchain does not relieve the system owners or operators of their responsibilities to implement security and maintenance measures. Ultimately, a blockchain—and its use within an electronic system—should reflect prudent Current Good Manufacturing Processes and solid procedural controls [25, 18].

2.2.2 Access Controls

The FDA describes the expected uses of internal and external access safeguards to protect data and system integrity. First, technical controls should limit access, which is essential for individual and role-based access controls (see 21 CFR §§ 11.10(d) and 11.30). User-access controls may include user name and password combinations, cards with security chips, or other biometrics such as the use of fingerprints or to establish identity [15].

Sponsors and clinical investigators must ensure that there are SOPs with descriptions of limiting unauthorized access (21 CFR §§ 11.10(d) and 11.30). As a starting point, sponsors should create a cumulative list of individuals authorized to access electronic systems that contain source data or case report forms [27]. These individuals' training and authorization should be appropriately documented (21 CFR § 11.10(i)). Similarly, there should be lists of study personnel, study roles, and details of their access privileges [16, 18]. Individuals should be assigned their own log-in access where that person's activities will be associated with their identity (21 CFR §§ 11.10(d) and 11.30), and access should be disabled when the individual no longer requires access to study materials [27]. Further, the system should limit the number of unsuccessful log-in attempts and record efforts at unauthorized access [16].

These identity procedures are also intended to segregate access roles. For example, a system administrator who has the authority to adjust files or system settings should be independent of the study staff responsible for entering or reviewing regulated records [18]. Also, original data entries and modifications cannot be made by anyone other than the clinical investigator and/or individuals on the investigator's staff [13].

Because blockchain is a backend software solution, these requirements may be met by other systems or software layers designed to administer access controls and monitor the access of authorized personnel [31]. Otherwise, blockchain technologies would need to offer access management.

Self-sovereign or decentralized identity capabilities that offer granular access controls based on an individual's verified credentials (e.g., [32]) can be incorporated into blockchain solutions. Other blockchain access control strategies have also been proposed to manage identity-based access controls. Di Francesco Maesa et al. [30] note that blockchain-based access controls often use smart contracts stored on the blockchain to execute code-based access policies and governance while maintaining event logs to record access. Adlam and Haskins [33] propose using attribute-based authentication that enables organizations to base authorization on narrow attributes, such as location or office. BurstIQ implemented blockchain-based granular controls and complex rules engines to manage access and governance for an Intermountain Healthcare program that managed surgical costs and health outcomes [34].

While blockchain-based controls can also be used to grant research participants access to health information collected during research participation, life sciences organizations have traditionally resisted allowing participants to access their health information—even information that would not unblind an individual's treatment assignment [21]. However, [35] adds that participant access to study-related health information could greatly improve study retention and increase trust among research participants in marginalized communities. A U.S. Government Accountability Office report [36] noted that there are no prohibitions preventing sponsors from providing research participants with their health information. However, the report also notes that there is also no economic incentive for sponsors to provide this information. In fact, the manual effort to curate this information for individual requests may be cost-prohibitive [37]. Therefore, blockchain-based research participant access solutions have been designed as efficient solutions to allow individuals to receive permissible health information (i.e., health information unlikely to compromise blinding procedures). With an access portal similar to an EHR patient portal, blockchain-based access controls can automate authorized permissions for each allowable variable or data point that an individual is permitted to access [21].

In addition to technological capabilities for access and identity management, the technical and procedural controls for identity and access management must be documented. These documents should outline the steps and processes for consistent operations and security, including procedures for removing access for individuals who leave the organization or are no longer part of the study. The list of authorized users should be maintained with study documentation, be accessible to applicable study personnel, and made available during an FDA inspection [10].

2.2.3 Audit Trails

The FDA defines an "audit trail" as "a secure, computer-generated, timestamped electronic record that allows for reconstruction of the course of events relating to the creation, modification, or deletion of an electronic record" (21 CFR § 11.10(e)). An audit trail allows for the reconstruction of electronic records to verify data for quality and integrity [10]. The following information must be captured in the audit trail:

- **The data originator**. The data originator could be a person, device, or instrument [27].
- **The date and time that data are added to the electronic case report form**. While the concept of a timestamp is straightforward, the audit trail recording begins when data enter the sponsor's EDC system [15]. For example, while wearable devices could capture important study-related information, the audit trail does not begin until this information is entered (manually or automatically) into the EDC.
- **The research participant about whom the data were collected** [27].
- **If changes are made to the data, the person making the change and why changes were made to the record** [16]. Many audit trails capture data corrections but may not capture why the changes are made. Further, any data corrections cannot obscure previous entries (21 CFR § 11.10(e); [27, 13]).

Audit trails may be created using different software programs or technologies but must also create logs that can be read by a human [27]. These audit trails cannot be altered or deleted and must be retained for the required duration and made readily available during an FDA inspection (21 CFR § 11.10(e); [27]).

If a blockchain is part of the electronic system used for the research, life sciences organizations should determine which electronic systems will create and preserve the audit trail. While blockchains create audit trails by design and default, the blockchain may or may not be used as the technology creating the audit trail.

Common Misunderstandings About Audit Trails

Life sciences research organizations are cautioned that some blockchain programmers or blockchain-as-a-service platforms may not understand components of FDA regulatory requirements for audit trails. First, some organizations describe the virtues of their audit trails as if the life sciences research organizations do not already use audit trails. Specifically, the concept of an audit trail is not new, and any blockchain company that hails its audit trail may be ignorant that the FDA has required secure audit trails since 1997 [14].

In addition, it is not clear whether a blockchain-based audit trail can provide any technical advantages over the audit trails already included in commercial off-the-shelf electronic systems. While some blockchain developers suggest that non-blockchain audit trails are easily modified by centralized services [31] or systems administrators [38], traditional Part 11 systems are tested and validated to ensure they are resistant to tampering—regardless of administrative privileges or centralized authority (21 CFR § 11.10(e)). However, it is possible that blockchain-based features could meet additional business and accountability purposes for authorized stakeholders.

Next, blockchain developers and operators are often unfamiliar with the regulated components required in an audit trail—confusing the concept of transaction logs with FDA-regulated audit trails (21 CFR § 11, Subpart B). Due to a lack of familiarity with FDA requirements, the specific elements of an FDA-regulated audit trail are often not met. In addition, some blockchain companies purport that transaction hashes constitute an audit trail. As an example of this misunderstanding, [39] wrote that the FDA could review transaction IDs in the audit log and compare this information to blockchain hashes to confirm that data were not altered. However, the FDA requirement that audit trails be "human readable" (21 CFR §§ 11.10(b) and 11.50(b)) means that an inspector should not have to decipher or cross-reference cryptographic hashes.

Second Audit Trail

Some life sciences organizations "anchor" private blockchains to a public blockchain (say, once per day or week). With anchoring, private clinical research information is stored off-chain or managed with a private blockchain accessible only to authorized parties. A public blockchain then captures periodic snapshots via hashes (e.g., [40, 41]). Organizations interested in a hybrid solution state that the public blockchain provides transparency that data were not altered and that the workflow was not modified in any way that could bias the results [39]. This arrangement is also intended to provide accountability, encourage honest behavior among collaborating parties [39], and promote public trust in research results because data integrity could be verifiable by anyone [40]. However, it is unclear how this solution would also maintain the FDA's requirements for the "confidentiality of electronic records" (21 CFR § 11.10). Even when there are requirements for the availability of data for public disclosure (21 CFR § 312.130) or data summaries (21 CFR § 812.38), these provisions still must provide for data confidentiality (21 CFR § 314.430).

While some organizations suggest that periodic anchoring to a public blockchain would also provide more confidence for regulatory purposes (e.g., [31, 38]), this argument may be tenuous for three primary reasons. First, organizations that utilize a snapshot to a public blockchain suggest that the public blockchain provides additional confidence that data were not changed. However, this argument lacks awareness of the validation requirements of existing electronic record systems that require testing and documentation to demonstrate that study data cannot be altered (21 CFR § 11.10(a); [18]). Next, concerning suggestions that a public blockchain would provide "additional trust" to regulatory agencies, it is unclear how additional trust would result from a public audit trail that was not designed, tested, and documented to meet the requirements of Part 11. Specifically, a list of hashes would not have meaning to an FDA inspector without the verification to support that the system is operating as intended (21 CFR § 11.10(a)). As one FDA Consumer Safety Officer explained to the author of this chapter, the FDA expects the primary audit trail to be designed and tested correctly. When the primary audit trail demonstrates trust, it is unclear why an organization would think a second audit trail would provide additional regulatory value. Last, if the public blockchain's audit trail is not human readable (21 CFR §§ 11.10(b) and 11.50(b)) or does not display the required audit trail elements [27], the public blockchain would not meet the regulatory criteria for an audit trail. Therefore, while anchoring a private blockchain to a public blockchain may serve some practical business purposes, it is unlikely that this arrangement would add value for regulatory purposes.

2.2.4 Data Storage and Retention

Federal agencies, such as the FDA, require data storage mechanisms to ensure the security and integrity of electronic records. Further, the data must be accessible to the FDA upon request to verify data quality and integrity [27]. Required records can be stored electronically with standard electronic file formats such as PDF or XML [42], and records must preserve the content and meaning.

Electronic records must be stored for the appropriate duration, depending on the investigational product and application process. For instance, the investigational drug regulations specify, "A sponsor shall retain the records and reports required by this part for 2 years after a marketing application is approved for the drug; or, if an application is not approved for the drug, until 2 years after shipment and delivery of the drug for investigational use is discontinued and FDA has been so notified" (21 CFR § 312.57(c)), which involves an average of 14 years [43]. The retention period for investigational devices is "for a period of 2 years after the latter of the following two dates: The date on which the investigation is terminated or completed, or the date that the records are no longer required for purposes of supporting a premarket approval application, a notice of completion of a product development protocol, a humanitarian device exemption application, a premarket notification submission, or a request for De Novo classification" (21 CFR § 812.140(d)), which takes an average of 9 years [43]. For the blockchain projects that purportedly support clinical trials, the author could not locate published methods by which these blockchains would manage all related records and logs for this duration. However, few blockchains plan for the future retention requirements if the current blockchains become obsolete and shut down [44]. Day [44] notes that there are few straightforward methods for blockchains to serve as archives, and an archiving solution would require a central authority. Therefore, Bhatia et al. [45] recommend migrating study data off blockchains to meet record archiving requirements.

Some blockchain-based systems have also not adequately addressed the ability to manage the long-term future of ledgers and whether data or logs could ever be deleted. Long-term planning is essential to manage data storage costs and risks after data sets have exceeded their useful lives. There have been several methods described for data deletion and chain pruning (e.g., [46, 47, 48]), but it is essential to consider that these methods rely on a governance structure and blockchain design that allows for some level of modification.

2.2.5 Backups and Disaster Recovery

Electronic systems and data storage remain vulnerable to downtime, server failures, data corruption, viruses, malware, and even malicious activities from users (e.g., [49, 50]). Therefore, it is critical to develop technology provisions for business continuity planning. Electronic systems used for life sciences research are expected to maintain backup and disaster recovery capabilities to protect against data loss [10]. The organization should also write SOPs and contingency plans that describe data and system recovery procedures.

It has been argued that the distributed ledger technologies utilized by blockchain provide the necessary redundancy for backups and should satisfy most organizations' data access needs (e.g., [31, 51]). However, ledger redundancy is not the same as a backup, and it is unlikely that "data redundancy" meets the FDA's requirements for "data backup." FDA notes that a "backup" is a true copy of the original in a format compatible with the original [18]. Unless the blockchain ledger can reproduce all data in their original formats (including files and images), a ledger would not qualify as a backup. As an additional consideration, while some blockchain platforms are technically capable of rolling back to an earlier set of blocks [52], blockchain operators should ensure that all recovery activities are documented and tested to ensure no data or audit trail activities are lost.

Next, it is unclear how the separate off-chain servers could produce backup data that are "exact, complete, and secure from alteration, inadvertent erasures, or loss" with the associated metadata [18, p. 2]. If the off-chain servers are storing FDA-regulated data, those servers must meet all applicable requirements of 21 CFR § 11, including access controls, maintenance, and backups. An audit trail of the network is not sufficient to demonstrate server-level data protections.

In preparation for an FDA inspection, an inspector is required to ask whether there are backups and disaster recovery plans and whether there were any system failures during the study period [13, 19]. The FDA also recommends that life sciences organizations maintain backup and recovery logs to enable assessments of system failures [16].

2.2.6 Recommendations

Blockchain technology offers tamper-resistant and tamper-evident data capabilities. However, the blockchain concept of decentralized storage raises questions regarding how all of the disparate, distributed storage locations would meet the FDA storage requirements. For example, many blockchains purported to be designed to manage clinical trials store only the blockchain hash and pointers on the ledger but maintain the actual data values in a separate server (e.g., [40, 53, 31, 38]). If regulated records are stored off-chain in a separate storage location (or several storage locations), each storage location involved in some aspect of "creating, modifying, maintaining, archiving, retrieving, and distributing" (21 CFR § 11.3(b)(6)) electronic records would have to meet the FDA requirements of Part 11. Specifically, each location would have to demonstrate and document the long checklist of Part 11 requirements for encryption, access controls, electronic signature standards, training, maintenance, and documentation, which is no small feat and no small expense. Simply having a blockchain-based audit trail that records after-the-fact activity is not acceptable if the electronic storage locations are not designed to prevent unauthorized access.

While some early stated goals for the role of blockchain in clinical trials are to "make the work of intermediaries and the data hosting systems used obsolete" [54, p. 3], intermediaries are often necessary to maintain oversight of a study. When a sponsor conducts a study, for example, the FDA expects the sponsor to know the location of all hardware storage locations to conduct a meaningful assessment of data security and ensure data can be retrieved during FDA inspections (21 CFR § 11.10(b); [15]). Further, there must be documented revision and change control procedures that describe how the audit trail is maintained, including "time-sequenced development and modification" (21 CFR § 11.10(k)(2)).

Because many blockchain projects and platforms have not appropriately planned for the end of blockchain's current capabilities, [44] recommends against using blockchain technologies for storing data that may extend beyond the useful life of the blockchain system. In general, decisions about record retention should be based on the ability to meet all applicable regulatory requirements [42].

2.3   Record and Signature Integrity

At the core, FDA regulations for electronic systems concentrate on the integrity of electronic records that serve as the basis for establishing whether a product is safe and effective. This section examines system design features that influence quality at the data level, although architecture and system integrity can also impact data integrity. While several blockchain-based data-level features can impact data integrity, this section will review the requirements for source data integrity and electronic consent.

2.3.1   Source Data Integrity

For Part 11, the concept of data integrity pertains to the trustworthiness of data managed by electronic systems—particularly those used for manufacturing, quality assurance, and safety [16]. Mechanisms for data entry require physical, logical, and procedural controls (21 CFR §§ 11.10 and 11.30). These controls are necessary to ensure that source data transferred to another system are not altered in value or meaning [15].

Among technical controls, the FDA recommends designing prompts for the nature of desired terminology, as well as alerts when data are missing or when values fall outside the acceptable range [16, 27]. When individuals perform manual data entry, the technology should prevent unauthorized changes before data are transferred to the sponsor's EDC system(s) [15].

As blockchain is increasingly proposed for regulated electronic systems, some proposals appear unfamiliar with the limitations of electronic systems. For example, some publications discuss the need for blockchain to prevent data falsification [55], reduce incorrect decisions [39], or perform source data verification [38]. However, these assertions do not address the "first-mile problem." The first-mile problem represents the discordance between the reality of a situation and the data digitally recorded [8]. Specifically, it is unlikely that any technology could prevent all data problems caused by human users, including data entry errors, misunderstandings, and/or non-compliance. Therefore, Alles and Gray [8] recommend that life sciences organizations continue to verify source data.

2.3.2 Electronic Consent

Electronic consent often includes components of a digital signature. The FDA defines a digital signature as "an electronic signature based upon cryptographic methods of originator authentication, computed by using a set of rules and a set of parameters such that the identity of the signer and the integrity of the data can be verified" (21 CFR § 11.3(b)(5)).

Life sciences organizations are demonstrating increasing utilization of electronic informed consent processes in place of paper documents. Many electronic consent programs offer audiovisual enhancements, including podcasts, websites, and/or graphics that allow for more individual engagement and interest [56]. Additional features may include quizzes or other methods to evaluate an individual's comprehension of the information, collect electronic signatures, and allow for better storage and retention of informed consent documents [56]. A copy of the consent form must be offered to the participant or his/her legally authorized representative signing the form, but this copy can also be electronic (21 CFR § 50.27(a)).

Part 11, Subpart C provides the criteria for accepting electronic signatures instead of paper-based handwritten signatures. Several pertinent issues within the requirements apply to blockchain-based systems.

Electronic Signatures

According to the FDA, "the technology authenticates the signer and prevents identity fraud" [15]. To meet the FDA's requirements for electronic records and electronic signatures, "electronic signatures should employ at least two distinct identification components such as an identification code and password" (21 CFR § 11.200(a)(1)).

Further, the electronic signature must be linked to the records to ensure the signatures cannot be transferred or falsified (21 CFR § 11.70). Further, the Part 11 regulations specify that after a user has signed into an electronic system with a password, all subsequent signatures can be executed using the password alone (21 CFR § 11.200(a)(1)(i)). Last, signatures should be executed under controlled circumstances, such as automatic inactivity disconnection [15]. These additional controls typically require additional programming in the user interface or other software systems, and it would be uncommon for this signature capability to be managed exclusively with a blockchain.

Key management is a critical issue for verifying that an electronic signature is unique to a single individual (21 CFR § 11.100(a)) and only used by their genuine owner (21 CFR § 11.200(a)(2)). With some blockchain-based systems, individuals are responsible for protecting their own keys. For example, if users lose their keys to cryptocurrency systems, they lose their currency [57]. However, in life sciences research, users cannot risk losing access to electronic systems. Therefore, organizations are required to implement loss management procedures (21 CFR § 11.300(c)) where alternate solutions are required to retrieve or restore keys. For example, a blockchain-based solution where individuals are responsible for their own keys is unlikely to meet the Part 11 requirement that "identification codes and password issuances are periodically checked, recalled, or revised (e.g., to cover such events as password aging)" (21 CFR § 11.300(c)). At the company where the author is employed, companies can create application layers on top of the blockchain to allow administrator oversight, link keys with usernames and passwords, or connect application programming interfaces to the organization's single sign-on. While it is argued that these are centralized solutions [58], such an approach appears necessary to maintain consistent user access.

Tracking Requirements

While a blockchain-based electronic signature is also associated with a timestamp on the ledger, blockchains may not be designed to include all components of an electronic signature (21 CFR § 11.50). Specifically, the signature must also contain the "the printed name of the signer" (§ 11.50(a)(1)) and "the meaning (such as review, approval, responsibility, or authorship) associated with the signature" (§ 11.50(a)(3)). To provide pseudonymity of blockchain-based hashes, most blockchains are unlikely to associate this information on the blockchain but may include additional software layers to capture this information. Last, the printed name, date/time, and meaning of the signature must also be available in a human-readable format for inspection, such as in an electronic display or printout (21 CFR § 11.50(b)). Further, electronic signatures must be linked to the respective records "to ensure that the signatures cannot be excised, copied, or otherwise transferred to falsify an electronic record by ordinary means" (21 CFR § 11.70; [15]). Blockchain-based hashes and audit trails are designed to associate users with their respective data-related actions, but few blockchains are designed to meet all electronic signature requirements.

Dynamic Consent

In addition to agreeing to participate at the beginning of a study, informed consent must involve ongoing communication between clinical investigators and research participants [59]. Research participants should receive new information that could affect their willingness to remain in a study [60], and participants are permitted to withdraw from the research at any time (21 CFR § 50.25(a)(8)). For participation in open-ended databases or specimen repositories, research participants should have the ability to modify their conditions for agreement when their preferences change [61]. Further, individuals should have the ability to review a consent form with family and friends, which may otherwise be difficult to achieve with limited user permissions in electronic consent platforms. Therefore, web-based dynamic consent mechanisms offer research participants the flexibility to update their preferences and remain more engaged in the research [59, 62]. Further, the blockchain-based granular access controls allow individuals to access information without imposing burdens on data administrators [63, 64]. This approach is also consistent with the stated approach of Dr. Sean Khozin, then-Director of the U.S. FDA's Oncology Center of Excellence [5]. The FDA has undertaken initiatives to empower patients and encourage them to access more of their health information.

In addition to allowing research participants to engage in dynamic and granular consent, blockchain-based mechanisms are also used for sharing data with research collaborators—allowing collaborators to receive only the variables necessary to complete analyses and for the prescribed amount of time [21]. As a mild caution, however, blockchain-based systems cannot control researchers' behavior. When a researcher receives data, there are no automated mechanisms to ensure that researchers follow research participant wishes [65] or use the data only for the stated purpose [21]. Taylor and Whitten [66] note that research plans often drift. While researchers with federal funding or are subject to FDA regulations, few mechanisms remain for detecting researchers' inappropriate study activity.

2.3.3 Submission to FDA

While it is unlikely that a blockchain-based system would manage all data nomenclature, any FDA-regulated electronic system needs to utilize the Common Data Interchange Standards Consortium (CDISC) standards, including use of the Medical Dictionary for Regulated Activities (MedDRA) for adverse events, Logical Observation Identifiers Names and Codes (LOINC) for laboratory tests, Systematized Nomenclature of Medicine—Clinical Terms (SNOMED-CT) for indications and usage, World Health Organization Drug Dictionaries, and several other requirements [67, 68]. A detailed description of the extensive terminology and nomenclature requirements are outside the scope of this chapter, but life sciences organizations are cautioned that newly introduced blockchain-based systems designed for clinical trials are unlikely to address these expectations.

2.3.4 Recommendations

There are extensive Part 11 requirements for electronic records and electronic signatures that may be difficult for some blockchain technologies to meet without adding additional integrated software layers and administrative controls. These controls also involve written procedures and training.

While blockchain offers promising technological features to enhance efficiencies with electronic records and electronic signatures, there are no technical replacements for an investigator's responsibility. Regardless of the technology used to capture informed consent, the investigator remains responsible for ensuring a thorough and voluntary informed consent process [56]. When the informed consent process is conducted electronically (using any technology), investigators must ensure that a member of the research team is available to answer questions, create processes so that the individual's signature is only used by the genuine owner, the individual receives a copy of the consent form and can withdraw from the study at any time [15]. This responsibility cannot be delegated to study personnel or electronic systems.

2.4 Verification and Validation

Thus far, this chapter has described the compliance of blockchain-based architecture, software-based system-level protection features, and record/signature level compliance considerations. These components can be designed as an integrated all-in-one software system that replaces current EDC or electronic consent systems or as separate but interconnected software layers or systems. Regardless of how the system is designed, if an electronic system is used for records subject to Agency regulations, the organization must ensure that electronic records and signatures are "trustworthy, reliable, and generally equivalent to paper records and handwritten signatures executed on paper" (21 CFR § 11.1(a)). Therefore, organizations are expected to conduct verification "to ensure accuracy, reliability, consistent intended performance, and the ability to discern invalid or altered records" (21 CFR § 11.10(a)). A thorough description of methods is outside the scope of this chapter, and only brief but pertinent details are addressed in this chapter.

2.4.1 Initial Validation

In many cases, the blockchain components will be outsourced to a vendor. When outsourcing the blockchain design or integration into an electronic system that processes electronic records, the same validation methodology applies to "customization in order to integrate with other software systems to address internal processes" [15, p. 6]. Therefore, it is critical to create a risk-based approach that considers the characteristics and intended use of the system to produce the records [15]. When validating a blockchain vendor's product, the same validation expectations would generally apply. A detailed plan should include installation qualifications, stress testing, dynamic testing, user acceptance testing, and any other testing deemed appropriate to ensure the system operates as intended [15].

Further, the FDA recommends that sponsors and other regulated entities obtain specific documentation from the vendor. Therefore, a blockchain vendor should provide SOPs, methods, and results to ensure that the technology functions as expected [15]. Of particular interest is the functioning of smart contracts designed into a blockchain-based system. As a refresher, smart contracts involve small programs that execute automatically when certain conditions are met [69]. Because smart contracts involve computer programming, they are unlikely to be devoid of mistakes [70]. Consequently, the code and branching logic should be tested to ensure accuracy and reliability.

2.4.2 Software Development Life Cycle

Responsibilities for verification and validation do not end with implementation but continue throughout the software development life cycle [16]. Blockchain systems also involve changes over time. Such changes typically involve software updates, upgrades, and maintenance. Therefore, processes and documentation should be in place to determine when changes require revalidation and the extent of testing to ensure the system functions within established operational limits (21 CFR § 11.10(a); [18]). All changes and testing should be documented [16] and available for review and copying for as long as required for review and copying for inspection (21 CFR §§ 11.10(b) and(c); 312.58(a); 812.140(b).

2.4.3 Recommendations

The FDA regulations do not prescribe any particular system, system design, or user interface, provided that the overall system (or integrated systems) can achieve the required technical and procedural controls in Part 11 [15]. As described earlier, blockchain is a backend technology that is not designed for direct interaction. There are layers of software typically needed for the complex user interface and data collection activities. Regardless of the number of software layers and systems, the sponsor is responsible for ensuring that systems are validated and perform as intended (21 CFR § 11.10(a); [15]. The validation results should be documented, and this documentation may be reviewed during an FDA inspection (21 CFR § 11.10(b); [15]).

Because blockchain vendors may not be familiar with regulatory requirements, these vendors may suggest that verification, validation and documentation are the sole responsibility of the life sciences organization. However, the vendor is expected to contribute to the process and should provide sufficient SOPs and documentation of initial and ongoing validation processes (21 CFR § 11.10(b)). Regardless of the vendor's role, the sponsor is ultimately responsible for ensuring the integrity of electronic systems used in clinical trials and typically performs user acceptance testing [13].

2.5 Training

Training is critical to ensure that individuals—including employees and contractors—who develop, use, or maintain computerized systems have the necessary education, training, and experience to perform their assigned duties (21 CFR § 11.10(i)). There are additional expectations that individuals who provide technical support will receive initial and ongoing training [19]. Therefore, if blockchain vendors "develop, maintain, or use electronic record/electronic signature systems" (21 CFR § 11.10(i)), the training requirement would apply to these developers as well. Training should be conducted continually to ensure consistent system operation and performance.

SOPs should describe the nature and frequency of training (21 CFR § 11.10(i); [16]), delegated roles and responsibilities for computer use (21 CFR § 11.10(j); [19]), and methods for ongoing training for technical support [13]. Such records should include evidence of initial and ongoing training [13]. If blockchain vendors are used, sponsors should determine the appropriate documentation of education, training, and experience (21 CFR § 11.10(i)) from the vendor.

3 Outsourcing

Throughout this chapter, there are references to the possible uses of vendors to develop and/or maintain blockchain components of an electronic system. The FDA recognizes that sponsors may outsource electronic services for cloud-based services and networks [15]. When services are used to process data for FDA-regulated electronic records, the sponsor should evaluate whether there are sufficient controls to ensure data consistency and security.

In the draft FDA document, *Use of Electronic Records and Electronic Signatures in Clinical Investigations Under 21 CFR § 11—Questions and Answers* (2017a), the FDA offers a suggested list of factors a sponsor should consider when determining whether an outsourced service provider is suitable.

- Validation documentation.
- Ability to generate accurate and complete copies of records.
- Availability and retention of records for FDA inspection for as long as applicable regulations require retention.
- Archiving capabilities.
- Access controls and authorization checks for users' actions.

- Secure, computer-generated, timestamped audit trails of users' actions and changes to data.
- Encryption of data at rest and in transit.
- Electronic signature controls.
- The performance record of the electronic service vendor and the electronic service provided.
- Ability to monitor the electronic service vendor's compliance with electronic security and data integrity controls [15, pp. 10–11].

The vendor should provide information about its ability to meet Part 11 requirements and appropriate security measures [15]. If the vendor cannot provide the safeguards offered in the list, sponsors are advised to consider the risks of using the vendor and the possible impact on data integrity [15]. This list appears to be a valuable starting point for evaluating potential blockchain vendors.

When establishing a relationship with a blockchain vendor, sponsors should obtain an appropriate service agreement that outlines the sponsor's requirements and the roles and responsibilities of the vendor [15].

The FDA notes the importance of performing vendor audits when using electronic systems integrated with other systems [15, p. 7]. Sponsors often perform audits of vendors' systems to verify appropriate design and development methods. Because blockchain technologies are often integrated with other systems, life sciences organizations should audit blockchain vendors using established risk-based approaches.

While the sponsor is responsible for providing requested records and data to the FDA, the FDA may also perform inspections of electronic service providers if the services include areas regulated by the FDA [15]. The determination depends on whether the records are not available from the sponsor and the criticality of the investigation. Therefore, blockchain vendors should be able to provide the necessary documentation in the event of an inspection. A sponsor should provide close oversight—and perform in-person or documentation audits, as appropriate—to ensure that the blockchain vendors' documentation is in order.

4   Future Directions

Blockchain technologies offer potential benefits in life sciences research; however, the technology cannot resolve all challenges faced in life sciences research. For example, blockchain cannot prevent "the first-mile problem" [8] described in an earlier section. There could be data entry errors, falsification, or impersonation before adding data to a blockchain [55]. This situation is ubiquitous with the "garbage in, garbage out" conundrum faced by most data management technologies [71]. Last, as this chapter has pointed out throughout many of the recommendations, blockchain vendors have been slow to learn and integrate technical and procedural controls into their blockchain designs and processes [2]. The following future directions are intended to provide awareness to life sciences organizations and vendors to create directions for advancing blockchain within regulated life sciences research.

4.1 Standards

While this chapter focused on FDA regulations, life sciences organizations should be aware of international regulatory requirements for electronic systems and emerging standards to create consistent blockchain design and infrastructure approaches. These standards are designed to improve blockchain-based data security and enhance interoperability with other systems [72] and blockchains [73].

The Institute of Electrical and Electronics Engineers (IEEE) [74] supports several standard projects to shape consistent terminology, governance structures, and inter-operability approaches for blockchain in healthcare and life sciences. The working groups also address the inclusion of Internet of Things devices, increased awareness of semantic variable and naming standards for regulatory agencies [75, 21], and more efficient mechanisms for data sharing [21]. These standards will continue to evolve as blockchain technologies offer increasingly advanced features.

Similarly, the International Organization for Standardization (ISO) supports technical committees and working groups to advance standards for blockchain and distributed ledger technologies. As of 2021, there are working groups to develop standards for identity, smart contracts, and data flow models [76]. Additional products are in development for governance and security best practices.

4.2 Blockchain Education

Many healthcare and life sciences organizations have tended to approach new technologies cautiously. De Filippi and Hassan [77] note that these organizations express concern whether new technologies would be acceptable to regulatory agencies or whether changes in regulations would require reprogramming. Because blockchain technology originated with Bitcoin, its applicability to recordkeeping in life sciences research may be questioned or misunderstood. Organizations' preconceived ideas about blockchain stifle innovation and reduce exploration of potential applications that could achieve more effective or efficient practices. To address this, educational sessions about blockchain technologies are sometimes included in industry conferences. Life sciences organizations should also conduct internal training sessions to increase their team's knowledge and awareness of blockchain technologies.

4.3   Research

Research is necessary to determine where emerging regulations may unintentionally inhibit progress for blockchain in life sciences research. An active area of research examines how on- and off-chain storage solutions and newer encryption capabilities that permit information verification without revealing identifiable information can address privacy statutes [78].

Additional research appears valuable for key management solutions that do not require centralized administration [57].

5   Conclusions

Blockchain projects often focus on software functions and features; however, few consider the depth of design and documentation required to meet regulatory requirements when blockchain is applied to life sciences research. The author has frequently espoused the message of "compliance by design" [2] to encourage blockchain organizations to learn about technological and procedural controls. Blockchain developers and life sciences organizations should work together to create the documentation evidence set describing design features, validation, and ongoing software maintenance throughout the software development life cycle. Most importantly, regulatory compliance is not a one-time event but requires ongoing testing and vigilance to ensure appropriate data integrity and respect for the human participants represented in the data sets.

5.1   Key Terminology and Definitions

Digital signature: A form of electronic signature where a set of rules and parameters allow the signatory's identity and the integrity of the data to be verified. "Signature generation uses a private key to generate a digital signature; signature verification uses a public key that corresponds to, but is not the same as, the private key. Anyone can verify the signature by employing the signatory's public key. Only the user that possesses the private key can perform signature generation" [79, p. i].

Electronic signature: "A computer data compilation of any symbol or series of symbols executed, adopted, or authorized by an individual to be the legally binding equivalent of the individual's handwritten signature" (21 CFR § 11.3(b)(7)).

Hashing: "A method of applying a cryptographic hash function to data, which calculates a relatively unique output (called a message digest, or just digest) for an input of nearly any size (e.g., a file, text, or image). It allows individuals to independently take input data, hash that data, and derive the same result—proving that there was no change in the data. Even the smallest change to the input (e.g., changing a single bit, such as adding a comma) will result in a completely different output digest" [80, p. 7]).

Institutional Review Board (IRB): "Any board, committee, or other group formally designated by an institution to review biomedical research involving humans as subjects, to approve the initiation of, and conduct periodic review of such research" (21 CFR § 50.3(i)).

Interoperability: "The ability of two or more products, technologies, or systems to exchange information and to use the information that has been exchanged without special effort on the part of the user" [80, p. 4].

Protected health information (PHI): "Individually identifiable health information transmitted or held by a covered entity or its business associate, in any form or medium, whether electronic, on paper, or oral" (45 CFR § 160.103).

Smart contract: "A collection of code and data (sometimes referred to as functions and state) that is deployed using cryptographically signed transactions on the blockchain network. The smart contract is executed by nodes within the blockchain network; all nodes must derive the same results for the execution, and the results of execution are recorded on the blockchain" [80, p. 32].

References

1. Efanov D, Roschin P (2018) The all-pervasiveness of the blockchain technology. Elsevier, Ltd. https://doi.org/10.1016/j.procs.2018.01.019
2. Charles WM, Marler N, Long L, Manion ST (2019) Blockchain compliance by design: Regulatory considerations for blockchain in clinical research. Front Blockchain 2(18). https://doi.org/10.3389/fbloc.2019.00018
3. Khozin S, Kim G, Pazdur R (2017) From big data to smart data: FDA's INFORMED initiative. Nat Rev Drug Discov 16(5):306. https://doi.org/10.1038/nrd.2017.26
4. Khozin S, Pazdur R, Shah A (2018) INFORMED: an incubator at the US FDA for driving innovations in data science and agile technology. Nat Rev Drug Discov 17(8):529–530. https://doi.org/10.1038/nrd.2018.34
5. Dorsey ER (2017) Digital footprints in drug development: a perspective from within the FDA. Digit Biomark 1(2):101–105. https://doi.org/10.1159/000481274
6. FDA takes new steps to adopt more modern technologies for improving the security of the drug supply chain through innovations that improve tracking and tracing of medicines (2019) U.S. Food and Drug Administration. Retrieved August 30, 2021, from https://www.fda.gov/news-events/press-announcements/fda-takes-new-steps-adopt-more-modern-technologies-improving-security-drug-supply-chain-through
7. U.S. Food and Drug Administration (2019b) Pilot project program under the Drug Supply Chain Security Act; Program announcement. Fed Regist 84(27):2879–2883. Article FDA–2016–N–0407. https://www.govinfo.gov/content/pkg/FR-2019-02-08/pdf/2019-01561.pdf
8. Alles M, Gray GL (2020) The first mile problem: deriving an endogenous demand for auditing in blockchain-based business processes. Int J Account Inf Syst 38:100465. https://doi.org/10.1016/j.accinf.2020.100465
9. U.S. Food and Drug Administration (2019a) FDA's Technology Modernization Action Plan (TMAP). Retrieved January 23, 2020, from https://www.fda.gov/about-fda/reports/fdas-technology-modernization-action-plan

10. Food and Drug Administration Amendments Act (2007) Pub L, 110–85, 121 Stat. 823 (September 27, 2007). https://www.govinfo.gov/content/pkg/PLAW-110publ85/pdf/PLAW-110publ85.pdf#page=82
11. U.S. Food and Drug Administration (2018c) Impact of certain provisions of the Revised Common Rule on FDA-regulated clinical investigations: Guidance for sponsors, investigators, and institutional review boards. Retrieved June 8, 2019, from https://www.regulations.gov/contentStreamer?documentId=FDA-2018-D-3551-0001&attachmentNumber=1&contentType=pdf
12. U.S. Food and Drug Administration (2018d) Regulations: good clinical practice and clinical trials. Retrieved June 8, 2019, from https://www.fda.gov/science-research/clinical-trials-and-human-subject-protection/regulations-good-clinical-practice-and-clinical-trials
13. U.S. Food and Drug Administration (2017b) Program 7348.810: Chapter 48—Bioresearch monitoring program. Sponsors, contract research organizations and monitors. Retrieved February 4, 2020, from https://www.fda.gov/media/75916/download
14. U.S. Food and Drug Administration (1997) Electronic records; electronic signatures. Fed Regist 62(54):13430–13466, Article RIN 0910–AA29. https://www.govinfo.gov/content/pkg/FR-1997-03-20/pdf/97-6833.pdf
15. U.S. Food and Drug Administration (2017a) Guidance for industry: Use of electronic records and electronic signatures in clinical investigations under 21 CFR Part 11—questions and answers (draft). Retrieved October 11, 2020, from https://www.fda.gov/media/105557/download
16. U.S. Food and Drug Administration (2007) Computerized systems used in clinical investigations. Retrieved June 14, 2019, from https://www.fda.gov/media/70970/download
17. Electronic Signatures in Global and National Commerce Act (2000) Pub L, 106–229, 114 Stat. 464 (June 30, 2000). https://www.govinfo.gov/content/pkg/PLAW-106publ229/pdf/PLAW-106publ229.pdf
18. U.S. Food and Drug Administration (2018a). Data integrity and compliance with drug CGMP: Questions and answers guidance for industry. Retrieved June 19, 2021, from https://www.fda.gov/regulatory-information/search-fda-guidance-documents/data-integrity-and-compliance-drug-cgmp-questions-and-answers-guidance-industry
19. U.S. Food and Drug Administration (2020a) Program 7348.811: Chapter 48—bioresearch monitoring program. Clinical investigators and sponsor-investigators. Retrieved January 11, 2021, from https://www.fda.gov/media/75927/download
20. U.S. Food and Drug Administration (2018b) Guidance for industry: use of electronic health record data in clinical investigations. Retrieved June 8, 2019, from https://www.fda.gov/media/97567/download
21. Charles WM (2021a) Accelerating life sciences research with blockchain. In: Namasudra S, Deka GC (eds), Applications of blockchain in healthcare, vol 83. Springer Nature, pp 221–252. https://doi.org/10.1007/978-981-15-9547-9_9
22. Benchoufi M, Porcher R, Ravaud P (2018) Blockchain protocols in clinical trials: Transparency and traceability of consent. F1000Res 6. https://doi.org/10.12688/f1000research.10531.5
23. Wong DR, Bhattacharya S, Butte AJ (2019) Prototype of running clinical trials in an untrustworthy environment using blockchain. Nat Commun 10(1):917. https://doi.org/10.1038/s41467-019-08874-y
24. Choudhury O, Sarker H, Rudolph N, Foreman M, Fay N, Dhuliawala M, Sylla I, Fairoza N, Das AK (2018) Enforcing human subject regulations using blockchain and smart contracts. Blockchain Healthc Today 1(10). https://doi.org/10.30953/bhty.v1.10
25. U.S. Food and Drug Administration (2011) Process validation: General principles and practices. Retrieved June 20, 2021, from https://www.fda.gov/regulatory-information/search-fda-guidance-documents/process-validation-general-principles-and-practices
26. WHO Expert Committee on Specifications for Pharmaceutical Preparations. (2019). Annex 3: Good manufacturing practices: Guidelines on validation. In: WHO Technical Report Series No 1019, 53rd edn. World Health Organization, pp 119–201. http://digicollection.org/whoqapharm/documents/s23430en/s23430en.pdf

27. U.S. Food and Drug Administration (2013) Guidance for industry: Electronic source data in clinical investigations. Retrieved June 14, 2019, from https://www.fda.gov/media/85183/download

28. Hanson-Heine MWD, Ashmore AP (2020) Calculating with permanent marker: how blockchains record immutable mistakes in computational chemistry. J Phys Chem Lett 11:6618–6620. https://doi.org/10.1021/acs.jpclett.0c02159

29. Gordon WJ, Catalini C (2018) Blockchain technology for healthcare: facilitating the transition to patient-driven interoperability. Comput Struct Biotechnol J 16:224–230. https://doi.org/10.1016/j.csbj.2018.06.003

30. Di Francesco Maesa D, Mori P, Ricci L (2019) A blockchain based approach for the definition of auditable access control systems. Comput Secur 84:93–119. https://doi.org/10.1016/j.cose.2019.03.016

31. Sadu I (2018) Auditing blockchain: Internal auditors need to focus on new risks and opportunities posed by blockchain technologies. Intern Audit 75(6):17–18. https://iaonline.theiia.org/2018/Pages/Internal-Audit-and-the-Blockchain.aspx

32. St. Clair J, Ingraham A, King D, Marchant MB, McCraw FC, Metcalf D, Squeo J (2020) Blockchain, interoperability, and self-sovereign identity: trust me, it's my data. Blockchain Healthc Today 3(122). https://doi.org/10.30953/bhty.v3.122

33. Adlam R, Haskins B (2020) A permissioned blockchain approach to electronic health record audit logs. Assoc Comput Mach 10(1145/3415088):3415118

34. Charles WM (2021b) Blockchain innovations in healthcare. PECB Insights (33):6–11. https://insights.pecb.com/pecb-insights-issue-33-july-august-2021/#page6

35. Rahimzadeh VN (2021) Pros and cons of prosent as an alternative to traditional consent in medical research. J Med Ethics 47:251–252. https://doi.org/10.1136/medethics-2020-106443

36. U.S. Government Accountability Office (2019) Artificial intelligence in health care: Benefits and challenges of machine learning in drug development (GAO-20–215SP). https://www.gao.gov/assets/710/703558.pdf

37. U.S. Government Accountability Office (2018) Urgent actions are needed to address cybersecurity challenges facing the nation (GAO-18–622). (Report to Congressional Committees, Issue. https://www.gao.gov/products/gao-18-622

38. Wang Y, Li J, Yan Y, Chen X, Yu F, Zhao S, Yu T, Feng K (2021) A semi-centralized blockchain system with multi-chain for auditing communications of wide area protection system. PLoS ONE 16(1):e0245560. https://doi.org/10.1371/journal.pone.0245560

39. Steinwandter V, Herwig C (2019) Provable data integrity in the pharmaceutical industry based on version control systems and the blockchain. PDA J Pharm Sci Technol 73(4):373–390. https://doi.org/10.5731/pdajpst.2018.009407

40. Benchoufi M, Altman DG, Ravaud P (2019) From clinical trials to highly trustable clinical trials: blockchain in clinical trials, a game changer for improving transparency? Front Blockchain 2(23). https://doi.org/10.3389/fbloc.2019.00023

41. Dai H, Young HP, Durant TJS, Gong G, Kang M, Krumholz HM, Schulz WL, Jiang L (2018) TrialChain: A blockchain-based platform to validate data integrity in large, biomedical research studies (1807.03662) [Preprint]. National Center for Cardiovascular Disease. https://arxiv.org/abs/1807.03662

42. U.S. Food and Drug Administration (2003) Guidance for industry: Part 11, electronic records; electronic signatures—scope and application. Retrieved January 18, 2020, from https://www.fda.gov/media/75414/download

43. Van Norman GA (2016) Drugs, devices, and the FDA: Part 1: an overview of approval processes for drugs. JACC Basic Transl Sci 1(3):170–179. https://doi.org/10.1016/j.jacbts.2016.03.002

44. Day MS (2019) The shutdown problem: how does a blockchain system end? (1902.07254) [Preprint]. Massachusetts Institute of Technology. https://arxiv.org/abs/1902.07254

45. Bhatia S, Wright de Hernandez AD (2019) Blockchain Is already here. What does that mean for records management and archives? J Arch Organ 16(1):75–84. https://doi.org/10.1080/15332748.2019.1655614

46. Dutta R, Das A, Dey A, Bhattacharya S (2020) Blockchain vs GDPR in collaborative data governance. Springer International Publishing. https://doi.org/10.1007/978-3-030-60816-3_10
47. Politou EA, Casino F, Alepis E, Patsakis C (2019) Blockchain mutability: challenges and proposed solutions [Preprint]. University of Piraeus, Greece. https://arxiv.org/abs/1907.07099
48. Radinger-Peer W, Kolm B (2020) A blockchain-driven approach to fulfill the GDPR recording requirements. In: Treiblmaier H, Clohessy T (eds) Blockchain and distributed ledger technology use cases. Springer Nature Switzerland AG, pp 133–148. https://doi.org/10.1007/978-3-030-44337-5_7
49. Gu J, Sun B, Du X, Wang J, Zhuang Y, Wang Z (2018) Consortium blockchain-based malware detection in mobile devices. IEEE Access 6:12118–12128. https://doi.org/10.1109/ACCESS.2018.2805783
50. Ibarra Jimenez J, Jahankhani H, Kendzierskyj S (2019) Cyber-physical attacks and the value of healthcare data: facing an era of cyber extortion and organised crime. In: Jahankhani H, Kendzierskyj S, Jamal A, Epiphaniou G, Al-Khateeb HM (eds) Blockchain and clinical trial: securing patient data. Springer Nature Switzerland AG, pp 115–137. https://doi.org/10.1007/978-3-030-11289-9_5
51. Venkatesan S, Sahai S, Shukla SK, Singh J (2021) Secure and decentralized management of health records. In: Namasudra S, Deka GC (eds) Applications of blockchain in healthcare, vol 83. Springer Nature, pp 114–139. https://doi.org/10.1007/978-981-15-9547-9_5
52. Juneja A, Marefat M (2018) Leveraging blockchain for retraining deep learning architecture in patient-specific arrhythmia classification. IEEE Eng Med Biol Soc. https://doi.org/10.1109/BHI.2018.8333451
53. Curbera F, Dias DM, Simonyan V, Yoon WA, Casella A (2019) Blockchain: an enabler for healthcare and life sciences transformation. IBM J Res Dev. https://doi.org/10.1147/JRD.2019.2913622
54. Goossens M (2018) Blockchain and how it can impact clinical trials. ICON. Retrieved December 18, 2018, from http://www2.iconplc.com/blog/blockchain
55. Hirano T, Motohashi T, Okumura K, Takajo K, Kuroki T, Ichikawa D, Matsuoka Y, Ochi E, Ueno T (2020) Data validation and verification using blockchain in a clinical trial for breast cancer. J Med Internet Res 22(6):e18938. https://doi.org/10.2196/18938
56. U.S. Department of Health and Human Services (2016) Guidance for institutional review boards, investigators, and sponsors: Use of electronic informed consent in clinical investigations—questions and answers. Retrieved June 14, 2019, from https://www.fda.gov/media/116850/download
57. Zhao H, Bai P, Peng Y, Xu R (2018) Efficient key management scheme for health blockchain. CAAI Trans Intell Technol 3(2):114–118. https://doi.org/10.1049/trit.2018.0014
58. Beckstrom K (2019) Utilizing blockchain to improve clinical trials. In: Metcalf D, Bass J, Hooper M, Cahana A, Dhillon V (eds) Blockchain in healthcare: innovations that empower patients, connect professionals and improve care. CRC Press, Taylor and Francis Group, pp 109–121. https://www.routledge.com/Blockchain-in-Healthcare-Innovations-that-Empower-Patients-Connect-Professionals/Dhillon-Bass-Hooper-Metcalf-Cahana/p/book/9780367031084
59. Kaye J, Whitley EA, Lund D, Morrison M, Teare HJA, Melham K (2015) Dynamic consent: a patient interface for twenty-first century research networks. Eur J Hum Genet 23(2):141–146. https://doi.org/10.1038/ejhg.2014.71
60. Custers B (2016) Click here to consent forever: expiry dates for informed consent. Big Data Soc 3(1):2053951715624935. https://doi.org/10.1177/2053951715624935
61. Ballantyne A (2020) How should we think about clinical data ownership? J Med Ethics 46(5):289–294. https://doi.org/10.1136/medethics-2018-105340
62. Leon-Sanz P (2019) Key points for an ethical evaluation of healthcare big data. Processes (Basel) 7(8):493. https://doi.org/10.3390/pr7080493
63. Albanese G, Calbimonte J-P, Schumacher M, Calvaresi D (2020) Dynamic consent management for clinical trials via private blockchain technology. J Ambient Intell Humaniz Comput. https://doi.org/10.1007/s12652-020-01761-1

64. Porsdam Mann S, Savulescu J, Ravaud P, Benchoufi M (2021) Blockchain, consent and prosent for medical research. J Med Ethics 47:244–250. https://doi.org/10.1136/medethics-2019-105963

65. Shabani M (2019) Blockchain-based platforms for genomic data sharing: a de-centralized approach in response to the governance problems? J Am Med Inform Assoc 26(1):76–80. https://doi.org/10.1093/jamia/ocy149

66. Taylor MJ, Whitton T (2020) Public interest, health research and data protection law: establishing a legitimate trade-off between individual control and research access to health data. Laws 9(1):6. https://doi.org/10.3390/laws9010006

67. U.S. Food and Drug Administration (2014) Providing regulatory submissions in electronic format—standardized study data: Guidance for industry. Retrieved June 14, 2019, from https://www.fda.gov/media/82716/download

68. U.S. Food and Drug Administration (2020b) Study data technical conformance guide: Technical specifications document. Retrieved July 19, 2019, from https://www.fda.gov/media/136460/download

69. McKinney SA, Landy R, Wilka R (2018) Smart contracts, blockchain, and the next frontier of transactional law. Wash J Law Technol Arts 13(3):313–347. http://hdl.handle.net/1773.1/1818

70. Abdullah T, Jones A (2019) eHealth: Challenges for integrating blockchain within healthcare. IEEE. https://doi.org/10.1109/ICGS3.2019.8688184

71. Learney R (2019) Blockchain in clinical trials. In Metcalf D, Bass J, Hooper M, Cahana A, Dhillon V (eds), Blockchain in healthcare: innovations that empower patients, connect professionals and improve care. CRC Press, Taylor and Francis Group, pp 87–108. https://www.routledge.com/Blockchain-in-Healthcare-Innovations-that-Empower-Patients-Connect-Professionals/Dhillon-Bass-Hooper-Metcalf-Cahana/p/book/9780367031084

72. Bittins S, Kober G, Margheri A, Masi M, Miladi A, Sassone V (2021) Healthcare data management by using blockchain technology. In: Namasudra S, Deka GC (eds) Applications of blockchain in healthcare, vol 83. Springer Nature, pp 1–27. https://doi.org/10.1007/978-981-15-9547-9_1

73. Anjum A, Sporny M, Sill A (2017) Blockchain standards for compliance and trust. IEEE Cloud Comput 4(4):84–90. https://doi.org/10.1109/MCC.2017.3791019

74. IEEE Standards Association (2021) P2418.6—Standard for the framework of distributed ledger technology (DLT) use in healthcare and the life and social sciences. IEEE. Retrieved September 25, 2021, from https://sagroups.ieee.org/2418-6/

75. Ethier J-F, Curcin V, McGilchrist MM, Choi Keung SNL, Zhao L, Andreasson A, Bródka P, Michalski R, Arvanitis TN, Mastellos N, Burgun A, Delaney BC (2017) eSource for clinical trials: implementation and evaluation of a standards-based approach in a real world trial. Int J Med Inform 106:17–24. https://doi.org/10.1016/j.ijmedinf.2017.06.006

76. International Organization for Standardization (2021) ISO/TC 307: Blockchain and distributed ledger technologies. Standards catalogue. International Organization for Standardization. Retrieved October 14, 2021, from https://www.iso.org/committee/6266604/x/catalogue/p/0/u/1/w/0/d/0

77. De Filippi P, Hassan S (2016) Blockchain technology as a regulatory technology: from code is law to law is code. First Monday 21(12). https://doi.org/10.5210/fm.v21i12.7113

78. Tomaz AEB, Nascimento JCD, Hafid AS, De Souza JN (2020) Preserving privacy in mobile health systems using non-interactive zero-knowledge proof and blockchain. IEEE Access 8:204441–204458. https://doi.org/10.1109/ACCESS.2020.3036811

79. National Institute of Standards and Technology (2013) Digital Signature Standard (DSS) (FIPS PUB 186–4). https://csrc.nist.gov/publications/detail/fips/186/4/final

80. Yaga D, Mell P, Roby N, Scarfone K (2018) Blockchain technology overview (NISTIR 8202). (NIST Interagency/Internal Report, Issue. https://www.nist.gov/publications/blockchain-technology-overview

Dr. Wendy Charles has been involved in clinical trials from every perspective for 30 years, with a strong background in operations and regulatory compliance. She currently serves as Chief Scientific Officer for BurstIQ, a healthcare information technology company specializing in blockchain and AI. She is also a lecturer faculty member in the Health Administration program at the University of Colorado, Denver. Dr. Charles augments her blockchain healthcare experience by serving on the EU Blockchain Observatory and Forum Expert Panel, HIMSS Blockchain Task Force, Government Blockchain Association healthcare group, and IEEE Blockchain working groups. She is also involved as an assistant editor and reviewer for academic journals. Dr. Charles obtained her Ph.D. in Clinical Science with a specialty in Health Information Technology from the University of Colorado, Anschutz Medical Campus. She is certified as an IRB Professional, Clinical Research Professional, and Blockchain Professional.

The Art of Ethics in Blockchain for Life Sciences

Ingrid Vasiliu-Feltes

Abstract Blockchain is one of the emerging technologies with profound societal and economic disruptive potential. It can also act as a catalyst for a new era where boundaries between physical, biological, and digital worlds become increasingly blended. This impact will likely trigger a complex cascade of adaptive changes in how we live, work, and educate future generations. Although ethics and moral values have been in existence for centuries, the digital era and rapid large-scale adoption of emerging technologies such as blockchain are posing novel digital ethics challenges that need to be addressed from a philosophical, legal, and self-sovereignty perspective. This chapter highlights how we can design proactive digital ethics programs in life sciences that mitigate potential negative consequences of blockchain deployments. Further, design thinking methodology combined with ethics principles can assist with building a human-centered blockchain ecosystem in the life sciences industry that will protect human rights. Specific digital ethics nuances related to various domains within life sciences as well as cultural or socioeconomic differences that can impact our blockchain ethical design frameworks will be addressed, and topics for future research will be suggested.

Keywords Ethics · Life sciences · Blockchain · Data governance · Identity · Research

1 Introduction

Ethics has been a very important discipline for centuries. After decades of marginalization, we are currently witnessing a resurgence within the scientific and business community due to the complex ethical issues we face while deploying emerging technologies at a larger scale. The scientific and business communities, as well as numerous not-for-profit and government agencies, are appropriately concerned

I. Vasiliu-Feltes (✉)
Detect Genomix, Sunrise, FL, USA
e-mail: ivfeltes@miami.edu

Government Blockchain Association, Washington, DC, USA

© The Author(s), under exclusive license to Springer Nature Singapore Pte Ltd. 2022   267
W. Charles (ed.), *Blockchain in Life Sciences*, Blockchain Technologies,
https://doi.org/10.1007/978-981-19-2976-2_12

about ethical issues that impact all industries. Topics such as bias, discrimination, data privacy, data ownership, transparency, and trust are making the headlines daily.

Industry leaders wish to be prepared for entering the next industrial revolution. Successful management of emerging technologies, such as distributed ledger technologies (DLTs, most notably blockchain), on all domains within the life sciences ecosystem will be required to display a complex armamentarium of novel skills, such as technology literacy and environmental, social, and governance (ESG) consciousness, as well as mastery of digital and applied ethics [1]. Furthermore, it has become evident that novel technologies like blockchain will also demand versatility in foundational ethical concepts. It is recommended to design proactive ethics programs to avoid negative consequences, and leaders that understand this imperative are poised to be successful. Ethical leaders of our digital era will be defined by upholding moral values, complementing state-of-the-art strategic planning, revising our education system, and embarking on an arduous, complex digital transformation journey.

Blending boundaries between physical, digital, and biological worlds will likely continue at an exponential pace. Emerging technologies such as DLTs, artificial intelligence (AI), Internet of Things (IoT), or next-generation computing have the potential to make a profound disruptive global impact. Many experts consider DLTs—and specifically blockchain technologies—to have a transformative impact across multiple industry sectors (e.g., [2, 3]). The life sciences industry is one of the most significantly affected post-pandemic and will demand unique ethics, business, and leadership challenges.

Deloitte's latest global life sciences outlook report highlights accelerated digitization, a new remote workforce, new customer-centric solutions, shortening of the research and development cycles, cross-border reliance via supply chain optimization [4]. One of the most significant challenges leaders face is the ethical and mindful deployment of emerging technologies. The life sciences industry is represented by a broad business ecosystem. Life sciences are also at the top of the agenda for most digital ethics experts concerned about potential negative consequences during deployments of emerging technologies (e.g., [5, 6]). Blockchain technologies have sparked numerous passionate debates among experts that emphasize the numerous opportunities they bring to the life sciences industry and experts who caution about all potential risks associated with their deployment.

The life sciences industry is undoubtedly experiencing tremendous growth. Several trends demand attention from key stakeholders: a rise in genomics-powered personalized and precision medicine, a rise of in silico trials, a reinvigorated focus on specific specialties such as immunology, pathology, imaging, as well as a remarkable increase in funding for some of the disinclines such as oncology or neurosciences [7].

The ethical aspects of blockchain deployment are complex for any industry. However, there are additional unique challenges related to the life sciences industry that must be addressed proactively. There are essential nuances in the ethical deployment of blockchain, which include societal and individual perspectives.

Among all domains that represent the life sciences, research is one of the most important to emphasize when considering blockchain deployments due to the exponential and long-term impact on all other sciences, healthcare, and the global business ecosystem.

At a basic level, we must ensure that blockchain deployments in life sciences uphold the basic ethical principles such as justice, beneficence, non-maleficence, confidentiality, integrity, and autonomy. A well-planned application of blockchain in life sciences must meet the impartiality and equality conditions, as well as ensure equal access and safeguard ownership of all data generated. The cryptography-based security offered by blockchain technology can contribute to our quest to offer maximum protection for the data stored and protect against unintended breaches, as well as malicious cyber-attacks. The life sciences industry generates massive datasets, numerous products, and solutions that are extremely difficult to safeguard. However, in some situations, a blockchain's attributes can offer a better solution for confidentiality, fidelity, and integrity than traditional technology architectures [8].

Perhaps one of the most convincing arguments for blockchain can be made for upholding the principle of autonomy. Self-sovereign identity has the potential to solve one of the major power dynamics and allow an optimal solution by offering individuals the right to their own digital identity and digital footprint [9]. Blockchain is the technology that can offer the necessary infrastructure to achieve a scalable, secure, decentralized model. In a recently published article about the use of blockchain in e-health, [9] provides a detailed overview of a centralized user-centric self-sovereignty model. The authors illustrate how the model gives users full control and provides the necessary steps for a successful implementation, such as decentralized identifiers, decentralized identifier documentation, and verifiable claims.

In addition to upholding the fundamental ethical principles, experts have called for the creation of a new Code of Ethics and Code of Conduct for Blockchain. Neitz [10] emphasizes the pros and cons of decentralization, as well as the dangers of human bias and conflict of interest for blockchain developers and other agents of interest. The author posits that while having a code of conduct would not eliminate challenges and ethical dilemmas for blockchain deployments, it could provide basic guidance to key stakeholders in the blockchain ecosystem [10].

Australia has taken the lead by drafting a Blockchain Code of Conduct that can serve as a blueprint for other countries. While it certainly offers opportunities for improvement, its content focuses on reputation, respect for rules, honesty, confidentiality, privacy, fairness, competence, self-improvement, conflicts of interest, and responsibility to others [11].

2 Digital Ethics Programs Design for Blockchain in Life Sciences

Digital ethics is a discipline that describes and addresses how we can translate classical ethics principles into the digital and virtual realms, such as beneficence, maleficence, autonomy, justice, and the values we desire to uphold as a society. Furthermore, applied ethics also aims to provide ethical guardrails that can assist us in maintaining trust, respect, responsibility, fairness, and citizenship.

Business ethics include governance, social and fiduciary responsibilities, as well as discrimination, fraud, abuse, or bribery [12]. Ethical life sciences leaders are expected to display a high regard for moral values such as honesty, fairness, respect for others. By striving to demonstrate ethical leadership in this digital era, leaders can greatly improve a Global Life Sciences Ethics Culture [13]. This section provides the overviews of digital ethics codes and summarizes the relevant literature.

2.1 General Application of Digital Ethics Across the Life Sciences Continuum

When evaluating the key elements that constitute a state-of-the-art proactive digital ethics program, we identify a need for a new code of digital ethics, a new code of digital conduct, new digital data governance, and a new digital bill of rights in addition to the traditional components. Gloria [14] forecasts a different future for digital rights, and Neitz [10] has questioned if we need a blockchain-specific code of ethics given "the libertarian origins of blockchain." Neitz expresses concerns about a potential backlash from blockchain developers "who embrace the libertarian ideal" and foresees that they would likely argue that implementation of a common standard goes against the very freedoms that make blockchain a revolutionary technology.

A recent systematic review of the blockchain literature reveals that most research had initially focused on cryptocurrencies. Only lately, a transition has been observed towards the ethical deployment of blockchain and the need for practical tools that can be utilized by industry experts, practitioners, and scholars [15]. The authors note that the spectrum of blockchain ethics research covers sustainability, greater societal good versus the needs of individual citizens, impact on law and democracy, the potential for digital twins and converging technologies, and the transformative power of blockchain for all industries in the digital era. Several publications call for the creation of international frameworks that can address the ethical considerations of blockchain technology infrastructure development and blockchain applications [15].

2.2 Research

Digital ethics has application for all types of research and all stages within the research lifecycle. There are numerous benefits of blockchain technologies in any research enterprise spanning across all domains: IRB review, audits, compliance, reporting, waste reduction, fraud prevention, informed consent, staff certification, patient recruitment, data privacy, addressing conflicts of interest, and advanced financial management [5, 6].

Enhancing the quality and safety in research is paramount to upholding the principles of beneficence and non-maleficence. Deploying blockchain for pharma research could not only reduce errors, reduce adverse events, improve outcomes but also aid with drug traceability, which has led to expanded use of blockchain in pharmaceutical supply chain management [16, 17]. A blockchain-powered pharma industry ecosystem could leverage smart contracts in a secure private permissioned distributed network of stakeholders and could lead to enhanced safety, improved integrity, and efficiency by reducing intermediaries.

Whether we aim to enhance study design, study implementation, study tracking, preparation for audits, or monitoring long-term impact, some of the unique benefits of blockchain can prove to be highly beneficial when deployed mindfully and with a strong data governance program [18]. Furthermore, the enhanced access, decentralized features, automation, and scalability can optimize efficiencies for all types of studies such as analysis of data and specimens, observational studies, interventional studies, case-control studies, cohort studies, cross-sectional studies, or qualitative studies [5]. Randomized controlled studies are often more complex and can serve as s excellent illustrative example of how blockchain technology can be deployed ethically by embedding guardrails and checkpoints during every process that ensures efficiency and compliance.

A comprehensive blockchain-powered digital ethics program can facilitate internal and external audit preparation. Additionally, many of blockchain's characteristics, such as proof of ownership and authority or its practical immutability, can reduce the overhead burden for staff, reduce waste, minimize or eliminate fraud [19].

A proactive robust data governance program requires transparency regarding data controls. The transparency afforded by blockchain technology ensures that all decisions and processes are auditable and confirms adequate data stewardship. Through blockchain's cryptographically backed-up infrastructure, we also achieve improved accountability, and its consent-based features allow seamless cross-disciplinary and inter-organizational collaboration without jeopardizing data sharing standards [20, 21].

Some of the most promising benefits of blockchain in data governance are highly desirable for any research enterprise. However, when deploying blockchain technologies for research, we must also mitigate some potentially negative aspects such as cost, limited lifespan of encryption, or network maintenance breakdowns [5, 9].

In life sciences research, multiple key stakeholders from a variety of public or private organizations are involved, and a reliance on private keys is often required. Therefore, a careful feasibility analysis of the specific type of blockchain technology to be deployed is essential. Furthermore, deciding what data need to be stored on and off-chain is also crucial and needs to occur early in the design phase [22]. For most life sciences projects, a hybrid design that enhances privacy by storing specific data elements on the chain and preserves some off-chain may prove to be an optimal solution.

Another potential barrier that needs to be overcome for life sciences research is the "zero state challenge." Specifically, the provenance of many records used for a specific research trial will require validation [23]. As described eloquently by La Pointe and Fishbane in the Blockchain Ethical Design Framework [23], an intentional design is essential to achieve optimal results. Specifically, the rules that govern human interaction must be prioritized and decided early in the process. Decision-makers will need to make tradeoffs that ensure the highest effectiveness of blockchain deployment. These tradeoffs can also impact inclusion, diversity, and enterprise return on investment [23].

As described above, the successful deployment of blockchain technologies applies to all domains within the life sciences continuum [16, 17]. However, several nuances are worth highlighting for a few high-impact domains that require a higher degree of customization for successful implementations, such as Genomics, Precision Medicine, Pharma, Biopharma, Biotech, or Biomed. The customization would ensure operational effectiveness and efficiency, as well as uphold ethical principles.

2.3   Genomics and Precision Medicine

Advanced genomic sequencing has opened a new world of opportunities in life sciences, from direct to consumer testing to novel scientific discoveries and the development of new personalized genomics-informed medical solutions. These solutions can include new molecules, new pharmaceutical agents, new medical devices, and new therapeutic pathways. All will require a safe, trusted method to access, store, share, and analyze the massive genomic data sets generated globally. Several publications are highlighting the numerous benefits of blockchain platforms in genomics-powered precision medicine. Most of them emphasize participatory access and distributed data stewardship (e.g., [7, 21, 24, 25]), while others highlight the enhanced security and self-sovereignty characteristics [9].

While there are clear opportunities for blockchain in genomics medicine, we must also overcome several challenges. Thiebes [24] determined that there are 17 technological advantages. The author also outlined the opportunities blockchain brings for increased flexibility, allowing dynamic access to various stakeholders and interdependent privacy. This dynamic consent process enables blood relatives to give data sharing permissions via smart contracts [24].

2.4 Digital Identity

Digital identity is a foundational element to successful ethical blockchain deployments in any industry and is crucial for life sciences. Digital identity can be represented by a person, organization, application, or device and includes electronic signatures, seals, website authentication, and registered delivery [9]. From an ethical perspective, we must reflect on all expressions and understand the impact of digital identity categorizations when deploying blockchain across the life science spectrum. Cameron's landmark publication [26] outlined identity principles, and blockchain is conducive to attaining all of them: user control and consent, minimal disclosure for a constrained use, justifiable parties, directed identity, pluralism of operators and technologies, human integration, and consistent experience across contexts.

Another essential article by Allen [27] describes four models of online identity, and each requires different digital ethics guardrails: centralized identity, federated identity, user-centric identity, and self-sovereign identity. He also drafted novel principles of self-sovereign identity, which should be foundational to those developing digital ethics blockchain playbooks: user-centricity, control, access, transparency, longevity, portability, interoperability, consent, minimized data disclosure, and protection.

Bouras et al. [9] provide a comprehensive review and overview of the impact of identity management and its importance in e-healthcare. The seven criteria of identity management they outline fully apply to life sciences research; autonomy, authority, availability, approval, confidentiality, tenacity, and interoperability. They also represent crucial elements of success in life sciences research and can be delivered via blockchain technology. The authors provide a helpful comparison of identity management models and how each type impacts the seven identity management criteria. Their findings suggest that decentralized models are the only ones offering autonomy, as well as the highest authority, availability, and confidentiality. They also highlight challenges with using centralized, federated, or user-centric identity models, such as lack of autonomy and interoperability in centralized models or lack of the approval feature in either centered or federated identity models [9].

3 Cultural, Legal, and Socioeconomic Influences

There is a complex and dynamic interplay between cultural factors, the legal landscape, and socioeconomic factors in each country or region that deeply influences the adoption of emerging technologies and their ethical deployment. There are marked differences in digital literacy and fluency that impact key stakeholders' ability to assess, design, develop, deploy, and monitor the deployment of all emerging technologies. However, blockchain has caused a marked cultural, legal, and socioeconomic divide that must be addressed globally. For research in the life sciences industry to thrive from leveraging blockchain technologies, we must develop new regulatory frameworks and legislative clarity.

A recent book, Future Law, eloquently highlights the challenges we encounter when developing legislation for emerging technologies, as well as some of the main regulatory and ethical intricacies lawmakers need to consider [28]. The authors also emphasize that arts and culture play a mediating role between technology and law. Mittelstadt and Floridi [29] also identify key societal issues and approaches that rule the debate on the ethical deployment of new and emerging technologies while calling for international collaboration to develop information governance policies. The authors caution against exceptionalism, parochialism, and adventitious ethics in life sciences research. While written to address ethical issues in big data management, the fundamental problems, main conclusions, and recommendations can be easily extrapolated and applied to blockchain technologies.

Another intriguing opinion highlights the convergence of ethics, law, and governance and the impact technology deployments in life sciences on significant decisions in the healthcare, military, defense, and space industries [18]. The authors also highlight how traditions and values in various global communities that share religious beliefs markedly impact the ability to draft laws for emerging technologies. They also point to the significant governmental bias, outdated regulations, and bureaucratic burdens existent in many geographic markets that preclude the development of legislation or policies that can assist with deploying emerging technologies such as blockchain [18]. The book calls out the tension between promoting innovation and entrepreneurship that stimulates economic growth and the regulatory hurdles. Examples are provided from various countries where the political process interferes with appropriate assessment of the benefits and risks associated with emerging technologies such as blockchain. Safety, privacy, responsibility, and public health are often crucial topics in the passionate debates, and key stakeholders within the life sciences and blockchain industries often find themselves caught in the middle of the polemic.

Carnevale and Occhipinti [30] pose several questions to all digital ethics advocates: Who is authorized to make decisions in a decentralized system? What about the mechanism for deciding? Authorized by whom? With what kind of consensus? To which principles must the decision-making mechanism respond? Answering these initial questions to optimize all aspects of the life sciences research industry is only the beginning of the digital ethics odyssey. It constitutes a moral imperative for all decision-makers [30].

Dierksmeier and Steel [12] forecast some of the moral dilemmas business leaders will have to solve before and during blockchain technology deployments. The authors share their views on the application of Habermasian corporate social responsibility theory in blockchain applications. The life sciences research industries are particularly amenable to data transparency to authorized stakeholders. State-of-the-art ethics programs will be required to navigate the numerous sources of ambivalence caused by those who endorse a utilitarianist, contractarianist, deontological, or virtue ethics approach [12].

The impact of ethical deployment of blockchain in life sciences research will inevitably also cause a recalibration of the educational and business processes within life sciences and, therefore, a novel emphasis on blockchain business ethics and educational ethics-related aspects. Other authors (e.g., [10]) caution about the ethical challenges with decision-making in all types of blockchain technologies that influence state and governing regulatory bodies.

Zatti [31] calls attention to how the pandemic has highlighted the need to share relevant biobank data and the benefits blockchain technologies offer while safeguarding intellectual property rights. The authors also echo other experts' calls to enhance legislation and more explicit regulatory guidelines that can facilitate large-scale adoption of blockchain.

Several governments worldwide have acknowledged the need for new laws and regulatory guidelines and already adopted blockchain. Europe and Asia are leading the way. However, there are promising efforts in North America, South America, Australia, and Africa. Lawmakers, policymakers, and ethicists will have to collaborate closely to align their new bodies of work with the global digital ethics frameworks. At a global level, we have a few universal opportunities that can drive successful blockchain deployments in life sciences and other industries, such as increased digital ethics advocacy, sustainability, and inclusion.

4 Blockchain Ethics and Purpose in Life Sciences

Life sciences leaders have the opportunity to shape the future by fostering a culture of digital ethics and contribute to the development of a Global Digital Ethics Framework for the life sciences research industry. This global framework can facilitate the attainment of the United Nations Sustainable Development Goals (SDGs) and further validate the existing sense of purpose in life sciences research.

While the deployment of blockchain technologies can have a large-scale impact on all United Nations SDGs, a few SDGs are more directly impacted by blockchain solutions where a lack of ethical deployment can have devastating circumstances on society. Blockchain technologies can augment and amplify sustainability efforts related to reducing poverty, reducing hunger, improving access to quality education, optimizing gender equality, promoting decent work and economic growth, building a robust infrastructure, reducing inequality, and creating sustainable cities [1]. Undoubtedly, blockchain technologies deeply influence the health and wellness ecosystem and specifically the life sciences industry through enhanced capabilities across various essential domains such as clinical trials, supply chain management, contract management, financial transactions, credentialing, and safety. At a global level, blockchain deployments can also accelerate research and development efforts, as well as act as an enabler for the large-scale adoption of other emerging technologies [6].

Perhaps one of the most important ethical aspects is blockchain's impact in ensuring appropriate assent, prosent, and consent in human research, as it transcends ethics and elicits legal, social, and philosophical considerations. Blockchain-enabled platforms also have a crucial potential to facilitate corporate ESG consciousness by becoming a foundational technology for data standardization, asset performance assessments, and compliance with ESG mandates or standards [1].

Ethical deployment of blockchain can only be successful with strong ethical leadership. We currently live in a globalized society that has become hyperconnected, with a high degree of automation and digitization embedded in our daily lives. Business leaders that wish to be successful in this new world must add a whole set of novel skills to their portfolio, such as ability to translate ethical concepts into daily practice, understand the basic methodologies defined by design thinking, enhance their digital acumen and become global digital citizens [13]. When we develop a state of the art enterprise digital ethics roadmap, it is recommended to align it with other key strategic initiatives and to embed all elements that are included in an ethics portfolio: social consciousness, concerns about climate impact, ethical use of cybersecurity software, as well as a customized digital code of conduct for the organization and its employees.

The exponential adoption of blockchain in life sciences will require a robust, sustainable digital ethics culture to avoid potential data breaches, optimize privacy and ensure ownership in this highly virtualized and digitized era. Digital ethics conscious leaders should be appropriately concerned about upholding core foundational ethical values, as well as those unique to the life sciences research ecosystem.

5 Future Directions: Disruption, Innovation, Evolution

The life sciences research industry has faced perhaps one of the highest pressures for digital transformation and disruption during and in the current post-pandemic era. The research enterprise has been disrupted by the global pandemic demands and has continued to evolve to meet the demand of a highly volatile, high-risk environment. From meeting novel regulatory and legislative guidelines, revising pricing structures in the face of economic downturn, and increasing efficiency, effectiveness, and safety while deploying the latest emerging technologies are just a few of the items life sciences leaders have to consider. Blockchain technologies have proven themselves feasible during the pandemic crisis and are now adopted at an accelerated pace within the life sciences disciplines and particularly in research [6]. However, enterprises must embark on a journey of continuous improvement, innovation, and disruption to remain competitive and ensure sustainability. Having a contours improvement mindset can facilitate the long-term success of digital ethics programs even in this highly volatile and high-risk post-pandemic era.

Industry experts forecast that blockchain technologies will continue to promote innovation and entrepreneurship while driving a new digital economy. For life sciences research, a few potential trends are emerging that can all benefit from blockchain deployments. These include novel use cases in various disciplines such as Psychology and Behavioral Sciences, Endocrinology, Immunology, Embryology, Neurobiology, as well as the emergence of new disciplines such as those that study the medical applications of brain–computer interfaces, human cloning, and bionic humans [32].

These disciplines pose unique ethics challenges that require innovative ethics approaches and a state of the art ethics governance. Organizations would be well advised to seek ethics counsel and create a robust ethics governance model to avoid or mitigate potential ethical breaches [33].

For all new use cases of blockchain technology deployment in life sciences, we have also noticed an exponential increase of converging technologies to optimize their impacts, such as the smart use of blockchain with AI, IoT, advanced computing methodologies such as quantum computing to create new concepts that can enhance development, quality, and safety such as digital health [34]. Designing state-of-the-art digital ethics programs that can accommodate the exponential ethical challenges brought upon by deploying multiple emerging technologies will become a moral imperative for leaders in this digital era.

For example, the combined deployment of AI & DLTs leverages the benefits of both technologies to optimize public health efforts, as well as facilitate the prevention, treatment and management of diseases. Large-scale adoption of converging emergent technologies such as blockchain AI, nanotechnology, and IoT can disrupt the current health care ecosystem and lead to improve global population health. To achieve long-term success, we must encourage and attain inter-and cross-disciplinary collaboration. There is a need to redesign the current life-sciences and healthcare delivery ecosystems to allow never paradigms such as precision and personalized medicine to fully develop. A completely redesigned AI and DLT-powered global health and life sciences ecosystem would be characterized by enhanced access to precision medicine solutions for patients worldwide. Last but not least, it would be essential to wisely and ethically deploy genomics-based precision medicine and further stimulate life-sciences research.

Evangelatos et al. [20] described how the unique combination of open source code software and blockchain technology could prove to be a viable solution for public biobanks' data governance. Building research ecosystems using decentralized blockchain technology that addresses the free-riding problem in the research community can lead to sustainability and aligns with free-market models.

By creating a virtual environment embodied as a digital twin, we can significantly enhance our ability to exchange valuable information with other stakeholders, enhance safety testing and optimize our data processing capabilities. Digital twins are designed and deployed to enable virtual collaboration, absorb and process big data, and assist us with managing the physical world more efficiently and safely [35]. The pandemic impact and disruption caused to the global economy have accelerated the pace and adoption of digital twins globally [35]. The design and deployment of digital twins are complex and intimately connected to other digital technologies such as blockchain, cloud computing, AI, IoT, 5G networks, virtual, augmented, or mixed reality. By maximizing the use of digital virtual replicas, we can exponentially accelerate our efforts in research and development, optimize quality assurance and safety testing, reduce waste, decrease operational inefficiencies and increase the return on our investments [35].

Life sciences and healthcare are examples where digital health twins could potentially solve several of the major challenges we are facing globally and have a profound disruptive effect. A global blockchain-powered precision medicine data exchange supporting research enterprises would allow us to derive meaningful and actionable insights exponentially and shorten the research and development lifecycle for novel drugs, devices, and treatment pathways [34].

Futurists and emerging technologies' advocacy groups are also envisioning blockchain technologies as a gateway technology for smart cities due to their ability to enable safer, more reliable, and transparent transactions among multiple stakeholders involved in the governance of smart cities. Smart research, smart health care, smart hospitals, smart research will hopefully become a golden standard for upcoming generations.

Overall, industry experts estimate that we will witness the increased incorporation of blockchain in the life sciences strategic planning process within the next few years [6]. To be successful, leaders ready to embark on this journey must address all stages from redesigning research processes, developing proofs of concept, deploying pilots, demonstrating the ability to scale, and creating an ethics culture mindset for the enterprise [15, 23].

A state-of-the-art digital ethics program for life sciences would have to start with infusing core ethics values at all levels within the organization. Such a program would require building an ethics mindset at the board level, including the C-suite, as well as middle management, employees, and patients. This program would also require developing a new vision and mission statement that emphasizes digital ethics, new policies and guidelines, new operating procedures, and embedding digital ethics guardrails into all relevant daily processes [36].

6 Conclusions

Beasley [37] questions if ethical leadership is an art. This author agrees and adds that implementing digital ethics programs in any organization requires ethical leadership and a proactive approach. Moral identity and moral imagination are not often included in a leadership skills list, yet they are crucial in successfully navigating some of the significant challenges leaders face, such as conflict management, ethical dilemmas, and uncertainty. Emerging technologies such as blockchain are perfect examples that showcase the complexity and need for inter-disciplinary collaboration of key stakeholders to be successful. Another key takeaway from this chapter is the need to develop and nurture a culture of digital ethics, encourage a continuous improvement mindset, and develop key digital ethics performance indicators to measure the impact of blockchain deployments in life sciences. Lastly, this author hopes that increased attention will be given to ethical deployments of blockchain as a sizable blockchain divide must first be overcome [38].

Digital ethics could and should become an integral part of our global education ecosystem and deeply embedded into the DNA of any life sciences research enterprise. Ideally, we would like to live and work in a world where we have designed, adopted a new Hippocratic Oath customized for the Digital Era and a New Code of Blockchain Ethics.

Key Terminology and Definitions

Applied ethics: Applied ethics is a branch of ethics devoted to treating moral problems, practices, and policies in personal life, professions, technology, and government.

Biobank: An extensive collection of biological or medical data and tissue samples amassed for research purposes.

Bionic humans: A human being whose body has been taken over in whole or in part by electromechanical devices.

Brain-computer interface (BCI): A system that measures the activity of the central nervous system (CNS) and converts it into artificial output that replaces, restores, enhances, supplements, or improves natural CNS output, and thereby changes the ongoing interactions between the CNS and its external or internal environment.

Contractarianism: A theory stemming from the Hobbesian line of social contract thought specifying that persons are primarily self-interested and that a rational assessment of the best strategy for attaining the maximization of their self-interest will lead them to act morally.

Cyberethics: The study of ethics pertaining to computers, covering user behavior and what computers are programmed to do, and how this affects individuals and society.

Digital ethics: The branch of ethics that applies to digital media, for example, in online contexts, how users interact with each other, both in representing themselves and controlling data about themselves in the platforms and technologies that they use and in their respect for other users and other users' rights to self-determination and privacy.

Digital twin: A digital representation of a real-world entity or system.

DLT: Distributed ledger technologies.

Environmental, social, and governance (ESG): Criteria are a set of standards for a company's operations that socially conscious investors use to screen potential investments.

Genomics: The branch of molecular biology concerned with the structure, function, evolution, and mapping of genomes.

Habermasianism: The theory by Jurgen Habermas, Sociologist, and Philosopher.

Human cloning: The creation of a genetically identical copy (or clone) of a human.

Open source code: Software for which the original source code is made freely available and may be redistributed and modified according to the requirement of the user.

Neurobiology: The branch of the life sciences that deals with the anatomy, physiology, and pathology of the nervous system.

Self-sovereignty: A feature of an ID or identity system, whereby individual users control when, to whom, and how they assert their identity.

Smart city: A smart city uses information and communication technology (ICT) to improve operational efficiency, share information with the public and provide a better quality of government service and citizen welfare.

Sustainable development goals (SDGs): A set of goals adopted by the United Nations in 2015 as a universal call to action to end poverty, protect the planet, and ensure that by 2030 all people enjoy peace and prosperity.

Utilitarianism: The doctrine that an action is right insofar as it promotes happiness, and that the greatest happiness of the greatest number should be the guiding principle of conduct.

Virtue ethics: Currently, one of three major approaches in normative ethics.

References

1. Singh R, Dwivedi AD, Srivastava G (2021) Blockchain for united nations sustainable development goals (SDGs). Front blockchain. https://www.frontiersin.org/research-topics/18154/blockchain-for-united-nations-sustainable-development-goals-sdgs
2. Heister S, Yuthas K (2021) How blockchain and AI enable personal data privacy and support cybersecurity. In: Blockchain potential in AI [Online First]. IntechOpen. https://doi.org/10.5772/intechopen.96999
3. Tan E (2021) A conceptual model of the use of AI and blockchain for open government data governance in the public sector (No. B2/191/P3/DIGI4FED). DIGI4FED. https://soc.kuleuven.be/io/digi4fed/doc/d-3-2-1-a-conceptual-model-of-the-use-of-ai-and.pdf
4. Levy V (2021) Global life sciences sector outlook. Deloitte. https://www2.deloitte.com/global/en/pages/life-sciences-and-healthcare/articles/global-life-sciences-sector-outlook.html
5. Benchoufi M, Ravaud P (2017) Blockchain technology for improving clinical research quality. Trials 18(1):1–5. https://doi.org/10.1186/s13063-017-2035-z
6. Charles WM (2021) Accelerating life sciences research with blockchain. In: Namasudra S, Deka GC (eds) Applications of blockchain in healthcare. Springer Singapore, pp 221–252. https://doi.org/10.1007/978-981-15-9547-9_9
7. Scott M (2019) Feature interview with David Koepsell, CEO of EncrypGen. Blockchain Healthc Rev. https://blockchainhealthcarereview.com/charting-the-blockchain-of-dna-feature-interview-with-david-koepsell-of-encrypgen/
8. Lo SK, Staples M, Xu X (2021) Modelling schemes for multi-party blockchain-based systems to support integrity analysis. Blockchain Res Appl 100024:100024. https://doi.org/10.1016/j.bcra.2021.100024

9. Bouras MA, Lu Q, Zhang F, Wan Y, Zhang T, Ning H (2020) Distributed ledger technology for ehealth identity privacy: state of the art and future perspective. Sensors 20(2). https://doi.org/10.3390/s20020483
10. Neitz MB (2020) Ethical considerations of blockchain: Do we need a blockchain code of conduct? The FinReg Blog. https://sites.law.duke.edu/thefinregblog/2020/01/21/ethical-consid erations-of-blockchain-do-we-need-a-blockchain-code-of-conduct/
11. Blockchain Australia Code of Conduct (2021). https://blockchainaustralia.org/codeofconduct/
12. Dierksmeier C, Seele P (2019) Blockchain and business ethics. Bus Ethics 29(2):348–359. https://doi.org/10.1111/beer.12259
13. Vasiliu-Feltes I (2020a) Ethical leadership in the fintech era. Xpertsleague. https://www.xperts league.com/ethical-leadership-in-the-fintech-era/
14. Gloria K (2021) Power and progress in algorithmic bias. Aspen Inst. https://www.aspeninst itute.org/publications/power-progress-in-algorithmic-bias/
15. Hyrynsalmi S, Hyrynsalmi SM, Kimppa KK (2020) Blockchain ethics: A systematic literature review of blockchain research. In: Cacace M, Halonen R, Li H, Phuong T, Chenglong O, Widén L, Suomi R (eds) Well-being in the information society. Fruits of respect. Springer, Cham, pp 145–155. https://doi.org/10.1007/978-3-030-57847-3_10
16. Uddin M (2021) Blockchain Medledger: hyperledger fabric enabled drug traceability system for counterfeit drugs in the pharmaceutical industry. Int J Pharm 597:120235. https://doi.org/10.1016/j.ijpharm.2021.120235
17. Uddin M, Salah K, Jayaraman R, Pesic S, Ellahham S (2021) Blockchain for drug traceability: architectures and open challenges. Health Inform J 27(2). https://doi.org/10.1177/146045822 11011228
18. Marchant GE, Wallach W (eds) (2017) Emerging technologies: Ethics, law and governance. Taylor and Francis. https://www.routledge.com/Emerging-Technologies-Ethics-Law-and-Gov ernance/Marchant-Wallach/p/book/9781472428448
19. Ingraham A, St. Clair J (2020) The fourth industrial revolution of healthcare information technology: key business components to unlock the value of a blockchain-enabled solution. Blockchain Healthc Today 3(139). https://doi.org/10.30953/bhty.v3.139
20. Evangelatos N, Upadya SP, Venne J, Satyamoorthy K, Brand H, Ramashesha CS, Brand A (2020) Digital transformation and governance innovation for public biobanks and free/libre open source software using a blockchain technology. OMICS 24(5):278–285. https://doi.org/10.1089/omi.2019.0178
21. Shabani M (2018) Blockchain-based platforms for genomic data sharing: a decentralized approach in response to the governance problems? J Am Med Inform Assoc 26(1):76–80. https://doi.org/10.1093/jamia/ocy149
22. Miyachi K, Mackey TK (2021) hOCBS: a privacy-preserving blockchain framework for health-care data leveraging an on-chain and off-chain system design. Inf Process Manag 58(3):102535. https://doi.org/10.1016/j.ipm.2021.102535
23. LaPointe C, Fishbane L (2018) The blockchain ethical design framework. Beeck center for social impact + innovation, Georgetown University. https://beeckcenter.georgetown.edu/wp-content/uploads/2018/06/The-Blockchain-Ethical-Design-Framework.pdf
24. Thiebes S, Kannengießer N, Schmidt-Kraepelin M, Sunyaev A (2020) Beyond data markets: opportunities and challenges for distributed ledger technology in genomics. In: Bui TX (ed) Proceedings of the 52nd Hawaii international conference on system sciences, pp 3275–3284. https://hdl.handle.net/10125/64142
25. Thiebes S, Schlesner M, Brors B, Sunyaev A (2019) Distributed ledger technology in genomics: a call for Europe. Eur J Hum Genet 28(2):139–140. https://doi.org/10.1038/s41431-019-0512-4
26. Cameron K (2005) The laws of identity. Microsoft corporation. http://myinstantid.com/laws.pdf
27. Allen C (2016) The path to self-sovereign identity. Life with alacrity. http://www.lifewithalac rity.com/2016/04/the-path-to-self-soverereign-identity.html
28. Edwards L, Schafer B, Harbinja E (eds) (2021) Future law: emerging technology, ethics and regulation. Edinburgh University Press. https://books.google.com/books/about/Future_Law.html?hl=&id=JheztAEACAAJ

29. Mittelstadt BD, Floridi L (eds) (2016) The ethics of biomedical big data. Springer International Publishing Switzerland. https://doi.org/10.1007/978-3-319-33525-4
30. Carnevale A, Occhipinti C (2019) Ethics and decisions in distributed technologies: A problem of trust and governance advocating substantive democracy. In: Bucciarelli E, Chen SH, Corchado JM (eds) Decision economics. Complexity of decisions and decisions for complexity. Springer, Cham, pp 300–307. https://doi.org/10.1007/978-3-030-38227-8_34
31. Zatti F (2021) Blockchains and dynamic consent in biobanking. SSRN. https://doi.org/10.2139/ssrn.3853352
32. Bar-Cohen Y (2004) Bionic: bionic humans using EAP as artificial muscles reality and challenges. Int J Adv Robot Syst 50(2):217–223. https://doi.org/10.1097/00002480-200403000-00188
33. Ishmaev G (2019) The ethical limits of blockchain-enabled markets for private IoT data. Philos Technol 33(3):411–432. https://doi.org/10.1007/s13347-019-00361-y
34. Popa EO, van Hilten M, Oosterkamp E, Bogaardt M-J (2021) The use of digital twins in healthcare: socio-ethical benefits and socio-ethical risks. Life Sci Soc Policy 17(1):1–25. https://doi.org/10.1186/s40504-021-00113-x
35. Vasiliu-Feltes I (2020b) Digital health twins—the great enablers of new healthcare ecosystems? LinkedIn. https://www.linkedin.com/pulse/digital-health-twins-great-enablers-new-healthcare-ingrid
36. Tandon A, Dhir A, Islam AKMN, Mäntymäki M (2020) Blockchain in healthcare: a systematic literature review, synthesizing framework and future research agenda. Comput Ind 122:103290. https://doi.org/10.1016/j.compind.2020.103290
37. Beasley B (2021) Is ethical leadership an art? Notre Dame Deloitte center for ethical leadership. https://ethicalleadership.nd.edu/news/is-ethical-leadership-an-art/
38. Tang Y, Xiong J (2019) Blockchain ethics research: a conceptual model. SIGMIS-CPR '19. In: Proceedings of the 2019 computers and people research conference, pp 43–49. https://doi.org/10.1145/3322385.3322397

Further Readings

39. Castellanos S (2021) Quantum computing scientists call for ethical guidelines. Wall Str J. https://www.wsj.com/articles/quantum-computing-scientists-call-for-ethical-guidelines-11612155660
40. Lemieux VL, Hofman D, Hamouda H, Batista D, Kaur R, Pan W, Costanzo I, Regier D, Pollard S, Weymann D, Fraser R (2021) Having our omic cake and eating it too? Evaluating user response to using blockchain technology for private and secure health data management and sharing. Front Blockchain 3:558705. https://doi.org/10.3389/fbloc.2020.558705
41. Parry G, Collomosse J (2021) Perspectives on good in blockchain for good. Front Blockchain 3:609136. https://doi.org/10.3389/fbloc.2020.609136

Dr. Vasiliu-Feltes is a healthcare executive, futurist, and globalist who is highly dedicated to digital and ethics advocacy. She is a passionate educator and entrepreneurship ecosystem builder, an expert speaker, board advisor, and consultant. Throughout her career, she has received several awards for excellence in research, teaching, or leadership. This past year, she has been named one of the Top 25 Leaders in Digital Twins, Top 50 Health Tech Global Thought Leaders, 100 Global Women Leaders, 100 Global Healthcare Leaders, Top 100 Global Finance Leaders, and Top 100 Women in Crypto. Additionally, she received the 2021 Excellence in Education Award, World Women Vision Award for Technology and innovation, serves as an Expert Advisor to the EU Blockchain Observatory Forum, and was appointed to the Board of UN Legal and Economic

Empowerment Network. Most recently, she also received the WBAF World Excellence Award for The Best Businesswoman Role Model Demonstrating Social Entrepreneurship. She is an active supporter of the UN SDGs illustrated in several global collaborations and through her contribution as Global Chairwoman for GCPIT and the Global SDG Summit.

During her academic tenure, she taught several courses while on faculty at the Miller School of Medicine, as well as for the combined MD/Ph.D. and MD/MPH programs. She currently teaches the Business Technology-Digital Transformation Course at the University of Miami Herbert Business School, as well as Innovation and Digital Transformation at the WBAF Business School. Throughout her career, Dr. Vasiliu-Feltes held several leadership positions and is a member of numerous prestigious professional organizations. She holds several certifications, such as Bioethics from Harvard, Artificial Intelligence and Business Strategy from MIT Sloan, Blockchain Technology and Business Innovation from MIT Sloan, Finance from Harvard Business School, Negotiation from Harvard Law School, Innovation and Entrepreneurship from Stanford Graduate School of Business, Certified Professional in Healthcare Risk Management, Fellow of the American College of Healthcare Executives, Patient Safety Officer by the International Board Federation of Safety Managers, Master Black Belt in Lean and Six Sigma Management, Professional in Healthcare Quality by the National Association of Healthcare Quality, Manager for Quality and Organizational Excellence, by the American Society for Quality, and Certified Risk Management Professional by the American Society for Healthcare Risk Management.

Cybersecurity Considerations in Blockchain-Based Solutions

Dave McKay and Atefeh Mashatan

Abstract Blockchain technology has a reputation for providing a higher level of assurance and security than other information systems. However, many design decisions, implementation realities, limitations, and trade-offs can create underlying vulnerabilities in blockchain-based solutions that are exploitable by malicious attackers. This chapter discusses some of the most common vulnerabilities in blockchain-based solutions that can arise in the context of life sciences research. For each of these vulnerabilities, mitigating strategies are proposed to address the identified risk. These mitigating strategies reduce the likelihood and impact of the occurrence and, thereby, bring the cybersecurity risk to an acceptable level. Like any other information system, securing a blockchain-based solution requires a holistic, contextual, and risk-based approach that investigates all possible attack vectors and contextually evaluates their potential harm.

Keywords Defense-in-depth · Threats · Vulnerabilities · Risk · Life sciences · Blockchain technology · Smart contracts

1 Introduction

Cybersecurity, or information system security and privacy, can be defined as the protection of information systems and their resources, processes, data, and people from unauthorized and malicious access or manipulation [1]. Effective cybersecurity employs a variety of techniques in a defense-in-depth strategy to ensure data confidentiality, integrity, and availability against a multitude of vulnerabilities, threats, and attack scenarios [2]. The defense-in-depth layered approach puts redundant security controls in place so that if the adversary successfully bypasses one, they still face

D. McKay (✉) · A. Mashatan
Cybersecurity Research Lab, Ted Rogers School of Management, Ryerson University, Toronto, ON, Canada
e-mail: dave.mckay@ryerson.ca

A. Mashatan
e-mail: amashatan@ryerson.ca

the remaining security controls [3]. First, cybersecurity professionals analyze each information system in a comprehensive manner, both as an individual component and in the context of the surrounding technological and organizational environment. A comprehensive risk assessment starts with identifying the assets, such as data, software, hardware, and communication technologies, that need protection. Next, they investigate how the identified assets are being threatened, e.g., unauthorized access to data, software or hardware modification, and communication disruption. Once the threat scenarios are identified, the risk assessors look for potential weaknesses and vulnerabilities in the system and its surrounding environment that the adversary can potentially exploit in an attack. They try to determine the likelihood of an attack, i.e., the probability of an adversary successfully exploiting a vulnerability, as well as the impact of the attack, i.e., what a successful attack would mean to the affected stakeholders (e.g., the organization, its employees, its clients, and customers). From here, cybersecurity professionals proceed to implement the appropriate mitigating controls. This reduces the likelihood and impact of the cybersecurity risk to an acceptable level for the organization, which is inevitably a very contextual decision-making process. While it is possible to make a system very secure, factors such as usability, cost, and efficiency must also be considered to make informed decisions that are contextually optimal. Blockchain-based solutions are no exception. Before delving deeper into discussing the cybersecurity of blockchain-based solutions for life sciences research, a brief review of the main concepts is presented herein.

Distributed ledger technology (DLT) [4, 5] is a ledger where multiple copies of a record are held in a decentralized manner eliminating the possibility of a single point of failure. Blockchain technology is a type of DLT designed to be append-only to make them tamper-resistant [6–8]. It was originally developed for financial transactions in the form of cryptocurrencies [9, 10]. In its most generic form, a blockchain is a record of transactions where the state and order are held in a consensus distributed peer-to-peer across many computers [11]. Blockchains use a variety of processes to enable properties that make them systems of shared trust. Theoretically, blockchains are very secure systems. In practice, a poorly configured blockchain, improperly designed smart contracts, insecure endpoints, coordinated attacks on consensus, and poorly protected private keys can disrupt the integrity of a blockchain opening it up to cybersecurity threats. Cybersecurity threats to Blockchain-based Solutions refer to the possibility of malicious attacks by individuals and organizations that aim to gain unauthorized access to the solution's systems, disrupt its normal activity, damage its systems, and corrupt, steal, or delete assets related to the solution [12]. The system assets of a blockchain-based solution include the collection of data, hardware, software, firmware, and communications, all encompassed in the deployed solution [13].

For any blockchain-based system, it is essential to understand where the system may contain vulnerabilities. A vulnerability is an underlying flaw or weakness in the information system that an attacker could exploit [14]. Whether it is introduced during the design, development, or operation of the system, it is critical to consider the potential of harm caused by an attack as a threat. An attack is when one or more of the threats to the assets are actually carried out. The attack surface of a system involves the collection of all reachable and exploitable vulnerabilities in the system [13]. It is imperative to understand the risks involved with running a blockchain-based system and how to mitigate them. The risk is measured as a function of the impact and the likelihood of the attack happening. Reducing that risk is a valuable contribution to a blockchain-based solution's design, development, and operation. This can be done using mitigation strategies. Mitigating controls, or countermeasures, are actions or techniques that are put in place to reduce the risk of an attack on an asset [13]. Even with mitigating controls in place, there may be some leftover residual risk to the asset. It is important to understand that the attackers need only to be successful once, but the solution providers need to be successful all the time to prevent all possible attacks. While it is theoretically possible to make near-perfect systems, in reality, it is too expensive and impractical to do so. Therefore, there will always be a balancing game between usability and efficiency on one side and cost and cybersecurity on the other side [15].

Life sciences research relies on data sharing and collaboration among various medical institutions and research organizations [16]. The patient-centricity of modern life sciences research has added a new dimension of distributed involvement to this discipline's already distributed nature of research collaboration. Remote participation has facilitated a greater level of access to patients and researchers [17]. These trends make blockchain technology a great fit for life sciences research. As in any other blockchain-based solution, the integrity of life sciences data and the validity of life sciences research drawn from that data are critically dependent on the protection of the three pillars of cybersecurity: confidentiality, integrity, and availability (CIA). CIA provides assurance of protection from unauthorized access, unauthorized modifications, and incidents that may threaten the availability of service or information to the authorized entities [2]. A properly implemented blockchain can provide integrity and availability but does not ensure confidentiality [11]. Confidentiality must be built on top of the blockchain, and because of that, it is the least secure part of using a blockchain-based solution for life sciences data.

To an attacker, a life sciences system may offer several opportunities to exploit vulnerabilities for gain. Life sciences data can hold value in the inherent expense of acquiring the data [18]. This can be exploited by selling that data to a competitor or denying access to that data—holding it for ransom. If the data hold personally-identifying information, there may be an incentive to acquire data on individuals if they are celebrities or politically compromised individuals. If the blockchain-based solution uses cryptocurrency to incentivize patient participation, the cryptocurrency holders could become the target. An attacker may be motivated to access the blockchain-based solution to use the assets as free resources. The motivations are varied, but the risk remains the same.

This chapter presents common blockchain-based solution patterns in network and system designs that may result in vulnerabilities. Each solution option is described and its most common vulnerabilities are discussed. For each vulnerability, mitigation strategies are presented. The chapter provides a survey of the common vulnerabilities found in blockchain-enabled solutions and offers guidance on mitigating the resulting risks.

2 Blockchain Solution Architecture

Blockchain technology offers a wide variety of architectural choices for developing solutions. The various components that make up a blockchain solution can be adjusted to better meet the requirements for the use case at hand. A solution architecture can make the blockchain central to what it is doing, or it can use the limited properties of an existing blockchain to enhance certain aspects of the solution. The different design choices that are made can have varying implications on the cybersecurity vulnerabilities of the system.

2.1 Network and Architecture Types

Blockchain networks and the solutions built with them and on top of them can be categorized to understand their major differences and implications on cybersecurity [11]. Most blockchain-based solutions for life sciences research have been using a *private*, as opposed to *public*, network and a *centralized/hybrid*, as opposed to *decentralized*, blockchain architecture. However, there are some recent initiatives where a public network and decentralized blockchain architecture are being considered as well [19, 20]. These different network types are described next, followed by their common vulnerabilities and mitigating strategies.

2.1.1 Public/Private Network

The initial blockchains were designed as public blockchains, where the network was publicly accessible, and anyone could join and interact with transacting parties without the need for permission [21]. The transparency of the system was not seen as a detriment but as a major benefit. Having the balance of all accounts and the contents of all transactions available to everyone to reach a consensus ensured that the double-spend problem with digital currencies was resolved. The original blockchains were developed to support the transfer of cryptocurrencies [10]. The driving force behind the work was to get government and financial institutions out of the financial transactions of individuals—they wanted a digital version of cash. To increase the system's adoption, it was generally recognized that there would be the main network, or *mainnet*, that allowed anyone in the world to participate. Mainnets for Bitcoin, Ethereum, and dozens of other blockchain networks are now developed and publicly available [22].

Blockchain systems only support a limited number of transactions per second. Having a mainnet for the entire world means that one must share that transaction throughput with all of the other applications on that network. Another drawback of having a mainnet is that all of the parameters for the blockchain are fixed. Many options can increase throughput and make the blockchain more suitable for certain types of data. To address these and other concerns, open-source blockchain projects can support private networks. A private network may run the exact same code as the mainnet, but it can have a different setup [22]. For example, transaction fees can be adjusted, the initial issuance of tokens can be set to favor the parties setting up the network, the block sizes can be optimized to the problem at hand, the consensus mechanism can be swapped out from Proof of Work to something like a Proof of Authority or Proof of Stake [11]. Therefore, a private blockchain is a strong option for solutions that do not necessarily need the worldwide consensus of a mainnet.

Private networks are being used for cases such as government or industry registries. Tamper-resistance and transparency are the desired aspects; however, there is no need for widely available interactions with other solutions on the network.

Vulnerabilities and Mitigation Strategies in Private Blockchains

Public blockchains have highly reviewed security, and the security settings are not controlled by a central authority. Instead, they are decided upon by collective action, usually under the guidance of a decentralized governance organization that draws from the participation of a large community. Private blockchains are set up by a select few organizations and by Information Technology staff who are not necessarily equipped with extensive experience in the systems they are tasked with configuring [21, 23]. A private blockchain must provide its own *genesis* file, a set of rules, and parameters determining how the blockchain will behave. This genesis file is used to build the first block in the blockchain (called a genesis block) and is passed on to all the participating nodes that join the network [24]. The genesis file for the Ethereum blockchain, the second largest blockchain platform, includes settings that identify the type of consensus, the level of difficulty for solving Proof of Work consensus problems, and the initial allocation of ETH coins to accounts [25]. Manipulations to this file before starting the network have implications for how the network will function, how strong the tamper-resistance protections are, and who has received the initial issuance of coins. If the participating parties are not technical, they can be misled about the initial supply of coins in the network.

Private networks may specify the software that is running on their nodes. They may provide an image that is to be used by all of the node operators. In turn, this can cause problems if the private network falls behind the public network in node software versions. Known vulnerabilities may exist in older versions. A network where the participants have not prepared for eventual updates in their node software is open to attacks on those vulnerabilities.

Private networks are typically smaller and are more susceptible to majority (51%) attacks [26], where taking control over more than half of the network allows the controller(s) to manipulate the whole system [27]. The consensus mechanism of a blockchain depends on the distribution of multiple different parties to agree on the truth of a set of transactions. A single node operator or one operator that controls a majority of the nodes removes the condition that needs to be present to have a trusted blockchain ledger history. The private network operators should avoid having a custom node software image. Typically, a custom node software image becomes a target for malicious software changes or injection. Networks are prone to malicious attacks if over 50% of the nodes work in conjunction to alter the transactions or transaction history on the blockchain. Attacking the software image allows the attacker to gain the advantages of a 51% attack without operating the nodes themselves.

Mitigation strategies for private blockchain vulnerabilities involve a variety of one-time and ongoing actions. An audit of the private blockchain settings should be completed by a third party who can explain the implications that the proposed settings pose to future blockchain participants. The genesis file is a fully transparent file that any joining node can see, but understanding what those settings mean is not an ability that can be expected of most private network members [28]. A similar audit should be made on any custom node software. Better yet, the participants in the network that are running nodes should be allowed to select their own blockchain compatible node software that is available for operating the public blockchain. Having multiple options will reveal malicious or erroneous nodes very quickly.

The private network is still a decentralized system, and it should have decentralized governance. One of the tasks that fall on the governance body is to stay on top of the public version of the blockchain. If there is a fork in the blockchain (i.e., when there is no complete consensus about what the next block should be [29]), the governance body should be aware of what caused it and its implications. If there is a cut-over to a new version of the node software standard, the governance organization should consider doing the same cut-over at a similar time. Falling out of sync with the public blockchain leaves the private blockchain open to know vulnerabilities that were the reason for the update on the public blockchain.

It is necessary for the private blockchain participants to have a complete understanding of the implications of node control. If one organization in the system is running all of the nodes, then the other parties are not protected by the properties of a fully decentralized blockchain. By participating in a blockchain and having part of an organization's operations secured by this blockchain, a participant must take the responsibility to run their own node or have a party that is not running other nodes on this identical blockchain run one or more for them.

2.1.2   Permissioned/Permissionless Blockchains

Public blockchains balance the requirement of transparency of transactions and balances with anonymous accounts. Nodes can see all of the activity on any account, but not who controls the account. Anyone can join and create transactions and participate in the consensus. No permission was required—these are referred to as permissionless blockchains [11, 22]. Some of the properties of these early blockchains appealed to businesses who were not concerned with removing government or financial institutions from their transactions. They appreciate the tamper-evident ledger shared between organizations, transactions' fast settlement times, and the low fixed fees for transfers. However, anonymous accounts do not work with generally accepted accounting principles, investor confidence, or financial reporting compliance. Also, the business model that private companies work under does not support the complete transparency of all transactions. Some data need to be kept private or confidential.

A new type of blockchain was created to overcome the problems with permissionless public blockchains [22]. Permissioned blockchains require permission from at least some sort of plurality of the parties in the network to join. Accounts are associated with an organization that was permitted to participate. Channels can be set up between the parties involved so that other organizations cannot see the details. Even within a channel, data can be marked as private so that only the two parties in the transaction can see the full record of the transaction. Everyone else may see a hash of the transaction data.

Permissioned blockchains are well suited for use cases like life sciences research [17, 23]. All of the parties can connect with each other. Information can be kept confidential from competitors and malicious parties, while at the same time supporting valuable properties like the provenance of data records.

Vulnerabilities and Mitigation Strategies in Permissioned Blockchains

Permissioned Blockchains are similar to large enterprise applications. They expose plenty of internal options that allow the blockchain to be configured exactly as the business use case required. This requires a deep understanding of the blockchain software, decentralized system behavior, consensus mechanisms, cybersecurity, cryptography, and the business use case that the permissioned blockchain needs to support [15]. The sheer amount of knowledge required to set one up is daunting. Often what happens is that default settings are used. The default settings may come from example or test networks. Often those examples are simplified or have lowered security to facilitate running on developer machines that may have lower processing power than servers. The default settings may have default cryptography files, accounts, or passwords. A common attack is to test the default accounts, passwords, and private keys. If any of these have been retained from the example network, then the security of the permissioned blockchain has been compromised [30]. This same vulnerability can exist with each and every participant in the network.

Permissioned blockchains, by nature, become very segmented. Multiple channels emerge that represent blocks of organizations that are transacting with each other. "Highly segmented" means that the data are not highly distributed. At that point, data loss can occur. Instead of having thousands of copies on the network like a public blockchain would have, there may be as few as two organizations in a channel. If both suffer a node loss, then all of the data is gone. Within a channel, there may be private data. The private data is exposed on the nodes of the two parties that are part of the transaction. The other parties will only see a hash of that data (i.e., a unique string value calculated using a cryptographic function [31]). If one party loses their data, there is an opportunity for the other party to change the data if it is advantageous to them. Even if the new data do not match the hash, the one party's data are the only record left.

Permissioned blockchains share smart contracts and executable transaction protocols that self-enforce contract terms [32]. The governance of the network must be determined to allow consensus on changes to the smart contracts. With systems like Hyperledger Fabric, the smart contracts are signed but are not part of the tamper-resistant ledger [33]. A policy needs to be set for changes to smart contracts (called chaincode in Hyperledger Fabric [34]).

A permissioned blockchain architecture does not typically have the support and bandwidth of a large network of nodes. When this architecture is undertaken, all parties involved must be realistic about the cost of setting up and operating it. Each party that belongs to the network should be prepared to run at least four ledger nodes and multiple copies of other specialized nodes like Certificate Authorities (responsible for identifying and verifying the nodes involved in the network) or Orderer Nodes (responsible for generating and distributing blocks to all peer nodes [35]). Having the extra nodes can improve application performance, but the main value is the redundancy of one's own data. While an organization can, they should not depend on rebuilding their ledger nodes from channel partner data.

The blockchain does not remove human interaction. In fact, a blockchain is best served by having human governance. The people can inform each other, bring up and discuss issues, and agree on resolutions. Strong governance and cooperation between permissioned blockchain participants can provide for more secure operation and reduce the possibilities of data loss and the temptation to commit fraud.

2.1.3 Centralized/Hybrid/Decentralized Blockchain Architectures

How a solution is architected with blockchain can vary based on the requirements. The ideal requirement for using a blockchain is a decentralized resulting system. Blockchain and offshoot technologies such as DLT technologies are the only options for building decentralized systems [11]. Decentralized systems are characterized by avoiding central points of failure [36]. The only thing close to centralization would be the governance organization that sits on top of the solution. In this case, governance is where people agree on the directions that the solution will take. This could consider who gets to participate, how the rules and processes function, or who has been designated to make agreed-upon changes to the code or settings of the system. Decentralized systems do not favor one participant over another. Anyone can come and compete, realize the advantages, or influence future directions. Decentralized systems work well across organizational boundaries or for groups of people where there is a lack of trust between parties [37].

Decentralized systems are new, and therefore, have not been fully comprehended in terms of their value in business applications [15]. There are very few people who understand how to properly implement them and make a business case for them. In contrast, people are very used to working with centralized systems. Centralized systems are characterized by having one or more parties that control the experience for all parties. They may host the system, control onboarding to the system, set the terms and conditions, and be in charge of updating and setting future directions.

Despite being fully centralized, centralized systems can capitalize on some of the properties of a decentralized blockchain. Centralized systems can benefit from the fast settlement times of cryptocurrency transactions, formalization of payment rules from smart contracts, timestamping data claims, or anchoring data hashes to take advantage of tamper resistance without having to run a tamper-resistant system [23].

Solution developers who have extensive experience developing centralized systems have a hard time adapting to the limitations of decentralized systems. A natural option has been to develop hybrid systems that combine the strengths of both centralized and decentralized systems. Decentralized systems have limitations in data storage, throughput of transactions, and high transaction costs. Centralized systems have a central point of failure, lack trusted transparency, and introduce onboarding barriers. It is common for a hybrid system to store a unique identifier on the blockchain ledger and map that to an entry in a centralized database. This overcomes a lot of the storage and transaction costs that may hinder the adoption of a decentralized solution. A hybrid system can help provide a "friendlier" user interface to the blockchain. The user interface is usually a web or mobile interface hosted by the organization developing the system [38]. Hybrid systems can be seen as taking advantage of the best of both worlds or suffering from the drawbacks of both.

Vulnerabilities and Mitigations in Centralized Systems that Make Use of a Blockchain

A centralized system is most vulnerable at the point where it connects to a blockchain. It must use a Software Development Kit (SDK) and an Application Programming Interface (API) to connect to the blockchain. The SDK provides software development tools for creating applications that invoke transactions and interact with the ledger [39], while the API provides the methods and protocols for communicating between the user interface, applications, and the blockchain network [40]. The centralized system needs to connect to a network using a network node. Moreover, it must protect its private key, a unique randomly generated piece of code used for data integrity and user authentication [41]. Blockchain systems have solid security by nature of how they are designed. The interface between a centralized system and the blockchain will not have the same level of auditing and code review that the blockchain has. That makes it a more attractive target for hackers and more prone to coding or operational errors.

Centralized systems will run asynchronously to the blockchain. Transactions can be generated much faster in a centralized system that does not require consensus on each action. The centralized system needs to wait for transactions to finalize. That could mean a result of success or failure. The asynchronous mismatch could result in the blockchain and the centralized system being out of sync. The order of transactions is imperative in blockchain applications, and the system may need to transfer funds into an account before sending them out. The opposite order could cause an unexpected failure of the send transaction.

The centralized system depends on the node to which they are connecting. If the node is under control by an operator other than the centralized organization, then the node could report values that are not true to the centralized system. Consequently, this node or the node operator now becomes a point of failure that is out of the control of the centralized system.

Centralized systems must take extra precautions when using a blockchain. It needs to be considered as integrating into an external system instead of connecting to a trusted database. SDK API use should be reviewed with blockchain experts to reduce costs and vulnerabilities. The private keys need to be handled by secure key management practices. A Public Key Infrastructure (PKI) must sign transactions in a separate application than the connected application. A PKI system facilitates the generation of private and public key pairs and the secure distribution of public keys [42] (i.e., unique codes that identify the nodes in the network [43]). If the centralized application is running in the cloud, then the cloud services for managing the private key and handling transaction signing should be used.

A message queue can reduce problems with the asynchronous nature of blockchain transactions. The blockchain side of the message queue is the consumer, and the centralized application is the producer. The message queue can handle resubmissions of failed transactions and can manage the transaction numbering and ordering.

The organization that is running the centralized application should take on running its own node. This will give them a better understanding of the blockchain they are connecting to and remove a layer of vulnerability by depending on another node. If this is not feasible, an agreement to use another organization's node should be made along with availability, e.g., a service level agreement.

Vulnerabilities and Mitigation Strategies in Hybrid Blockchain Systems

Hybrid systems have vulnerabilities for the operators as well as the users. The user is not granted the same blockchain protections on their data that do not reside on the blockchain. Any data stored in a centralized system or any off-chain data processing are not tamper-evident or transparent. As a user, one cannot guarantee that the data will not be altered or deleted, and there is no way to recover it outside of the centralized system operator. Any data processing does not promise that the processing was done by code that one can inspect. There is no verifiable transparency on storage or processing, the availability of the data is not ensured, and the centralized system can be a central point of failure. Failure of that system to respond can lock the transactions one can do with the blockchain data. Lack of response can influence downstream effects in removing context data in other systems depending on the centralized system to provide the data associated with a database primary key.

A hybrid system must provide similar protections that a blockchain does. The onus is on the centralized system to provide redundancy and fail-over protections to ensure the high availability of their system. High availability will reduce downtime and knock-on effects. Any data processing by the hybrid system should accompany explanations of what the processing is doing, and what standards are involved; the code should be audited and made available for inspection in a public repository. There should be a way for users to reproduce the processing themselves to review and test the transformation that their data could undertake.

2.2 Design Decisions

When developing blockchain solutions, developers face several design decisions. The decisions are to be made around which technology to use, what the system and cloud architecture should look like, how to implement features, and what controls are in place. These decisions are made through a balance of requirements over ease of use, cost to implement or run, knowledge of the staff who will implement the solutions, and existing systems that need to integrate with the solution.

2.2.1 Application Design Decisions, Vulnerabilities, and Mitigating Strategies

When developing a blockchain solution, the application is the most visible part of the system. It is directly exposed to the internet and provides the user experience in interacting with the system. Often this is a web, mobile, or legacy application that has been developed or modified to use blockchain services.

A significant consideration for blockchain projects is what data are stored and where it is stored. Blockchains have limitations on what one can store. The data one stores are going to be copied to all nodes in the network. The data are tamper-evident, so it grows continuously. For example, the Ethereum blockchain uses gas fees to discourage application designers from storing large amounts of data on the network. Permissioned blockchains like Hyperledger Fabric do not have this limitation. However, all parties must agree on the smart contract storage ahead of time and have an idea of the storage requirements to which they have committed. These limitations drive application designers and architects to three solutions. They can keep the application to minimal storage, so it is feasible to run on the network, store the data in a decentralized data store, or use hybrid storage with identity keys on the blockchain and full storage in a centrally managed database or data store.

Traditional centralized software design looks at all the different types of people who will use the system and assigns roles accordingly. The roles provide authentication and authorization to access operations in the application. The roles are treated as groups, and users who are assigned a role become members of that group. A role-based access control (RBAC) system will often allow users to hold multiple roles [2]. The user is assigned the roles when they register and as they are granted access to more functions. The roles are usually administered by a superuser or other type of system administrator. The roles are assigned based on knowledge of who the person is and often will match their role concerning the central organization responsible for the system. This type of access management does not work well for blockchain. In public systems, anyone can join anonymously. It becomes difficult to assign meaningful roles when it cannot be determined to whom it is assigned. In a permissioned blockchain, organizations must trust other organizations to grant roles based on their agreed-upon criteria. There is no built-in way to verify. Application designers have several easy solutions to this. They can make very simple roles assigned at the time of registration and limit someone to one role. Specifically, they can use an RBAC pattern and use financial transactions to determine the role, or implement Verifiable Credentials (VC) and use external sources of truth to validate the role criteria.

Blockchain data on the network are protected and secure. However, if data refer to data stored in another place, then data are vulnerable to tampering. A common pattern is to move the storage to a distributed system like the Interplanetary File System (IPFS), a peer-to-peer distributed file system that creates a decentralized web for a faster and safer web [44]. IPFS takes the data file one is storing and calculates a hash of it. That hash is then used to retrieve the data. The data are broken into blocks, and those blocks are distributed across the system. Blocks may be duplicated in the distributed storage. When the file is retrieved from the hash, IPFS tries to find the closest copies to where the retrieval request happens to decrease the time to retrieve. The process of handling the blocks of data for distribution and retrieval falls on IPFS node software. Often an application will delegate the IPFS to a commercial node that offers pinning. That node acts as a gateway to IPFS. This is now a central point for hacking, failure, and fraud.

The hybrid system pattern for reduced blockchain storage stores a database index identifier on the blockchain that points to a database record in a centralized database. This presents a lot of vulnerabilities. The external access to the centralized database is now a centralized point of failure. All parties must trust that the record details that the identifier points have not been tampered with. If there are time-sensitive data in the record, the system becomes vulnerable to the quality of service that the system provides to the different parties accessing it. It is common to have duplicate or predictable information in database records. Suppose a hashing technique is used to try to prove the contents of a record. In that case, a malicious database operator could precalculate alternate record data that will still match the hash of the given database record. They could choose the more favorable alternative and still be compliant with the security rules. Databases can become corrupt, go offline, be deleted, suffer Structured Query Language (SQL) injection, be subject to distributed denial-of-service attack (DDoS), or any other techniques to poison, deny, or delay the data from the off-chain source. An SQL injection is an attack used in the SQL where malicious SQL code is injected or weaponized to seize and capture complete control of the application database [45, 46]. A DDoS attack targets the availability of its targeted victim by exhausting the network's resources and communication [13, 47].

In a hybrid or decentralized system, a role-based approach to access control is vulnerable to a variety of attacks. If the role represents a skill, identity, or resource, the system needs to verify the claim that the user who is granted that role has made to acquire the role. This claim will often be based on information that is off chain. A fraudulent claim or a fraudulent verification of that claim reduces the security of the RBAC. If a claim for role granting is verified and accurate, it is only accurate when the claim is verified. If a person can prove their legal standing to practice law in the required jurisdiction, that claim could become invalid the next day if their license to practice law is revoked. There is no way that the blockchain system could re-validate the claim, nor would there be a reasonable process to notify the revocation status. The same could be true for claims that have an expiry date.

Applications that make extensive and sensitive use of a decentralized or distributed data store need to include the node ability in the application itself. Relying on a third party for sensitive data is a dangerous vulnerability.

Hybrid storage requires a robust mechanism to prove that the contents retrieved from the centralized database represent what was intended to be stored. A hash of the data to be stored can be placed on the blockchain along with the identifier. That will allow a party to verify that the data that have been retrieved has not been modified. To strengthen that assurance, a nonce or random seed value can be included that adds some extra entropy to the hash. This prevents situations where alternate record contents can be generated in advance using brute force techniques that try all possibilities in the search space [48]. If duplicate record contents are stored, the nonce will ensure that the hash will be different. If the nonce is random or at least variant, then the search for alternate data contents that produce the same hash becomes extremely challenging, as all nonce values need to be taken into account.

Any hybrid storage system that uses a centralized database to pair with a blockchain should be created as a database cluster to match the high availability and redundancy of the blockchain. This will reduce the possibility of data loss, stave off the worst effects of a DDoS attack at the database level, and increase the availability due to increased system uptime.

RBAC systems are a mismatch with blockchain systems if the role is a proxy for an off-chain authority. A better option would be to use a VC and have credential-based access control with privacy-preserving and cryptographically verified authorship [49]. VCs are a World Wide Web Consortium (W3C) standard that uses signed data to prove a third-party claim about the subject of the credential [50]. For example, the VC could be a claim from the organization that governs who can practice law in a jurisdiction where the holder of the VC is currently a valid lawyer. The power of the VC is that it can be used to machine verify the claim, and it can do it any time it needs to prove it. VCs can also come with an expiry date and a revocation registry. Any time the VC is confirmed, it can see if the claim as made was valid and is still valid.

2.2.2 Cryptographic Key Management, Vulnerabilities, and Mitigations

By its very nature, decentralized technology requires the security of the private keys to be controlled at the network edges. The private key is used to prove that the key holder controls the account they are making transactions on [42]. It is used to provide cryptographic proof that the commands originated from that account, and it allows for complete transparency and verifiability on all transactions from that account.

The onus is on the account holder to protect their private key. There is no centralized service in a blockchain network to hold one's private key.

From a software design perspective, putting the onus of key management on the end-user may not be desired. The software requirements may be for ease of use and to avoid key loss. There are two ways to handle this: a custodial model, where the software application holds the private key for the customer, and a non-custodial model, where the application provides the user with the ability to backup and restore the keys.

Custodial key management is one of the most significant vulnerabilities in blockchain applications. Some of the largest thefts of cryptocurrency have come from crypto exchanges that held the private keys for their users. The users traded ease of use for reduced security and lost control of their tokens. Instead of the blockchain protecting the account access, the security moved to the centralized application holding the users' keys [51]. A system that contains multiple keys that represent millions of dollars is a very obvious target for hackers. A centralized application would not have the same scrutiny as an open-source project like a blockchain. It is also open to manipulation by the developers and operators of the system themselves. Relinquishing control of a private key to any other party is a huge risk.

Private keys should never leave the system in which they were generated. If the key has to move, it must be encrypted, then transferred, and then decrypted. Private keys should not be kept open in internet-addressable applications. A better solution would be to run a PKI system where the application can pass transaction data to be signed by the PKI system and passed back to the application. The application can then pass that signed transaction to the blockchain. The signed data prove that the application has control over the data without holding a copy of the private key.

2.2.3   Smart Contract Design Decisions, Vulnerabilities, and Mitigations

The discipline of smart contract engineering is still new. Decentralization and tamper resistance put many restrictions on smart contract capabilities. There are limits to the size, storage, and execution cycles imposed by the requirement to operate the smart contracts on multiple nodes and share the execution resources with all other smart contracts. Once a smart contract is deployed, there are issues with updating them when stored in a tamper-resistant manner.

In recent years, software developers have had the luxury of being able to write web applications. With these applications, updates can be pushed to the server, and the software user does not have to participate in any way with the upgrade. It is a very centralized way of updating the software that is very convenient to the developers. With smart contracts on public blockchains, one cannot change it once they deploy the smart contract. Instead, one would have to deploy a new contract and transfer all the data across from the old contract to the new one at great cost and an unpredictable length of time. It may not even be possible, based on the data structures that have been used. The onus on the software developer is to get it right the first time. There might also be a case where the requirements change. Fixing bugs and changing features is commonplace in centralized software development.

A possible way around tamper-resistant contracts is to build upgradeable smart contracts. These are based on the ability of smart contracts to call other smart contracts. The address of the contract to be called can be updated. This allows a stub contract to use a static contract address, and the contract to be called from the stub can change. The data storage structures would be in the stub contract.

Smart contracts can handle large amounts of value either stored in the contract or transferred in and out. A common requirement is to be able to halt activity when a flaw or suspicious activity is detected. Some common smart contract design patterns can allow for pausing or halting a smart contract.

A common use for smart contracts is to create a token, which is a digital asset issued on a blockchain [52]. A very common standard for tokens is the ERC20 [53] standard developed by the Ethereum blockchain community. The initial concept of a token was to have a fixed supply of tokens. This simplifies the tokenomics used to understand the value [54]. Some use cases require the total supply of tokens to increase and decrease. Patterns were developed for minting and burning tokens. For example, a fiat currency or commodity-backed derivative token may require minting and burning based on the underlying asset deposited or withdrawn.

A common pattern is to have a smart contract to make a call to another smart contract. An example might be when a person uses a smart contract requiring a single account to provide payment. There might be a special case where the payment transfer requires splitting the payments to multiple parties. If the original contract does not support split payments, then a split payment contract can be used in place of the account. In most blockchains, smart contracts and accounts are interchangeable.

The trust in blockchain data is only extended to data that were created and processed on the blockchain. However, there are many use cases where external data are required to make decisions or provide supporting documentation. For example, a supply chain smart contract that deals in perishable goods may need to track the temperature from a weather service. Alternatively, perhaps, a derivative smart contract would need snapshot information on market values for calculating valuations. In practice, these values can be drawn from an Oracle. An Oracle is a smart contract that is updated by a trusted third party that updates the data on a regular basis [55].

Decentralized programming imposes some restrictions on the data structures that can be used in smart contract software development. The use of unbounded arrays, large strings, and large binary objects that a person would use in other languages is very limited in a smart contract. Since the smart contract must be executed across every node in the network participating in the network consensus, the virtual machine executing the smart contract must limit the storage and number of instructions that can be executed in one call. This means that arrays need to be bound to a determined size to ensure no iterations across the array exceed the instruction limit. Mappings, too, are limited because they do not allow one to iterate across the key pairs. This introduces a change in mindset for centralized system developers who are used to having databases that they can access. Large strings and byte arrays are also discouraged. In particular, it is necessary to limit the size of strings that can be passed into an application. Even operations like a string comparison between two long strings can eat up precious instruction executions.

The upper bound to the number of instructions allowed to be executed shows up in Ethereum as a gas limit. A maximum number of instructions, that an individual can pay for, limits the scope of an application to one that can be executed in a shared limited resource environment. Smart contracts need to be small, have deterministic execution paths, and avoid patterns like iterators and recursion.

The practice of building an "upgradeable" smart contract introduces vulnerabilities to the owner of the contract and those that use it. For the smart contract user, there is a huge gap in the potential to trust the smart contract. If the contract can change, then the underlying promises of the contract can change. Trusting one's data and money in a contract that can change at any point is imprudent. The contract could be altered by the owner, someone who has gained control of the owner's private key, or a party that has found a vulnerability in the contract. An upgradeable smart contract involves a contract that calls another contract. This level of complexity introduces a variety of attack vectors (i.e., paths or means through which attackers gain unauthorized access to the system [56]) that can affect execution to which singular smart contracts may not be vulnerable. An upgradeable smart contract can be called from another smart contract that can manipulate the amount of gas available or the size of the call stack. These changes can force the contract to fail at an inopportune time. If the calling contract thinks that it is complete after passing to the upgradable contract, then the saved state data are vulnerable. A smart contract could be constructed to appear like the calling contract and trick the upgradable contract into using bad data or logic. The sheer complexity of having a mechanism where a contract can call another and have that other contract change introduces a lot of unintended vulnerabilities.

Smart contracts that have the ability for the owner to pause, halt, or stop the transferability of the contract make the user of the contract vulnerable to the owner's whims. Often these abilities are built into token smart contracts. This means that when the owner locks the contract, the value held in all of the tokens represented by that contract is locked until the owner deems it appropriate to unlock it. The token holders will not be able to transfer, trade, or exchange their tokens. New buyers will not be able to purchase the token, and editing holders will not be able to take advantage of the special features of the token. By removing all utility for the token, the value and liquidity of the token has been removed. If the halt on the token is too long or not justified, the token's value may drop to zero, wiping out all value in the token.

The ability to mint and change the supply of tokens leaves the token economy in an unpredictable state. The decision to mint or burn tokens rarely is decided by on-chain data. It is often due to external actions that may not be verifiable. There have been cases of stable coins that are supposed to mint and burn based on US dollar holdings [57], where the company behind the token has refused to release audit data that reconciles with the minting and burning activity. At this point, the trust in the token is erased and drops to less than the ensured value of the underlying asset.

On-chain data, and in particular data created due to auditable transactions, can be trusted. Data that come from other sources needs to be held with a lower level of assurance of accuracy. This applies to the data that come from an Oracle. Oracle data are vulnerable to any security issue or software flaw. Oracle data could also be maliciously manipulated to change the outcome of a smart contract transaction in favor of the malicious party.

Any smart contract that tries to use structures like large strings, unbounded arrays, or recursion has the potential to lock any value stored in the contract permanently. If the code path to transfer value out of the contract hits an execution gas limit, the contract will never be complete. It will never let any value held in the contract be transferred out. This may not show up in testing but may occur over time as data are added to the structures.

Smart contracts should never be designed to be "upgradeable"; instead, the contract should either be designed correctly, to begin with, or designed such that it can be retired and a new version deployed. A transfer strategy needs to be designed into the original contract so that existing data can be transferred without having to leave the blockchain when it is retired. Only smart contracts that use public variables can be migrated without using off-chain data. Furthermore, even with that restriction, mapping data cannot be recovered on-chain. The only way to manage that is to have recorded the data using emit notifications or going back through the ledger to replay all of the function calls to determine the state of the mapping. This is not a trustable exercise.

The value of a token must consider the ability to halt, mint, or burn tokens. These are actions that individual token holders cannot control. They are at the whim of the token contract owner. Adjust the calculated value based on the trust or reputation of the party in control of the private key of the owner account.

Smart contracts that depend on Oracle values should have delays on them that allow for corrections of data in case of an error in the data passed into the Oracle. There should be a voting or multiple signature governance method (i.e., distributing approval over multiple independent parties [58]) built into the smart contract to address situations where the Oracle data are demonstrably wrong.

A third party should review all smart contracts. If there is a significant value held or controlled by the smart contract, it should be audited by a third party that can grant assurance of the contract's validity and who holds insurance in case of malpractice.

2.2.4 Network Design Decisions, Vulnerabilities, and Mitigating Strategies

For an application to make a call to a smart contract, it needs to have an account on the blockchain and access to a node on the blockchain. Smart contract function calls on Ethereum are recorded just like transfer transactions. The gas fee and limit are set, the "transfer to" address is the smart contract address, and the extra data are stored by the function call and parameters. This special case of a transaction is then passed to a node on the network for processing. At this point, the application developers need to choose how the application will connect to the network to send transactions, check on the finalization, listen for events, and read state and network data. A mobile device will not be able to have its own network node. It will have to depend on a node in the cloud as an access point. A web application typically uses a node as an access point. The access point is now the application's only connection to the source of truth, the blockchain. The most popular access point for the Ethereum blockchain is Infura [59]. Infura provides free developer access and fee-based access for developers where it is cheaper to use Infura than the cost of running their own node. Most node operators limit who can access their nodes for read and write operations.

The consensus mechanism for a blockchain can be based on a majority of the nodes agreeing on the true state of the data in a block [60]. This could be all nodes, a select few authoritative nodes, or a random selection of nodes. This mechanism is used to remove malicious nodes from participating in the network. Anyone that wants to change the state values from what was submitted in signed transactions to something else will require a majority of the nodes in the network to agree with them. A healthy blockchain should have a wide distribution of entities that control the nodes.

Infura.io is now a point of centralization for many otherwise decentralized applications (DApps). If Infura is down, then a large number of DApps hosted by the peer-to-peer network [61] are taken down with it. A blockchain uses the wide distribution of control of nodes to ensure that the transactions are correctly recorded on the ledger. By using a service like Infura, an application can no longer be assured that the state data are protected. Infura can act like a 51% attacker. It can report erroneous data to the application, and the application cannot detect the attack.

Any application that depends on node access should run its own node and use a combination of commercial third-party services or have reciprocal agreements in place with other applications that run their own nodes. Having multiple nodes reduces the opportunity for a single node to falsify data, and it provides a way for an application to verify across other nodes. All nodes should be reporting exactly the same data. Having multiple nodes drops the centralization problem of lowered uptime.

2.2.5 State Data Design Decisions, Vulnerabilities, and Mitigations

The ledger of a blockchain is intended to be permanent. A copy of all data is available to anyone who wants to read the data. Copies are agreed upon and maintained across all participating nodes. An individual can change a value in a smart contract, but the history of the original value stays on the ledger. The information of what account set the original value, what that value was, who changed it, and what they changed it to is permanent. There are no deleting values from the blockchain or forgetting data.

There are business use cases that require a random number. It could be for generating unique identifiers, probabilistic prediction models, or games. A true random number generator uses entropy from a non-computer source to generate a random number. A blockchain cannot access external systems. Blockchains are limited to pseudo-random number generators. Pseudo-random number generators need a seed to help generate an unpredictable random number [62]. Often the hash of a block or transaction is used as the seed.

The consensus process of blockchains is complex and is handled by nodes. Most nodes are now run as a revenue stream by organizations colloquially referred to as miners [63]. In a Proof of Work consensus mechanism, miners set the order of transactions in a block. They must solve a problem to compete for a reward of a new token and/or the transaction fees included in the blocks' transactions [64].

Smart contracts are often treated the same way as accounts. For example, in the Ethereum blockchain, a smart contract can act as an account. It can receive and transfer ETH or tokens. When executing a send command, a person may be calling another smart contract.

Centralized programmers are trained to write large and complicated programs. Small and simple programs are dismissed as being trivial and not worthy of being developed. Smart contract development requires that the programs be as simple as possible, use as few instructions as possible, and be highly predictable in situations where the entity that called the contract does not control the environment in which the smart contract is executed. Smart contracts can help transfer large amounts of value from one account to another.

There are no sources of random numbers in a blockchain. Some numbers look like good candidates, like block times (i.e., the time to generate and add a new block to the blockchain [65]) or transaction hashes. Still, miners can manipulate these to select favorable outcomes from the random number generator for which they provide a seed. As a simple example, if a smart contract pays out based on a 50/50 coin flip calculation, a miner can change the order of transactions or delay the write time to garner a favorable result. This may only happen if the miner has won the ability to mine the block. However, that makes it all the harder to detect the vulnerability.

Applications should take into account that the order of the transactions coming into them or generated by them to another contract is not entirely under their control. Miners can manipulate first-come-first-served contract rewards for gain.

There are subtle differences in commands like send, call, and transfer. Each has a different use and behavior. The send, call, and transfer commands differ in how gas is handled and their behavior on encountering errors. The approve and transferFrom commands are used in cases where a third party is granted access to transfer the funds of an account. It allows a token holder to approve another account making a transfer from the holder's account.

If a smart contract has more than 300 lines of original code, it should be considered suspect and open for review. The industry average for code defects is between 15 and 50 errors for 1000 lines of code.

Transaction order dependencies should be removed or cause a failure of the contract if they are detected as out of order. Smart contracts should always check if things like transfer amounts are higher than account balances. For example, if the order of transactions where the deposit from one party arrives after a withdrawal from another, that may trigger an alert and/or contract failure.

Solidity developers need to be made aware of the implications of every built-in function available to them. They need to keep up to date on the latest exploits and make a practice of offering and accepting peer reviews of contracts and sending contracts for evaluation and auditing.

The send and transfer functions will automatically pass on a fixed amount of gas at 2300 [66]. This may not be enough to handle if the recipient is a smart contract like a payment split contract.

Smart contracts should be small, easy to understand, and only offer a few features. Wherever possible, existing vetted contracts like those from the OpenZepplin project [67] should be used as a starting point. Smaller contracts are safer contracts.

2.2.6 Human Vector—Social Engineering, Phishing, and Ransomware

Just like any other system, blockchain applications are open to social engineering hacks. Wherever there are user interfaces or humans making decisions about actions to take, there is the possibility of problems [68].

Blockchains push the management of private keys to the edge devices that provide entry points to the system. This puts the users in charge of their own security, backup, and transaction approvals. Corporate blockchain systems require users to transfer amounts to accounts. Programmers need to provide access to private keys to the programs that use them. These are all problem areas.

An important part of any defense-in-depth system is to provide cybersecurity training to the users of the system. They need to be reminded of how to identify phishing and spear-phishing attacks [69, 70]. Phishing is a form of cyberattack that attempts to steal sensitive information by imitating websites and legitimate emails to fool users into providing their confidential information. Spear-phishing is a type of phishing consisting of cleverly crafted fraudulent emails containing malware disguised as hyperlinks and attachments to deceive users into revealing their information. Corporations can use various techniques to reduce their employees' vulnerability to phishing [71]. At a minimum, the users require training on the accepted protocols of how the systems are managed not to allow anyone to override those without explicit authorization from management. End-users require tools to help them properly guard the backup of private keys. Hardware wallets, seed phrases (i.e., series of words that store the information needed to access the wallet [72]), or social key recovery methods (i.e., backing up the key by splitting it among one's social cycle [73]) help users from losing their private keys. Edge devices should be protected by biometric security or at least a limited access pin code. Lost or stolen devices should be considered compromised. Accounts should be moved, and parties notified just like if a wallet was stolen with credit cards in it.

Programmers need to be alert to security problems they may be introducing. Private keys should never be placed in code; instead, they should be held in an external environment or key file. Better yet, they should be held in a PKI system that is not internet accessible that can provide transaction signing services. Most of the cloud providers provide application-secret services or transaction signing services.

An application in the cloud should not have an easily accessible private key and account that a hacker could access.

3 Future Threats

The previous section dealt with existing threats to blockchain applications. These applications that are currently being written may be in place for several decades. The tamper-evident record may be accessible for even longer. Here are some future threats and possible ways to mitigate them.

3.1 Quantum-Based Attacks

A very real concern is that quantum computing may break the encryption techniques currently used to create the public/private key pairs that are the basis for blockchain trust systems [74]. A quantum computer uses quantum mechanics and, consequently, functions differently from classical computers, giving them advantages in solving certain types of problems. Early models of quantum computers currently exist, but they are not yet powerful enough to break any cryptographic technique. When a powerful-enough quantum computer is made, it will give a quadratic speed up against asymmetric cryptography such as Elliptic Curve Cryptography and Rivest-Shamir-Adleman [75]. While there are different estimates as to when a cryptographically-relevant quantum computer can be expected, most experts agree that one will be engineered in a decade with a non-negligible probability [76].

Currently, it would take vast amounts of computing power and a long time to work backward from a sufficiently large public key to find the private key. The nature of our current generation's architecture and foreseeable generations of computers ensures that these accounts are safe. A capable quantum computer can reverse a public key to get the associated private key. Therefore, any account currently protected by private keys is no longer in control just by the original holder of that key [77].

The best way to mitigate this is to move to quantum-resistant alternatives [78]. Although there are several good cryptographic candidates, they have not been standardized yet [79]. Once there is a better understanding of these new cryptographic techniques, mass migration can take place. The intention is for these migrations to take place before they are required as a result of a powerful-enough quantum computer.

It is also important to note that encrypted data stored on blockchains will not always stay encrypted. If that data hold any secrets that need to stand over a long period of time, then eventually, they will be broken. For that reason, it is never a good practice to store personally identifiable information on a blockchain, even if it is encrypted.

3.2 Forking

From time to time, there are severe problems found in the security of a blockchain. Sometimes the only solution to this is to fork the blockchain from when the flaw was discovered. In the transition from the original to the forked version, any transactions after the fork that are made on the original blockchain are lost in the fork. This is hugely disruptive and can put the value transfers from flaw discovery to fork in contention. The financial implications of a fork are serious enough that, in some cases, the old version continues to run and is renamed. Several blockchains split this way. For example, Ethereum and Ethereum Classic, as well as Bitcoin and Bitcoin Gold, are the result of forks [80].

Forking takes place by having the miners of the blockchain all simultaneously adopt a new protocol. A good way to stay alert to this type of change is to run a node and stay up to date on notifications of proposed changes [81].

3.3 Interoperability

As more decentralized systems are created on blockchains, new solutions will become available that will cater to one type of blockchain over another. Those systems may still find it advantageous to interact with each other. Perhaps, a permissioned blockchain insurance system wants to make payouts using a smart contract on a public blockchain. There needs to be a way for these systems to interoperate. In the sections above, off-chain and external data were identified as a suspect and an area for vulnerabilities. Interoperating between blockchains is a perilous proposition. Recently, a smart contract that mapped the Ethereum blockchain and the Poly Network, a cryptocurrency [82], was shown to have a vulnerability, and $600 million was stolen [83].

There is work on new types of distributed ledger protocols that will offer greater protections and a path for maintaining trust in data across systems. Currently, the Key Event Receipt Infrastructure (KERI) protocol is being developed. It contains concepts like built-in key rotation and single-party ledgers that may offer a path to secure blockchain interoperability [84].

3.4 Consensus Flaws in New Methods

The current Proof of Work consensus mechanism used in both Bitcoin and Ethereum is very energy-intensive and a slow mechanism for maintaining state data. There are other approaches like Proof of Stake and Proof of Authority to overcome these challenges. There are also new approaches like Proof of Space and Time, Proof of Accuracy, Proof of Benefit, and other alternative methods [85]. As a secure consensus mechanism, Proof of Work has never been broken in public implementations. This may not hold for some of these different proposed strategies.

A mitigation strategy is to weigh the return of being an early adopter to a blockchain using a new consensus mechanism against the complete loss of control over an individual's data or the value they have entrusted to the system. A risk assessment is required.

3.5 Collision Existence

Like asymmetrical encryption, the algorithms currently used for blockchain and decentralized storage could impose security vulnerabilities. Hashing algorithms are considered one-way algorithms. It is infeasible to go from the hash back to the original file that was hashed [42]. The flaw with a hash is that it is not assured that the hash that has been created for a file is unique and not actually a duplicate of a hash for another file. This is a hash collision. Adding more bits to the hashing algorithm lowers the probability of a collision. With the current hash algorithms, systems can be safe to assume that the hash is unique.

Deliberately finding an alternative version of a file that still follows a reasonable file schema resulting in a collision is currently nearly impossible. However, a mathematical method may be discovered that allows that calculation to fall within reasonable computer power and time limits. If that happens, then data protection schemes used in permissioned blockchain applications for private data, hybrid systems that store a hash and database identifier, and timestamping basic blockchain applications may no longer be trusted.

Data migration to a new hashing algorithm will be required. Documentation may make the difference in areas of contention between two parties contending over the original data represented by a hash. A good mitigation strategy for this type of possible scenario is keeping records of data off the chain and not trusting all data retention to the blockchain.

4 Conclusion

Cybersecurity vulnerabilities threaten all information systems alike, and blockchain-based solutions are not immune. A blockchain-based solution is a system composed of several blockchain-specific components such as blockchain protocols, network communications, smart contract virtual machines, and account structures. These blockchain components are strong and secure, with very few problems with these components of a blockchain system. However, blockchain systems are still required to be hosted in a variety of environments, access to the blockchain nodes may be controlled by other parties, use case realities may require deviation from pure blockchain implementations, and human interactions are required. All of these factors open up a blockchain system to possible attacks or disruptions. This chapter discussed some of the most common implementation weaknesses resulting in such attacks and disruptions, as well as mitigating strategies against them.

When building a blockchain-based solution, there are many choices to make. The design decisions outlined in this chapter are some of the most common areas to look out for the introduction of vulnerabilities. Still, they do not comprise an exhaustive list of all areas of concern. These are a new type of system, and assumptions should always be challenged. There is a constant stream of new information about vulnerabilities in these systems that should be taken into consideration when developing a new one or maintaining an existing system.

One assuring point from this chapter is that for each vulnerability identified, there is a mitigation strategy. There are ways to reduce the risk in the architectural and design decisions required to be made based on system requirements and limitations. As long as the system developers and operators are aware of the possible areas of vulnerability, the issues can be addressed.

A good defense-in-depth strategy can be used to ensure the confidentiality, integrity, and availability of a blockchain system. A mindset which always assumes that external systems and actors are capable of malicious activity is a good starting point in defending against attacks for any solution, not just blockchain-based ones.

References

1. Guttman B, Roback EA (1995) An introduction to computer security: the NIST handbook. Diane Publishing. https://doi.org/10.6028/NIST.SP.800-12r1
2. Stallings W (2018) Effective cybersecurity: a guide to using best practices and standards. Addison-Wesley Professional. https://www.pearson.com/us/higher-education/program/Stallings-Effective-Cybersecurity-A-Guide-to-Using-Best-Practices-and-Standards/PGM1835803.html
3. Mosteiro-Sanchez A, Barcelo M, Astorga J, Urbieta A (2020) Securing IIoT using defence-in-depth: towards an end-to-end secure industry 4.0. J Manuf Syst 57:367–378. https://doi.org/10.1016/j.jmsy.2020.10.011
4. Lesavre L, Varin P, Mell P, Davidson M, Shook J (2019) A taxonomic approach to understanding emerging blockchain identity management systems. National Institute of Standards and Technology. White Paper. https://doi.org/10.6028/NIST.CWSP.01142020
5. Green JS, Daniels S (2019) Digital governance: leading and thriving in a world of fast-changing technologies. Routledge, London, UK. https://doi.org/10.4324/9780429022371
6. Brühl V (2017) Bitcoins, blockchain und distributed ledgers. Wirtschaftsdienst 97(2):135–142. https://doi.org/10.1007/s10273-017-2096-3
7. Henninger A, Mashatan A (2021) Distributed interoperable records: The key to better supply chain management. Computers 10(7):89. https://doi.org/10.3390/computers10070089
8. Yaga DJ, Mell PM, Roby N, Scarfone K (2018) Blockchain technology overview. In: National Institute of Standards and Technology, Gaithersburg, MD, USA, Technical Report 8202. https://doi.org/10.6028/NIST.IR.8202
9. Mashatan A, Lemieux V, Lee SHM, Szufel P, Roberts Z (2021) Usurping double-ending fraud in real estate transactions via blockchain technology. J Database Manag 32(2):27–78. https://doi.org/10.4018/JDM.2021010102
10. Nakamoto S (2008) Bitcoin: a peer-to-peer electronic cash system. https://nakamotoinstitute.org/bitcoin/ or https://doi.org/10.2139/ssrn.3440802

11. Farouk A, Alahmadi A, Ghose S, Mashatan A (2020) Blockchain platform for industrial healthcare: Vision and future opportunities. Comput Commun 154:223–235. https://doi.org/10.1016/j.comcom.2020.02.058
12. Li X, Jiang P, Chen T, Luo X, Wen Q (2020) A survey on the security of blockchain systems. Future Gener Comput Syst 107:841–853. https://doi.org/10.1016/j.future.2017.08.020
13. Stallings W, Brown L, Bauer MD, Bhattacharjee AK (2012) Computer security: principles and practice. Pearson Education, Upper Saddle River, NJ, USA. https://doi.org/10.5555/2685921
14. Sahinoglu M (2005) Security meter: A practical decision-tree model to quantify risk. IEEE Secur Priv 3(3):18–24. https://doi.org/10.1109/MSP.2005.81
15. Demir M, Turetken O, Mashatan A (2020) An enterprise transformation guide for the inevitable blockchain disruption. Computer 53(6):34–43. https://doi.org/10.1109/MC.2019.2956927
16. Park J, Gabbard JL (2018) Factors that affect scientists' knowledge sharing behavior in health and life sciences research communities: differences between explicit and implicit knowledge. Comput Hum Behav 78:326–335. https://doi.org/10.1016/j.chb.2017.09.017
17. Charles WM (2021) Accelerating life sciences research with blockchain. In: Applications of blockchain in healthcare. Springer, Singapore, pp 221–252. https://doi.org/10.1007/978-981-15-9547-9_9
18. Manion ST, Bizouati-Kennedy Y (2020) Blockchain for medical research: accelerating trust in healthcare. Productivity Press. https://doi.org/10.4324/9780429327735
19. "The project," PharmaLedger. https://pharmaledger.eu/about-us/the-project/. Accessed 14 Nov 2021
20. "Ethereum-based solutions for healthcare & life sciences," ConsenSys Health. https://consensyshealth.com/. Accessed 14 Nov 2021
21. Morkunas VJ, Paschen J, Boon E (2019) How blockchain technologies impact your business model. Bus Horiz 62(3):295–306. https://doi.org/10.1016/j.bushor.2019.01.009
22. Xu X, Weber I, Staples M, Zhu L, Bosch J, Bass L, Pautasso C, Rimba P (2017) A taxonomy of blockchain-based systems for architecture design. In: 2017 IEEE international conference on software architecture (ICSA), IEEE, pp 243–252. https://doi.org/10.1109/ICSA.2017.33
23. Ruoti S, Kaiser B, Yerukhimovich A, Clark J, Cunningham R (2019) Blockchain technology: What is it good for? Commun ACM 63(1):46–53. https://doi.org/10.1145/3369752
24. Ncube T, Dlodlo N, Terzoli A (2020) Private blockchain networks: a solution for data privacy. In: 2nd international multidisciplinary information technology and engineering conference (IMITEC). IEEE, pp 1–8. https://doi.org/10.1109/IMITEC50163.2020.9334132
25. Li Z, Hou J, Wang H, Wang C, Kang C, Fu P (2019) Ethereum behavior analysis with NetFlow data. In: 2019 20th Asia-Pacific network operations and management symposium (APNOMS), IEEE, pp 1–6. https://doi.org/10.23919/apnoms.2019.8893121
26. Hwang GH, Chen PH, Lu CH, Chiu C, Lin HC, Jheng AJ (2018) InfiniteChain: a multi-chain architecture with distributed auditing of sidechains for public blockchains. In: International conference on blockchain. Springer, pp 47–60. https://doi.org/10.1007/978-3-319-94478-4_4
27. Zhang P, Zhou M (2020) Security and trust in blockchains: Architecture, key technologies, and open issues. IEEE Trans Comput Soc Syst 7(3):790–801. https://doi.org/10.1109/tcss.2020.2990103
28. Pimentel E, Boulianne E, Eskandari S, Clark J (2021) Systemizing the challenges of auditing blockchain-based assets. J Inf Syst 35(2):61–75. https://doi.org/10.2308/ISYS-19-007
29. Liu B, Qin Y, Chu X (2019) Reducing forks in the blockchain via probabilistic verification. In: 2019 IEEE 35th international conference on data engineering workshops (ICDEW), IEEE, pp 13–18. https://doi.org/10.1109/ICDEW.2019.00-42
30. Schneier B (2019) There's no good reason to trust blockchain technology. Wired Mag. https://www.wired.com/story/theres-no-good-reason-to-trust-blockchain-technology/.
31. Antonopoulos AM (2014) Mastering Bitcoin: Unlocking digital cryptocurrencies. O'Reilly Media Inc. https://www.oreilly.com/library/view/mastering-bitcoin/9781491902639/
32. Yu XL, Al-Bataineh O, Lo D, Roychoudhury A (2020) Smart contract repair. ACM Trans Softw Eng Methodol 29(4):1–32. https://doi.org/10.1145/3402450

33. Aswin AV, Kuriakose B (2019) An analogical study of Hyperledger Fabric and Ethereum. In: Intelligent communication technologies and virtual mobile networks, Springer, pp 412–420. https://doi.org/10.1007/978-3-030-28364-3_41

34. Liang X, Zhao J, Shetty S, Liu J, Li D (2017) Integrating blockchain for data sharing and collaboration in mobile healthcare applications. In: IEEE 28th annual international symposium on personal, indoor, and mobile radio communications (PIMRC). IEEE, pp 1–5. https://doi.org/10.1109/PIMRC.2017.8292361

35. Hyperledger, "Hyperledger Fabric docs documentation, release master," 2021. https://hyperledger-fabric.readthedocs.io/_/downloads/en/release-2.0/pdf/.

36. Rathore S, Kwon BW, Park JH (2019) BlockSecIoTNet: Blockchain-based decentralized security architecture for IoT network. J Netw Comput Appl 143:167–177. https://doi.org/10.1016/j.jnca.2019.06.019

37. Beck R (2018) Beyond Bitcoin: The rise of blockchain world. Computer 51(2):54–58. https://doi.org/10.1109/MC.2018.1451660

38. Zikratov I, Kuzmin A, Akimenko V, Niculichev V, Yalansky L (2017) Ensuring data integrity using blockchain technology. In: 2017 20th conference of open innovations association (FRUCT), IEEE, pp 534–539. https://doi.org/10.23919/FRUCT.2017.8071359

39. Hyperledger, "Fabric SDK for node.js," 2018. https://fabric-sdk-node.github.io/index.html.

40. Mackey TK, Miyachi K, Fung D, Qian S, Short J (2020) Combating health care fraud and abuse: Conceptualization and prototyping study of a blockchain antifraud framework. J Medical Internet Res 22(9):e18623. https://doi.org/10.2196/18623

41. Jesus EF, Chicarino VR, De Albuquerque CV, Rocha AADA (2018) A survey of how to use blockchain to secure internet of things and the stalker attack. Secur Commun Netw 2018, Art. no. 9675050. https://doi.org/10.1155/2018/9675050

42. Stinson DR (2005) Cryptography: theory and practice. Chapman and Hall/CRC. https://doi.org/10.1201/9781420057133

43. Kaushik A, Choudhary A, Ektare C, Thomas D, Akram S (2017) Blockchain—literature survey. In: 2017 2nd IEEE international conference on recent trends in eElectronics, information & communication technology (RTEICT), IEEE, pp 2145–2148. https://doi.org/10.1109/RTEICT.2017.8256979

44. Nyaletey E, Parizi RM, Zhang Q, Choo KKR (2019) BlockIPFS-blockchain-enabled interplanetary file system for forensic and trusted data traceability. In: 2019 IEEE international conference on blockchain (Blockchain), IEEE, pp 18–25. https://doi.org/10.1016/j.bcra.2021.100032

45. Su G, Wang F, Li Q (2018) Research on SQL injection vulnerability attack model. In: 2018 5th IEEE international conference on cloud computing and intelligence systems (CCIS), IEEE, pp 217–221. https://doi.org/10.1109/CCIS.2018.8691148

46. Jang YS (2020) Detection of SQL injection vulnerability in embedded SQL. IEICE Trans Inf Syst 103(5):1173–1176. https://doi.org/10.1587/transinf.2019EDL8143

47. Martinasek Z (2015) Scalable DDoS mitigation system for data centers. Adv Electr Electron Eng 13(4):325–330. https://doi.org/10.15598/aeee.v13i4.1531

48. Juels A, Ristenpart T (2014) Honey encryption: security beyond the brute-force bound. In: Annual international conference on the theory and applications of cryptographic techniques, Berlin, Springer, Heidelberg, pp 293–310. https://doi.org/10.1007/978-3-642-55220-5_17

49. Alzahrani B (2020) An information-centric networking based registry for decentralized identifiers and verifiable credentials. IEEE Access 8:137198–137208. https://doi.org/10.1109/access.2020.3011656

50. World Wide Web Consortium (2019) Verifiable credentials data model 1.0: expressing verifiable information on the web. https://www.w3.org/TR/vc-data-model/?#core-data-model.

51. Bruschi F, Tumiati M, Rana V, Bianchi M, Sciuto D (2020) A decentralized system for fair token distribution and seamless users onboarding. In IEEE symposium on computers and communications (ISCC). IEEE, pp 1–6. https://doi.org/10.1109/ISCC50000.2020.9219642

52. Wieninger S, Schuh G, Fischer V (2019) Development of a blockchain taxonomy. In: 2019 IEEE international conference on engineering, technology and innovation (ICE/ITMC), IEEE, pp 1–9. https://doi.org/10.1109/ICE.2019.8792659

53. "EIP 20: ERC-20 Token standard," Ethereum improvement proposals. https://eips.ethereum.org/EIPS/eip-20. Accessed 14 Nov 2021
54. Au S, Power T (2018) Tokenomics: The crypto shift of blockchains, ICOs, and tokens. Packt Publishing Ltd. https://doi.org/10.5555/3306877
55. Alruwaili A, Kruger D (2020) Hybrid-trusted party contract agrees on clients input. In: IEEE 3rd 5G World Forum (5GWF). IEEE, pp 127–132. https://doi.org/10.1109/5GWF49715.2020. 9221388
56. Simmons C, Ellis C, Shiva S, Dasgupta D, Wu Q (2014) AVOIDIT: a cyber attack taxonomy. In: 9th annual symposium on information assurance. pp 2–12. https://doi.org/10.22937/IJC SNS.2021.21.8.1
57. Lee S (2018) Explaining stable coins, the holy grail of cryptocurrency. Forbes. https://www. forbes.com/sites/shermanlee/2018/03/12/explaining-stable-coins-the-holy-grail-of-crytpocur rency/?sh=289155cc4fc6
58. Aitzhan NZ, Svetinovic D (2016) Security and privacy in decentralized energy trading through multi-signatures, blockchain and anonymous messaging streams. IEEE Trans Dependable Secure Comput 15(5):840–852. https://doi.org/10.1109/TDSC.2016.2616861
59. "Ethereum API | IPFS API gateway | ETH nodes as a service," Infura. https://infura.io/. Accessed 14 Nov 2021
60. Zheng Z, Xie S, Dai H, Chen X, Wang H (2017) An overview of blockchain technology: architecture, consensus, and future trends. In IEEE international congress on Big Data (BigData Congress). IEEE, pp 557–564. https://doi.org/10.1109/BigDataCongress.2017.85
61. Cai W, Wang Z, Ernst JB, Hong Z, Feng C, Leung VC (2018) Decentralized applications: the blockchain-empowered software system. IEEE Access 6:53019–53033. https://doi.org/10. 1109/ACCESS.2018.2870644
62. Gentleman R, Ihaka R (2000) Lexical scope and statistical computing. J Comput Graph Stat 9(3):491–508. https://doi.org/10.2307/1390942
63. Kannengießer N, Lins S, Dehling T, Sunyaev A (2020) Trade-offs between distributed ledger technology characteristics. ACM Comput Surv 53(2):1–37. https://doi.org/10.2307/1390942
64. Gervais A, Karame GO, Wüst K, Glykantzis V, Ritzdorf H, Capkun S (2016) On the security and performance of proof of work blockchains. In: Proceedings of the 2016 ACM SIGSAC conference on computer and communications security. pp 3–16. https://doi.org/10.1145/297 6749.2978341
65. Doku R, Rawat DB, Garuba M, Njilla L (2019) LightChain: on the lightweight blockchain for the Internet-of-Things. In: 2019 IEEE international conference on smart computing (SMARTCOMP), IEEE, pp 444–448. https://doi.org/10.1109/SMARTCOMP.2019.00085
66. Ethereum, "Solidity—Solidity 0.8.10 documentation," 2021. http://solidity.readthedocs.io.
67. "OpenZeppelin," OpenZeppelin. https://openzeppelin.com/. Accessed 14 Nov 2021
68. Krombholz K, Hobel H, Huber M, Weippl E (2015) Advanced social engineering attacks. J Inf Secur Appl 22:113–122. https://doi.org/10.1016/j.jisa.2014.09.005
69. Jansson K, von Solms R (2013) Phishing for phishing awareness. Behav Inf Technol 32(6):584–593. https://doi.org/10.1080/0144929X.2011.632650
70. Arachchilage NAG, Love S, Beznosov K (2016) Phishing threat avoidance behaviour: An empirical investigation. Comput Hum Behav 60:185–197. https://doi.org/10.1016/j.chb.2016. 02.065
71. Roghanizad M, Choi E, Mashatan A, Turetken O (2021) Mindfulness and cybersecurity behavior: a comparative analysis of rational and intuitive cybersecurity decisions. In AMCIS 2021 proceedings (ASAC 2020), August 9–13, 2021. https://aisel.aisnet.org/amcis2021/info_s ecurity/info_security/13.
72. "What is a seed phrase?," Coinbase. https://www.coinbase.com/learn/crypto-basics/what-is-a-seed-phrase. Accessed 14 Nov 2021
73. Wang F, De Filippi P (2020) Self-sovereign identity in a globalized world: Credentials-based identity systems as a driver for economic inclusion. Front Blockchain 2:28. https://doi.org/10. 3389/fbloc.2019.00028

74. Mashatan A, Heintzman D (2021) The complex path to quantum resistance. Commun ACM 64(9):46–53. https://doi.org/10.1145/3466132.3466779
75. Mashatan A, Turetken O (2020) Preparing for the information security threat from quantum computers. MIS Q Exec 19(2):157–164. https://aisel.aisnet.org/misqe/vol19/iss2/7/
76. Mosca M (2018) Cybersecurity in an era with quantum computers: will we be ready? IEEE Secur Priv 16(5):38–41. https://doi.org/10.1109/MSP.2018.3761723
77. Gheorghiu V, Gorbunov S, Mosca M, Munson B (2017) Quantum proofing the blockchain. Blockchain Research Institute: University of Waterloo. https://evolutionq.com/quantum-safe-publications/mosca_quantum-proofing-the-blockchain_blockchain-research-institute.pdf
78. Buchmann J, Lauter K, Mosca M (2018) Postquantum cryptography, part 2. IEEE Secur Priv 16(5):12–13. https://doi.org/10.1109/MSP.2018.3761714
79. Chen L, Jordan S, Liu YK, Moody D, Peralta R, Perlner R, Smith-Tone D (2016) Report on post-quantum cryptography. U.S. Department of Commerce, National Institute of Standards and Technology. https://doi.org/10.6028/NIST.IR.8105
80. Spurr A, Ausloos M (2021) Challenging practical features of Bitcoin by the main altcoins. Qual Quant 55(5):1541–1559. https://doi.org/10.1007/s11135-020-01062-x
81. Andersen JV, Bogusz CI (2019) Self-organizing in blockchain infrastructures: Generativity through shifting objectives and forking. J Assoc Inf Syst 20(9):11. https://doi.org/10.17705/1jais.00566
82. "PolyNetwork," PolyNetwork. https://poly.network/. Accessed 14 Nov 2021
83. Fung B (2021) $600 million gone: The biggest crypto theft in history, CNN. https://www.cnn.com/2021/08/11/tech/crypto-hack/index.html.
84. "KERI," KERI. https://identity.foundation/keri/. Accessed 14 Nov 2021
85. Kaur S, Chaturvedi S, Sharma A, Kar J (2021) A research survey on applications of consensus protocols in blockchain. Secur Commun Netw 2021, Art. no. 6693731. https://doi.org/10.1155/2021/6693731

Dave McKay is a Fractional CTO for Healthcare and Blockchain companies. He is also a professor for the Blockchain Development Program at George Brown College and a Research Professional at the Cybersecurity Research Lab of the Ted Rogers School of Management at Ryerson University. Dave has made contributions to the technology and adoption of Self-sovereign Identity systems. He is the Co-chair of the Innovation Experts Committee of the Digital Identity and Authentication Council of Canada and sits on the Technology Stack Working Group and the Utility Foundry Working Group of the Trust Over IP Foundation. Dave is a reviewer for the CIO Strategy Council of Canada and reviews and comments on specifications surrounding healthcare, identity, and verifiable credentials.

Dr. Atefeh Mashatan (Associate Professor and Director, Cybersecurity Research Lab, Ryerson University) holds a BMath (Carleton University, 2002), an MMath (University of Waterloo, 2003), and a PhD in Combinatorics and Optimization (University of Waterloo, 2009). She holds the Canada Research Chair (Tier II) in Quality of Security for the Internet of Things. Her research is focused on the development of novel cybersecurity designs based on emerging technologies such as IoT, Blockchain, and Quantum Computing. Her expertise at the frontlines of the global cybersecurity field was recognized by SC Magazine in 2019, when she was named one of the top five Women of Influence in Security. She was recognized as one of Canada's Top 19 of 2019 Tech Titans at IBM CASCON Evoke conference. In 2020, she received the Enterprise Blockchain Award in the category of New Frontiers in Blockchain Academic Research by Blockchain Research Institute for developing the Mosaïque Digital Wallet. Most recently, she received the recognition of Top Women in Cybersecurity in Canada. Prior to joining Ryerson University, she was a Senior Information Security Consultant and a Solutions Architect at CIBC

(Canadian Imperial Bank of Commerce), in 2012–2016, with a focus on cryptography and enterprise architecture. Prior to that, she was a Scientific Collaborator with the Security and Cryptography Laboratory, School of Computer and Communication Sciences, Swiss Federal Institute of Technology Lausanne (EPFL), Lausanne, in 2009–2012. She is also a Certified Service Oriented Architect with Honors. She received the Certified Information Systems Security Professional Certification from the International Information Systems Security Certification Consortium.

The Future of Blockchain

Wendy M. Charles

Abstract Blockchain's current uses demonstrate potential for enhancing efficiencies and patient-centered solutions in life sciences research. For blockchain to continue to present new features and remain relevant in life sciences research, it is critical for blockchain capabilities to evolve and integrate with newer technologies. This chapter introduces the role of blockchain technologies in smart data, quantum computing, digital twins, and the emergence of the metaverse. Additional predictions and recommendations for preparing for future blockchain needs are provided.

Keywords Blockchain · Smart data · Quantum computing · Artificial intelligence · Digital twins · Metaverse

1 Future of Blockchain

To accommodate future research needs, life sciences research organizations are reducing timelines and costs using artificial intelligence (AI) applied to real-world data and previous clinical trials [1]. As this book has demonstrated thus far, life sciences organizations have identified meaningful opportunities to use blockchain in genomics, governance, regulations, security, and legal realms as uses of blockchain are accelerating. Therefore, it is critical for life sciences research organizations to determine the best methods for sustaining this forward momentum. To be successful, organizations should take the lessons learned in this book to create new ways of collaborating and accelerating research advancements. This chapter describes trends within life sciences research and advances in new technologies that are further facilitated by blockchain.

W. M. Charles (✉)
Life Sciences Division, BurstIQ, Denver, CO, USA
e-mail: wendy.charles@cuanschutz.edu

© The Author(s), under exclusive license to Springer Nature Singapore Pte Ltd. 2022 315
W. Charles (ed.), *Blockchain in Life Sciences*, Blockchain Technologies,
https://doi.org/10.1007/978-981-19-2976-2_14

1.1 Predictions of Future Blockchain Trends

A Gartner report about top trends in 2021 predicts a movement toward "distributed everything" [2]. While the future of technology is unpredictable and subject to variations of technological advances and market forces, this section offers a few trends that will likely affect the direction and adoption of distributed ledger technologies.

1.1.1 Decentralized Clinical Trials

Within life sciences research industries, the uses of blockchain will likely increase to meet the needs of decentralized clinical trials (DCTs). The term "decentralized" is reminiscent of the distribution of nodes. However, within the context of DCT, "decentralized" refers to remote collection [3] and/or the use of remote/virtual technologies [1]. The primary goals of DCTs are to bring clinical studies to the research participants where they live and work [3]. These studies can be designed as pragmatic studies to capture individuals' real-world experiences or be highly structured as clinical research studies [3]. By allowing remote data capture capabilities using electronic technologies [1], DCTs can enroll and retain more research participants and capture data under real-world circumstances [3]. During the Covid-19 pandemic, there was an extensive movement toward more virtual data collection [1], and there are strong indications that the adoption of virtual trials will increase in the future [4].

Components of blockchain technologies appear necessary for the success of future DCTs. For example, Dr. Khozin [3]—the former Associate Director of the U.S. Food and Drug Administration (FDA) Oncology Center of Excellence—describes successful DCTs as involving "distributed networks of connected technologies" (p. 27), and he has advocated for uses of blockchain to foster innovation (e.g., [5, 6]). In fact, the FDA joined the Decentralized Trials and Research Alliance that includes more than 50 international life sciences organizations [7] and was co-founded by ConsenSys Health, a health-oriented blockchain company [8].

Blockchain capabilities are recognized for offering more security for DCTs than centralized data management and limiting the potential for data loss [7]. There is also potential for enhancing protections of connected devices along distributed channels [3] and the benefit of an inherent audit trail to promote data reliability and integrity [4].

When considering the prospect for blockchain-based DCTs implementations, it is valuable to recall that there are no one-size-fits-all solutions. Each DCT has unique needs, and the selected technologies must be suitable for the population studied [4]. Further, any technology used to process protected health information or data regulated by the FDA must meet applicable Health Insurance Portability and Accountability Act (HIPAA) regulations, FDA regulation 21 CFR § 11 for electronic records and electronic signatures, and/or Good Clinical Practice guidelines [9]. Last, because DCTs could involve telemedicine for health management of research participants, organizations must be vigilant about evolving telemedicine statutes and guidelines

Fig. 1 Components for data
objects that create smart data

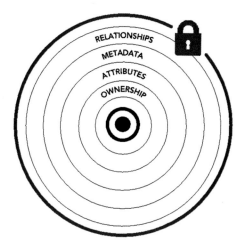

[4]. Overall, Dr. Gail [7] notes that DCTs have "arrived" (p. 387) and will likely change the nature of clinical trials. With the value added by blockchain technologies, blockchain will likely become an increasingly valuable technology for facilitating future success.

1.1.2 Evolving Data Features

Advances in life sciences research depend on increasing data value and establishing data networks that promote connections between disparate data sets [10]. Because previous book sections described the use of blockchain for connecting and sharing data sets, this section focuses on future predictions of using blockchain and related technologies to create greater value.

Smart Data

Data and analytic models are increasingly used to accelerate business intelligence and insights. While the concept of blockchain-based data integrity has been around for a long time, blockchain data structures are now designed to create deeper insights, sometimes referred to as "smart data" [11]. As shown in Fig. 1, blockchain-based data can be stored with components of data ownership, attributes, metadata, and relation-ships. Ownership could represent a person, university, company, lab, or anything else. The ownership is linked using blockchain methods to each data object in ways that cannot be modified, increasing the ability to trust data [11]. Life sciences researchers then connect data sources to create more enriched and insightful analytics.

Fig. 2 Roles of edges and
nodes in graph diagrams

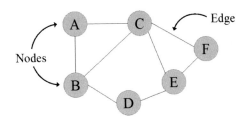

Blockchain Interoperability

A Gartner report predicts that "by 2023, 35% of enterprise blockchain applications will integrate with decentralized applications and services" ([12], p. 3). While considerable research and progress are required to achieve this level of interoperability, the IBM Blockchain group writes, "83% of organizations today believe assurance of governance and standards that allow interconnectivity and interoperability among permissioned and permissionless blockchain networks to be an important factor to join an industry-wide blockchain network, with more than one-fifth believing it to be essential" ([13], p. 3).

Successful interoperability strategies are needed to promote more intelligent health-oriented ecosystems. Health ecosystems require data management solutions that connect systems more efficiently, involving frameworks of users across healthcare facilities, research facilities, and academic institutions that need to transmit and store large volumes of data [14]. For example, blockchain-based frameworks increasingly connect ehealth technologies [15] and telehealth information systems [16]. These integrations may also require communication of organizations' business processes and models to ensure integrations address the desired value propositions [14]. Last, blockchain programmers are encouraged to learn healthcare and life sciences ontologies to connect to existing healthcare and research systems. Ultimately, progress in blockchain-based health and research ecosystems is predicted to lead to healthcare personalization, data intelligence, and autonomous systems [10].

Graph Technologies

While relational databases are most commonly used for large-scale data systems, they are limited by strict data schema and limitations regarding how data can be displayed and queried [17]. However, graph databases allow data to be represented by nodes, edges, and other properties, creating complex data relationships [18]. As shown in Fig. 2, "nodes" are people, places, or things that have roles in data relationships. "Edges" represent different types of connections between nodes and indicate connection strength [18]. Shifting data analytics to edge relationships allows opportunities for scaling capabilities and analytics where health-related data cannot move outside specific geographic boundaries [2]. Figure 3 shows how data relationships can be presented visually to enable the review of the interrelationships.

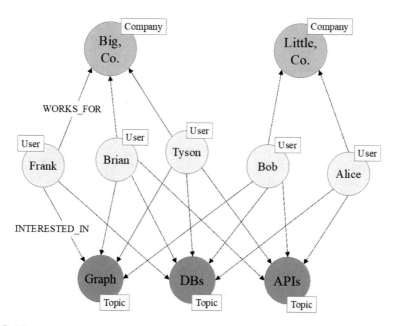

Fig. 3 Visual representation of graph relationships. Diagram inspired by [18]

Gartner predicts that by 2023, graph technologies will facilitate decision-making in 30% of organizations around the world [19]. By 2025, Gartner predicts that graph technologies will enable 80% of analytics innovations, resulting in faster decision-making [20]. Within life sciences research, this growth is driven by researchers' desire to uncover unexpected relationships that would be difficult to identify or analyze with traditional statistical programs [19]. Other scientists seek graph technologies to manage unstructured or semi-structured data more efficiently [20]. Therefore, graph technologies create new underlying data management technologies that can facilitate machine learning models and research collaborations [2].

Blockchain-based technologies store data more effectively for graph analytics. First, blockchain technologies store data on a ledger that manages data provenance, creating longitudinal records of individuals [19]. Blockchain technologies can also interconnect a complex network of collaborators and participants on a granular level [11] while providing a nearly immutable history for data management practices and security [17]. Last, blockchains are used for graphing technologies to track data assets, grant and revoke permissions, and involve strong encryption [21].

The following companies offer graph technologies and related visualization tools for blockchains. For a more comprehensive listing of blockchain visualization technologies with images, examine the systematic review published by [22].

BitExTract was designed by researchers at Hong Kong University of Science and Technology in 2018 [23]. This software provides a multi-view analytics tool that displays and compares bitcoin transaction relationships [23]. The Connection View creates node-edge diagrams that show relationships of transactional exchanges.

Each node is colored to represent the continent in which the exchange originated. The edge thickness shows the frequency or intensity of transactions between nodes [23]. BitExTract designed an "ego-network" graph visualization tool so that individuals can drag nodes of interest to the center of the display to view related transactions. As of November 2021, it does not appear that BitExTract is available as a commercial product.

Bitquery (https://bitquery.io/) offers a set of software products that index and query blockchain data. One of their products, Bitquery Explorer, is a client-side web application that connects an analytics explorer to query across more than 30 different blockchains [24]. Bitquery GraphQL allows for querying blockchains and can create actionable and insightful graphics [25]. Forensic data companies and government agencies use Bitquery's technology to track blockchain transactions and recover stolen funds [26]. Bitquery also advertises that its technology is used for scientific research [24].

Blockchain 3D Explorer (https://blockchain3d.info/) creates visualizations of blockchain transactions as 3D graphs and in virtual reality. The open-source software creates timeline-based 3D graphs that connect input and output addresses over time. The technology currently supports virtual reality systems for Google Cardboard as an immersive experience that allows individuals to view a history of transactions inside the blockchain [22].

BlockchainVis [27] is a blockchain forensic tool developed by the Italian Distributed Ledger Technology Working Group. The tool creates visual displays and queries of a transaction network. A filter panel can restrict specific nodes when examining connections between nodes [22]. As of November 2021, it is unclear whether BlockchainVis is a commercial product.

BurstIQ, Inc. (https://www.burstiq.com/) offers a graph technology called LifeGraph® that combines blockchain with machine learning methods designed for the secure handling of personally identifiable information. The network turns digital health assets into smart data that enforce data ownership, control, and security [11]. The smart data are integrated into a network model that enriches AI algorithms to make solutions more personalized and optimized [11]. For example, this technology was utilized in collaboration with the National Center for Advancing Translational Sciences (NCATS), a division of the National Institutes of Health. The NCATS team sought graph technology to predict the feasibility of creating synthetic molecular reactions [21]. The BurstIQ LifeGraph® network was integrated with the NCATS computational infrastructure so that researchers could collaborate while maintaining traceability and ownership [11]. The solution demonstrated that a collaborative research network could reduce the cost and risks of collaborative research while accelerating the pace of discovery [11].

Databricks is a San Francisco-based company (https://databricks.com) that offers graph analytic platforms and visualization tools for blockchain transaction data. A transaction can be associated with any detail created on the blockchain, including name, ID, or unit [28]. Databricks uses Apache Spark and GraphFrame coupled with graph visualization libraries to identify significant patterns in blockchain transactions [28]. Using Graph APIs, the technology analyzes data for users' incoming and outgoing transactions. The GraphFrames product creates vertices and edges from the transaction data to form directed graphs. The edges are shown with arrows and thickness to represent traffic volume [28].

Dan McGinn and his colleagues from the **Data Science Institute at Imperial College London** designed an unnamed blockchain visualization tool built with the Neo4j graph database [29]. This tool creates blockchain node activity profiles that display connections between nodes in cryptocurrency networks [30]. The goal is to reveal temporal transactions patterns along the entire graph as an edge-weighted adjacency matrix [30]. As of November 2021, it is unclear whether this tool is available as a commercial product.

While the uses of graph technologies in blockchain are still evolving, life sciences research leaders are encouraged to explore opportunities for integrating graph technologies into their analytics solutions. A blockchain may offer advantages for connecting data and AI/ML algorithms to improve these initiatives.

1.2 Quantum Computing

Modern connected networks rely on cryptography to protect our identities, communication, and financial transactions [31]. Blockchain security depends on one-way (asymmetric) cryptography for digital signatures and to validate transactions on the ledger [32]. One-way cryptography can be run on conventional computers. However, efforts to reverse the encryption would require substantial computing resources [33], requiring many years to solve [32]. However, scholars caution that a newer type of technology, quantum computing, will create future risks for blockchain networks that run on traditional computers (e.g., [31, 32]).

Traditional computing uses bits to encode data as 0 or 1. However, quantum computing uses particles of light (photons) to encode quantum bits that similarly have two (basis) states (0 or 1) but could be manipulated in ways that can only be explained by quantum mechanics [33]. Very simply, the particles become "entangled" when the state of one component cannot be described without the others. This composite is a sum, or "superposition," that can be measured [33]. The system then collapses the superposition to one of the basis states to extract information [34]. Quantum computing can execute calculations much more efficiently because the system can simultaneously perform whole ranges of numbers and return only one result [34]. An image of a quantum computer is shown in Fig. 4.

For life sciences research, quantum technologies enable computing speed and complexity that conventional computing cannot achieve [34]. The improvements also include minimal storage and near-guaranteed security [33]. Quantum computing capabilities can accelerate life sciences research by creating simulations of molecular compounds to discover future medications, performing DNA sequencing, and optimizing personalized medicine [33]. For example, Boehringer Ingelheim partnered with Google to create a Quantum Lab [35]. Ryan Babbush, Google's head of quantum algorithms, noted that "extremely accurate modeling of molecular systems is widely

anticipated as among the most natural and potentially transformative applications of quantum computing" ([35], p. 2).

As a caution noted earlier, blockchain-based networks are designed with the premise that traditional computing would not be able to reverse encryption without extraordinary time and effort [36]. However, quantum computers are projected to calculate the cryptographic codes used by many blockchains within ten years [32]. As the most imminent threat, malicious actors could use quantum computing to deduce private keys from the published public keys with little effort [34]. Cryptocurrency owners are then at high risk of losing control of their cryptocurrency. Further, it is feared that the few cryptocurrency miners who gain access to quantum computing will monopolize future block generation and sabotage transactions, such as engaging in double-spending [32].

The threat of quantum computing, though, appears limited by the cost and complexity of quantum networks [32]. Fedorov et al. [32] state that quantum computers need a "quantum internet" to connect across computers in a communications network. Without an intermediary, each node would require fiber optic channels to connect to other nodes, resulting in a quantum blockchain [32].

Regardless of the time and scope of the emergence of quantum computers, life sciences research organizations should start planning for security threats introduced by quantum computing. Campbell [37] argues that cybersecurity should be a primary concern because organizations cannot afford to lose the protection of their data and intellectual property. This risk assessment should include devices and data storage vulnerabilities for the cyber-attacks that will inevitably arrive [33]. Cybersecurity measures for blockchain should also include extensive planning and testing of post-quantum-resistant cryptography [37]. Post-quantum cryptography involves newer methods of cryptography that utilize suites of algorithms proposed to be more secure than conventional algorithms [34]. Proposals involve quantum key distribution, where the technology creates unconditionally secure message authentication [38]. As alternate approaches, Yaqoob et al. [36] recommend replacing traditional digital signatures, while Fedorov et al. [32] advocate encrypting all peer-to-peer channels in a blockchain network.

It may take years of analysis before industries and individuals trust quantum-resistant security measures [37]. Therefore, longer-term protection measures require investments and guidance from governments. Countries currently leading research developments in quantum technologies include China, the U.S., and several members of the European Union [32]. Legislators should engage in honest discussions and methods to approach cybersecurity regulations in a manner that allows innovation and market forces to drive advancements [37]. Campbell [37] also recommends that countries collaborate toward designing global standards.

Until post-quantum security measures are standardized and established, life sciences organizations are encouraged to consider blockchain platforms that can change cryptographic algorithms or utilize flexible encryption functions [32]. Campbell [37] encourages organizations to ask blockchain vendors about technologies that could adapt to quantum computing to protect regulated data. Implementation plans and updates could be included in contractual obligations. Campbell [37] also reminds organizations to update their policies, procedures, and risk assessments with any modifications to cryptographic methods. Without sufficient planning for quantum computing, the threat to blockchains—and all technologies involving encryption—could be severe [32].

1.3 Digital Twins

Another emerging technology involves "digital twins." While various definitions of digital twins are available, unifying concepts involve software that takes real-world data to create a digital, cyber, or virtual representation (a "twin") that generates valuable insights about the real-world object [39]. First used in manufacturing to create digital representations of machinery or sensors [40], digital twins have evolved

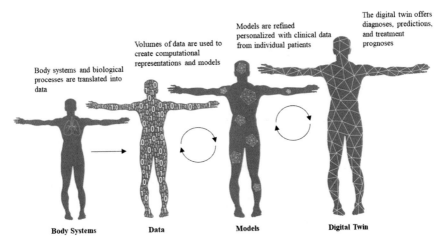

Fig. 5 Representation of methods for creating a digital twin. Substantially adapted from [44]

beyond digital models and now imply that a digital twin is connected in some way to the real-life individual or item (Kritzinger et al. 2018). More than a simulation or 3D model [41], a digital twin model must continually adapt to changing data and information to forecast future conditions [42]. The concept of digital twins has been named one of the top ten emerging strategies, with projected spending of nearly $11 billion in 2022 [42].

In medicine, "digital twins healthcare" (DTH) is an emerging discipline for creating digital twins with health information [43]. As shown in Fig. 5, DTH models involve three components: a physical object (such as a body part), a virtual object (such as a simulation or 3D print design), and healthcare data. Therefore, DTH should not be considered one technology but a cocktail of technologies [42].

In life sciences research, digital twins have been studied to create replicas of human body parts and influence therapy decisions or personalized medicine [40]. For example, Liu et al. [43] designed a digital twin for elder care that provides remote diagnosis health consultations and real-time monitoring. Oklahoma State University researchers created a digital twin of trachea models with individual alveolar sacs to simulate pulmonary oncology drugs delivered with inhalers [45]. Typical aerosol drugs can only reach 25% of the intended cancerous cells; however, Feng et al.'s digital twin model inspired aerosol molecular modifications that could reach 90% of cancerous cells ([45], p. 26). Clinical trials are critical for model modifications and validations, and it is necessary to perform longitudinal studies to characterize long-term DTH responses in various settings [44].

The adaptation of real-life health conditions to digital twins is facilitated by AI, cloud computing, and—in many cases—blockchain [39]. While blockchain is not required to create a digital twin, digital twin projects can innovate faster and more securely with blockchain features [39]. Blockchain-based smart contracts are used to manage the granular consent of multiple collaborators for the collective development of digital twins. At the same time, the audit trail provides accountability for any changes or updates to a model [46]. For example, Leng et al. [47] created a hybrid digital twin/blockchain model called ManuChain that adds a layer of digital twin models on top of a blockchain layer. The blockchain-based smart contracts automate individualized tasks for the twin model and perform a critical role in connecting the cyber and physical components of manufacturing. The blockchain also maintains multiple copies of digital twins on the ledgers for federated learning [47]. In a model designed by [48], the digital twin model is stored on a permissioned blockchain and records all changes and provenance of the model while the IoT data are stored off-chain in separate servers. The smart contracts then update the parameters of the digital twin accordingly [48]. Putz et al. [46] also utilize a hybrid on-chain/off-chain model for their DTH where the healthcare data are supplied by off-chain IoT data. These approaches reduce the computation and storage requirements for the permissioned blockchain [46, 48].

The development of digital twins in life sciences research, however, has been slow, and few models have reached clinical use [49]. First, researchers must model human biology [41]. Laubenbacher et al. [44] note that the ability to replicate the complexity of multi-system interactions, such as an immune response, is currently out of reach because of difficulties with model validation. A human body generates constant molecular changes and adaptations that make it challenging to model physiological processes [40]. Compounding these factors is the need for large volumes of data to perform comparisons of disease states to evolving and heterogenous definitions of "healthy" or "normal" states [40]. Kendzierskyj et al. [40] add that the concepts of "healthy" and "normal" can only be drawn from population statistics but often cannot inform individual digital twin models with sufficient precision. Tao and Qi [41] note it is too early in the development process to create "accepted standards" or "norms."

Progress with developing digital twins has also been slowed by the difficulty of obtaining and integrating data. Tao and Qi [41] relayed that there may be a need to aggregate data from thousands of sensors, and data providers may maintain data in different formats. If unable to obtain sufficiently representative data, the digital twin results will be distorted [41]. This data limitation is compounded by the need for laboratories worldwide to integrate and validate each other's work, requiring central coordination [44]. While blockchain could be used for data integration and protection, the more significant issue of sharing DTH knowledge and software involves the desire to maintain commercial secrecy [41]. Additionally, when DTH research is intended to treat, diagnose, or mitigate medical decisions, the software would be regulated as a medical device subject to FDA review and approval [50]. This level of evidence, quality control, and algorithmic performance is believed to hinder current adoption [49].

When designing DTH, life sciences research organizations should be aware of significant privacy risks. As a preliminary factor, data must be obtained from electronic health records and multiple other sources of health information that must be linked to each individual represented in the data set, creating concern about data theft and tampering [40]. As a more extensive consideration, when designing a DTH for an individual, an accurate digital model effectively becomes part of that person's identity [40]. In conclusion, DTH has made significant progress with modeling physiological and behavioral features, but accurate multi-system models still appear far from reality.

1.4 The Metaverse

The term metaverse was coined using the prefix "meta" and "universe" to create a virtual reality environment that can replicate aspects of the physical world [51]. In fact, in October 2021, Facebook changed its name to Meta, stating, "Our company's vision is to help bring the metaverse to life, so we are changing our name to reflect our commitment to this future" ([52], p. 1). Metaverses use high-speed networks and AI to create simulations where avatars represent humans. As shown in Fig. 6, avatars are similar to digital twins in that they can look and behave like humans to create an enhanced user experience [53]. These avatars can engage in various activities in the metaverse, including cultural, economic, and social interactions designed to mimic

Fig. 6 Example of a Second Life Avatar. *"My Second Life Avatar" by Lisa Tripp is licensed under CC-BY-SA 2.0 that allows reusers to distribute, remix, adapt, and build upon the material in any medium or format, so long as attribution is given to the creator. To view a copy of this license, visit* https://creativecommons.org/licenses/by/2.0/

the real world [51]. The connected ecosystems, including social norms and rules, are often analogous to existing norms and rules in the real world [53]. Lee et al. [53] suggest that the metaverse can support the production of intangible assets in order to become a potentially self-sustaining economic ecosystem.

The earliest metaverse developed was Second Life (https://secondlife.com/), founded in 2003 by Linden Research Inc., based in San Francisco, CA. There are thousands of sites to explore in this virtual environment, including markets that exchange a unique form of payment called a Linden dollar [53]. Other popular metaverses include gaming apps Minecraft (https://www.minecraft.net/) designed by Microsoft, Fortnite (https://www.epicgames.com/fortnite/) built by Epic Games, and Roblox (https://www.roblox.com/) created by Roblox Corporation. All involve avatars and digital economies. To date, the majority of published health-oriented research in metaverses focused on Second Life settings, but later research has studied a wide range of virtual reality environments.

The Covid-19 pandemic has been a significant driver for drawing greater participation in the metaverse. The mass closures and social distancing limitations generated tremendous interest in metaverse environments that offer virtual social interactions [54]. Minecraft was even used to create a virtual graduation for University of California, Berkeley students during the Covid 2020 shutdowns—complete with a speech by Chancellor Carol Christ, Pomp and Circumstance music, and flying mortarboards [55]. These metaverses are also increasingly used in place of videoconferencing and virtual conference attendance [56]. Microsoft announced in November 2021 that Microsoft Teams video conferencing software will offer a mixed-reality platform that combines real-world participation with a metaverse [57]. Microsoft Teams aims to create greater engagement and interaction during remote encounters.

1.4.1 Health Activities in the Metaverse

Metaverses can offer unique support and capabilities for health-oriented education and treatment, including life sciences research. Using avatars, individuals can visit multiple doctors without leaving their (real) homes, and blockchain-based systems have been designed to manage the storage of their health information [53]. To promote interpersonal reactions, avatars can be controlled with body-centric sensors that allow for subtle movements and facial expressions to create a more realistic presence and more profound connection in the virtual environment [53].

The realistic social interactions make the metaverses a viable venue for innovations that have implications in real life. First, metaverses create immersive environments that allow for more effective training simulations. For example, clinical nursing training has been conducted in Second Life to provide examples of simulated patients in high-risk situations [58]. Also, Schaffer et al. [59] found that this virtual reality setting provides an effective learning platform for preparing nursing students to manage clinical situations that rarely occur in real life. Last, Second Life was used to create education sessions to train medical students to review radiology images

[60]. Overall, this immersive training environment has improved decision-making and is believed to translate into more effective clinical practices.

Specialty sites within metaverses have been used to offer one-on-one meetings with physicians, nurses, and other healthcare providers [61]. When coupled with real-world biometric sensors attached to humans who visit a virtual healthcare clinic, the individuals/avatars can receive health monitoring and assessments [53]. In addition, certain health-oriented therapies have proven remarkably successful in the metaverse. Individual consultations are available (for a fee) in an anonymous manner that encourages individuals to ask questions and receive medical or psychological advice that they might not pursue in real life [61, 62]. Gorini et al. [62] describe how individuals have received successful treatment for specific phobias in the metaverse, such as claustrophobia, arachnophobia, and agoraphobia, using desensitizing simulations of fearful environments without the (real) individuals experiencing any physical danger.

Within the metaverse, some sites also facilitate virtual meeting places for patient support groups and community education (Fig. 7). Support groups conducted in virtual reality settings are often more comfortable for individuals seeking support for sexual abuse or other sensitive or stigmatizing conditions [62]. There are also themed lectures and education events for patient communities that discuss diseases, treatment plans [61], and offer a supportive environment for friends and family [62].

Fig. 7 Second Life group meeting place. *"Avatar-Based Marketing: What's the Future for Real-Life Companies Marketing to Second Life Avatars?" by John' Pathfinder' Lester, licensed under CC-BY-SA 2.0 that allows reusers to distribute, remix, adapt, and build upon the material in any medium or format, so long as attribution is given to the creator. To view a copy of this license, visit* https://creativecommons.org/licenses/by/2.0/

For life sciences research, the metaverse offers the opportunity to conduct research directly or indirectly. Within Second Life, there are laboratories and clinics where individuals (as avatars) participate in research and receive tokens for participation [61]. These Second Life research sites also actively recruit for research participation within Second Life and real-world settings.

1.4.2 Blockchain and the Metaverse

Lee et al. [53] assert that blockchain is "expected to connect everything in the world in the metaverse" (p. 16). Therefore, several metaverse applications utilize a blockchain model of distribution where components of virtual spaces are synchronized amid connected users, and users' activities are recorded on the blockchain [56]. The data connectivity features of blockchain can connect computer vision, AI, and IoT into individual patient profiles for more accurate healthcare simulations [53]. Further, emerging blockchain-based graph technologies and edge relationships allow for high-quality queries and make the necessary data selectively available [51]. Blockchain also allows for highly scalable, flexible, and secure data storage [53].

To protect the avatars—and the individuals behind them—it is also necessary to ensure a trusted-based information system can manage identities [53]. van der Merwe [54] recommends implementing a gatekeeping function or levels of access restrictions in the metaverse, depending on the nature of the metaverse site and vulnerabilities. Blockchain technologies are used, then, to enforce specific rules. Ryskeldiev et al. [56] proposed a blockchain-based layer that creates unique identifiers (hashes) for each avatar and created space. For example, when a new space is created, a new block would be formed that contains its geographical coordinates within the metaverse, the URL to a 360° image, and a timestamp [56].

1.4.3 Metaverse Drawbacks

A virtual reality environment introduces unique and complicated ethical considerations. First, an avatar is a digital representation of an individual that creates new questions about what it means to be a person [53]. van der Merwe [54] wonders whether the friendships and romances formed in a metaverse environment are any less real. van der Merwe [54] further speculates about human rights and obligations. Specifically, does an avatar have the same legal rights associated with humans? Is there any recourse available against theft, violence, or harassment within a metaverse? Instead, Jeon et al. [51] argue that virtual people have no legal basis, and it is unlikely that they would have legal rights.

When a person engages in any virtual social interaction, there are questions about the degree to which social behavior in the metaverse reflects individuals' behavior in real life, creating significant cautions where adults could interact with children in the metaverse [53]. Jeon et al. [51] point out that individuals could engage in racial or gender discrimination under the protections of avatar anonymization. Of more significant concern, the avatar-based interactions do not indicate the individual's true identity behind the avatar. Anyone could enter the metaverse to create any avatar, creating possible misrepresentations during social interactions [51]. At the same time, van der Merwe [54] notes that an avatar is not required to portray a real person, so it is unrealistic to expect that the avatar's behavior would correspond with the real person.

When questions about the separations of avatars and humans are applied to healthcare, must a healthcare provider in the metaverse possess a clinical license in real life? Can a metaverse healthcare clinician provide treatment to a person (or specifically, an avatar) who resides in a different state or jurisdiction than permitted by that clinician's state-issued license (in real life)? Gorini et al. [62] advocate for international guidelines to govern the delivery of regulated services in a virtual environment.

Metaverses were not designed for healthcare or research, so virtual environments are not designed to protect individuals' privacy or confidentiality. In the metaverse, it would be necessary for clinicians and researchers to create protected environments where entry requires a secret code [62]. There are also questions about whether HIPAA or other healthcare regulations would apply to health information collected in a metaverse with no geographic boundaries. These environments collect extensive information about avatars, such as their locations and surroundings, which could involve privacy regulations [53]. However, it is unknown how privacy statutes and regulations would apply. Therefore, Lee et al. [53] recommend that these systems collect the minimum amount possible and only for as long as needed. Blockchain technology governance layers could also enforce specific rules [53]. Last, Lee et al. [53] advocate for autonomous agents in the metaverse to observe behaviors and expectations.

Overall, the metaverse is rapidly expanding, and blockchain-based technologies can facilitate some of the required protections and governance. However, there are many questions about the degree to which healthcare or research regulations apply to the virtual world. There are also many questions about managing the privacy of virtual encounters intended to be private. Many of these questions may require the cooperation of international governments to create determinations and guidelines.

1.5   The Carrier Wave Principle

When contemplating the many future directions of technology advancements, it is valuable to reflect on "the carrier wave principle." This principle was coined by Sinnreich and Gilbert [31] to raise concern that "as the cultural infrastructure has become increasingly reliant on computational processing, everyday users have become commensurately less capable of understanding the consequences of their actions and interactions" (p. 5818). Sinnreich and Gilbert [31] use the analogy of radio and television carrier waves to advise that we are constantly creating digital content—that we might not even be aware of—that could remain stored and searchable into perpetuity. Further, when carried forward by digital technologies that expand the scale and speed of knowledge, the cultural and social meaning of that information could be misinterpreted or exploited. Last, Sinnreich and Gilbert [31] explain that information is often taken out of context and then amplified through social platforms, resulting in incorrect AI algorithms that further distort the information's meaning.

In summary, the carrier wave principle raises a caution that "the growing reliance on computational processing as a foundation of knowledge production and social governance makes public oversight and development of best practices for data collection, management, and processing imperative for a functional civil society" ([31], p. 5831). Therefore, the authors advocate for cultural awareness and political reaction to the growing effects of the carrier wave principle as technologies evolve [31].

2   Future Research

Additional research is necessary to understand the best ways to adapt blockchain technologies to meet the future needs of life sciences research. First, it is valuable to consider that health-oriented information is now generated by an increasing number of smart devices [63], clinicians, researchers, and organizations that support life sciences industries [64]. While this book has positioned that this information can be managed successfully by blockchain, the technology must meet applicable regulations [9], and the data should only be used in accordance with individuals' permissions [65, 66]. Additional research is needed to create standards and best practices for evolving representations of individuals, including digital twin and metaverse environments.

Second, future research is needed to advance blockchain interoperability with life sciences applications and services. Organizations will benefit from testing integrations of their blockchains to compatible blockchains or other systems, including healthcare ecosystems [67, 68]. The interoperability should be tested in cross-national and cross-international contexts to create context-based solutions [14]. These solutions should include open standards such as Health Level 7 and Fast Healthcare Interoperability Resources [69].

As blockchain technologies become more diverse, there is an increasing threat of cyber-attacks [37]. Vulnerabilities may exist in underlying algorithms, side-channels, software integrations, and coding errors [37]. Additionally, organizations must prepare for the growing computational threat of quantum computing [33]. To address future technological threats, it is crucial to create long-term security strategies to mitigate emerging risks. Future research should address methods for managing cryptographic keys [14] and protocols for advancing quantum-resistant encryption [38]. Future research is also necessary to create appropriate blockchain cybersecurity standards and design industry-relevant cybersecurity programs for due diligence at protecting networks and organizations [37]. These research efforts should extend to safeguarding increasingly sensitive information with blockchains, such as biometrics and genetic information [14]. Organizations are also encouraged to review and implement appropriate policies and guidelines.

Last, research should determine the factors that may facilitate or hinder the adoption of blockchain technologies in life sciences research. This research should assess how blockchain could create or enhance value within life sciences research organizations [14]. Assessments should focus on the identification and development of strategic issues, such as performance [15], technical requirements [70], and resource limitations [71]. Research on these issues may advance blockchain architectures that offer more utility and efficiencies. For example, there are needs for more cost-effective node management [72], privacy management [73], authentication management [16], and patient-centered access controls [70]. This research will require a multi-disciplinary approach to address the roles of blockchain with evolving technology, regulatory, ethical, and legal considerations [65, 66].

3 Conclusions

As this book has described, blockchain technologies serve as an infrastructure layer of a holistic life sciences research ecosystem. As with any change, blockchain technologies may be perceived to pose threats to current paradigms [74]; however, blockchain technologies need not replace existing life sciences research technologies but can enhance their current capabilities [75]. Ideally, technological advances would allow research professionals to focus more on human interactions and other research tasks requiring human knowledge and facilitation. Therefore, as market forces drive innovation such as quantum computing, digital twins, and the metaverse, blockchain technologies can provide checks and balances on these systems [74]. Sinnreich and Gilbert [31] advise that "we can play a more proactive role in shaping that future by being more deliberate now about what kinds of media we build, what kinds of messages we send, and what kinds of laws and ethics we embrace to guide their development and deployment" (p. 5832).

Acknowledgements The author gratefully acknowledges the review and thoughtful feedback from Brooke Delgado, Leanne Johnson, and Hayley Miller.

References

1. Levy V (2021) 2021 Global life sciences outlook. https://www2.deloitte.com/global/en/pages/life-sciences-and-healthcare/articles/global-life-sciences-sector-outlook.html
2. Panetta K (2021) Gartner top 10 data and analytics trends for 2021. Gartner, Inc. https://www.gartner.com/smarterwithgartner/gartner-top-10-data-and-analytics-trends-for-2021. Accessed 23 Sept 2021
3. Khozin S, Coravos A (2019) Decentralized trials in the age of real-world evidence and inclusivity in clinical investigations. Clin Pharmacol Ther 106(1):25–27. https://doi.org/10.1002/cpt.1441

4. Datacubed Health (2020). Implementing solutions to virtualize and decentralize clinical trials. https://www.datacubed.com/wp-content/uploads/2020/10/Implementing-Solutions-to-Virtual-and-Decentralize-Clinical-Trials-Datacubed-Health.pdf
5. Khozin S, Kim G, Pazdur R (2017) From big data to smart data: FDA's INFORMED initiative. Nat Rev Drug Discovery 16(5):306. https://doi.org/10.1038/nrd.2017.26
6. Khozin S, Pazdur R, Shah A (2018) INFORMED: an incubator at the US FDA for driving innovations in data science and agile technology. Nat Rev Drug Discovery 17(8):529–530. https://doi.org/10.1038/nrd.2018.34
7. Van Norman GA (2021) Decentralized clinical trials. JACC Basic Transl Sci 6(4):384–387. https://doi.org/10.1016/j.jacbts.2021.01.011
8. Dalton B (2021) ConsenSys health joins decentralized trials & research alliance (DTRA) to democratize and accelerate clinical trials. ConsenSys Health. https://consensyshealth.com/news/consensys-health-joins-decentralized-trials-research-alliance-dtra-to-democratize-and-accelerate-clinical-trials/. Accessed 21 Oct 2021
9. Charles WM, Marler N, Long L, Manion ST (2019) Blockchain compliance by design: regulatory considerations for blockchain in clinical research. Front Blockchain 2(18). https://doi.org/10.3389/fbloc.2019.00018
10. BurstIQ (2021) Why smart data is the future of data security. https://www.burstiq.com/smart-data-white-paper/
11. Charles WM (2021) Blockchain innovations in healthcare. PECB Insights (33):6–11. https://insights.pecb.com/pecb-insights-issue-33-july-august-2021/#page6
12. Litan A (2021) Hype cycle for blockchain 2021; More action than hype. Gartner, Inc. https://blogs.gartner.com/avivah-litan/2021/07/14/hype-cycle-for-blockchain-2021-more-action-than-hype/. Accessed 15 Oct 2021
13. Schlapkohl K (2020, April 10) The future of blockchain. IBM. https://www.ibm.com/blogs/blockchain/2020/04/the-future-of-blockchain/. Accessed 28 Aug 2021
14. Tandon A, Dhir A, Islam AKMN, Mäntymäki M (2020) Blockchain in healthcare: a systematic literature review, synthesizing framework and future research agenda. Comput Ind 122:103290. https://doi.org/10.1016/j.compind.2020.103290
15. Hyla T, Pejaś J (2019) eHealth integrity model based on permissioned blockchain. Futur Internet 11(3):76. https://doi.org/10.3390/fi11030076
16. Ji Y, Zhang J, Ma J, Yang C, Yao X (2018) BMPLS: blockchain-based multi-level privacy-preserving location sharing scheme for telecare medical information systems. J Med Syst 42(8):147. https://doi.org/10.1007/s10916-018-0998-2
17. Ermolaev V, Klangberg I, Madhwal Y, Vapper S, Wels S, Yanovich Y (2020) Incorruptible auditing: blockchain-powered graph database management. IEEE. https://doi.org/10.1109/icbc48266.2020.9169431
18. Robinson I, Webber J, Eifrem E (2015) Graph databases, 2nd edn. O'Reilly Media, Inc. http://bit.ly/dl-neo4j
19. Goasduff L (2020, October 19) Gartner top 10 trends in data and analytics for 2020. Gartner, Inc. https://www.gartner.com/smarterwithgartner/gartner-top-10-trends-in-data-and-analytics-for-2020. Accessed 22 Sept 2021
20. Adrian M, Jaffri A, Feinberg D (2021) Market guide for graph database management solutions (G00737853). https://info.cambridgesemantics.com/graph-database-management-solution-market-guide-gartner
21. Warr WA (2021) National Institutes of Health (NIH) workshop on reaction informatics. https://chemrxiv.org/engage/api-gateway/chemrxiv/assets/orp/resource/item/611cf1a6ac8b499b36458d19/original/national-institutes-of-health-nih-workshop-on-reaction-informatics.pdf
22. Tovanich N, Heulot N, Fekete J-D, Isenberg P (2019) Visualization of blockchain data: a systematic review. IEEE Trans Vis Comput Graph 27(7):3135–3152. https://doi.org/10.1109/tvcg.2019.2963018
23. Yue X, Shu X, Zhu X, Du X, Yu Z, Papadopoulos D, Liu S (2019) BitExTract: interactive visualization for extracting bitcoin exchange intelligence. IEEE Trans Vis Comput Graph 25(1):162–171. https://doi.org/10.1109/TVCG.2018.2864814

24. Bitquery (2021) Bitcoin analysis: track bitcoin transactions and address. https://bitquery.io/blog/bitcoin-analysis. Accessed 11 Nov 2021
25. Bitquery (2021) Blockchain GraphQL APIs. https://bitquery.io/labs/graphql. Accessed 14 Nov 2021
26. Brown D (2021, Sep 22) Tracking stolen crypto is a booming business: How blockchain sleuths recover digital loot. The Washington Post. https://www.washingtonpost.com/technology/2021/09/22/stolen-crypto/
27. DLT Group (2020) Italian Distributed Ledger Technology Working Group. http://dltgroup.dmi.unipg.it/tools.php. Accessed 14 Nov 2021
28. Mahapatra A, Gieseke E (2021) Analyzing algorand blockchain data with databricks delta (Part 2). Databricks. https://databricks.com/blog/2021/03/03/analyzing-algorand-blockchain-data-with-databricks-delta-part-2.html. Accessed 11 Nov 2021
29. McGinn D, Birch D, Akroyd D, Molina-Solana M, Guo Y, Knottenbelt WJ (2016) Visualizing dynamic bitcoin transaction patterns. Big Data 4(2):109–119. https://doi.org/10.1089/big.2015.0056
30. McGinn D, McIlwraith D, Guo Y (2018) Towards open data blockchain analytics: A Bitcoin perspective. R Soc Open Sci 5(8):180298. https://doi.org/10.1098/rsos.180298
31. Sinnreich A, Gilbert J (2019) The carrier wave principle. Int J Commun 13:5816–5840. 1932-8036/20190005
32. Fedorov AK, Kiktenko EO, Lvovsky AI (2018) Quantum computers put blockchain security at risk. Nature 563(7732):465–467. https://doi.org/10.1038/d41586-018-07449-z
33. Farouk A, Alahmadi A, Ghose S, Mashatan A (2020) Blockchain platform for industrial healthcare: vision and future opportunities. Comput Commun 154:223–235. https://doi.org/10.1016/j.comcom.2020.02.058
34. Stewart I, Ilie DI, Zamyatin A, Werner S, Torshizi MF, Knottenbelt WJ (2018) Committing to quantum resistance: a slow defence for bitcoin against a fast quantum computing attack. R Soc Open Sci 5(6):180410. https://doi.org/10.1098/rsos.180410
35. Hale C (2021) JPM: boehringer partners with Google to bring quantum computing to biopharma R&D. Fierce Biotech. https://www.fiercebiotech.com/medtech/boehringer-partners-google-to-bring-quantum-computing-to-biopharma-r-d. Accessed 11 Nov 2021
36. Yaqoob I, Salah K, Jayaraman R, Al-Hammadi Y (2021) Blockchain for healthcare data management: opportunities, challenges, and future recommendations. Neural Comput Appl. https://doi.org/10.1007/s00521-020-05519-w
37. Campbell RE (2019) Transitioning to a hyperledger fabric quantum-resistant classical hybrid public key infrastructure. J Br Blockchain Assoc 2(2):4. https://doi.org/10.31585/jbba-2-2-(4)2019
38. Sun X, Kulicki P, Sopek M (2020) Lottery and auction on quantum blockchain. Entropy (Basel) 22(12):E1377. https://doi.org/10.3390/e22121377
39. Raj P (2021) Empowering digital twins with blockchain. Adv Comput 121:267–283. https://doi.org/10.1016/bs.adcom.2020.08.013
40. Kendzierskyj S, Jahankhani H, Jamal A, Ibarra Jimenez J (2019) The transparency of big data, data harvesting and digital twins. In: Jahankhani H, Kendzierskyj S, Jamal A, Epiphaniou G, Al-Khateeb HM (eds) Blockchain and clinical trial: securing patient data. Springer Nature Switzerland AG, pp 139–148). https://doi.org/10.1007/978-3-030-11289-9_6
41. Tao F, Qi Q (2019) Make more digital twins. Nature 573:490–491. https://doi.org/10.1038/d41586-019-02849-1
42. Popa EO, Van Hilten M, Oosterkamp E, Bogaardt M-J (2021) The use of digital twins in healthcare: socio-ethical benefits and socio-ethical risks. Life Sci Soc Policy 17(1). https://doi.org/10.1186/s40504-021-00113-x
43. Liu Y, Zhang L, Yang Y, Zhou L, Ren L, Wang F, Liu R, Pang Z, Deen MJ (2019) A novel cloud-based framework for the elderly healthcare services using digital twin. IEEE Access 7:49088–49101. https://doi.org/10.1109/access.2019.2909828
44. Laubenbacher R, Sluka James P, Glazier James A (2021) Using digital twins in viral infection. Science 371(6534):1105–1106. https://doi.org/10.1126/science.abf3370

45. Feng Y, Chen X, Zhao J (2018) Create the individualized digital twin for noninvasive precise pulmonary healthcare. Significances Bioeng Biosci 1(2):26–30. https://doi.org/10.31031/SBB. 2018.01.000507

46. Putz B, Dietz M, Empl P, Pernul G (2021) EtherTwin: blockchain-based secure digital twin information management. Inf Process Manag 58(1):102425. https://doi.org/10.1016/j.ipm. 2020.102425

47. Leng J, Yan D, Liu Q, Xu K, Zhao JL, Shi R, Wei L, Zhang D, Chen X (2019) ManuChain: combining permissioned blockchain with a holistic optimization model as bi-level intelligence for smart manufacturing. IEEE Trans Syst Man Cybern Syst 50(1):182–192. https://doi.org/ 10.1109/tsmc.2019.2930418

48. Lu Y, Huang X, Zhang K, Maharjan S, Zhang Y (2021) Communication-efficient federated learning and permissioned blockchain for digital twin edge networks. IEEE Internet Things J 8(4):2276–2288. https://doi.org/10.1109/jiot.2020.3015772

49. Corral-Acero J, Margara F, Marciniak M, Rodero C, Loncaric F, Feng Y, Gilbert A, Fernandes JF, Bukhari HA, Wajdan A, Martinez MV, Santos MS, Shamohammdi M, Luo H, Westphal P, Leeson P, Diachille P, Gurev V, Mayr M et al (2020) The 'digital twin' to enable the vision of precision cardiology. Eur Heart J 41(48):4556–4564. https://doi.org/10.1093/eurheartj/eha a159

50. Charles WM (2021). Accelerating life sciences research with blockchain. In: Namasudra S, Deka GC (eds) Applications of blockchain in healthcare, vol 83. Springer Nature, Berlin, pp 221–252. https://doi.org/10.1007/978-981-15-9547-9_9

51. Jeon H-J, Youn H-C, Ko S-M, Kim T-H (2021) Blockchain and AI meet in the metaverse. In: Fernández-Caramés TM, Fraga-Lamas P (eds) Blockchain potential in AI [Working Title]. IntechOpen. https://doi.org/10.5772/intechopen.99114

52. Meta (2021, October 28) Connection is evolving and so are we. https://about.facebook.com/ meta. Accessed 13 Nov 2021

53. Lee L-H, Braud T, Zhou P, Wang L, Xu D, Lin Z, Kumar A, Bermejo C, Hui P (2021) All one needs to know about metaverse: a complete survey on technological singularity, virtual ecosystem, and research agenda. University of Helsinki. https://doi.org/10.13140/RG. 2.2.11200.05124/7

54. van der Merwe D (2021) The metaverse as virtual heterotopia. Diamond Scientific Publishing. https://www.dpublication.com/abstract-of-3rd-socialsciencesconf/41-20250/

55. Kell G (2020) Unforgotten: COVID-19 era grads to be celebrated virtually this Saturday. University of California, Berkeley. https://news.berkeley.edu/2020/05/14/unforgotten-covid-19-era-grads-to-be-celebrated-virtually-this-saturday/. Accessed 13 Nov 2021

56. Ryskeldiev B, Ochiai Y, Cohen M, Herder J (2018) Distributed metaverse: creating decentralized blockchain-based model for peer-to-peer sharing of virtual spaces for mixed reality applications. Association for Computing Machinery. https://doi.org/10.1145/3174910.317 4952

57. Roach J (2021, November 2) Mesh for Microsoft Teams aims to make collaboration in the 'metaverse' personal and fun. Microsoft. https://news.microsoft.com/innovation-stories/mesh-for-microsoft-teams/. Accessed 13 Nov 2021

58. Hudson K, Taylor LA, Kozachik SL, Shaefer SJ, Wilson ML (2015) Second life simulation as a strategy to enhance decision-making in diabetes care: a case study. J Clin Nurs 24(5–6):797–804. https://doi.org/10.1111/jocn.12709

59. Schaffer MA, Tiffany JM, Kantack K, Anderson LJW (2016) Second Life® virtual learning in public health nursing. J Nurs Educ 55(9):536–540. https://doi.org/10.3928/01484834-201608 16-09

60. Rudolphi-Solero T, Jimenez-Zayas A, Lorenzo-Alvarez R, Domínguez-Pinos D, Ruiz-Gomez MJ, Sendra-Portero F (2021) A team-based competition for undergraduate medical students to learn radiology within the virtual world Second Life. Insights Imaging 12(1). https://doi.org/ 10.1186/s13244-021-01032-3

61. Beard L, Wilson K, Morra D, Keelan J (2009) A survey of health-related activities on second life. J Med Internet Res 11(2):e17. https://doi.org/10.2196/jmir.1192

62. Gorini A, Gaggioli A, Vigna C, Riva G (2008) A second life for eHealth: prospects for the use of 3-D virtual worlds in clinical psychology. J Med Internet Res 10(3):e21. https://doi.org/10.2196/jmir.1029
63. Casado-Vara R, Corchado JM (2019) Distributed e-health wide-world accounting ledger via blockchain. J Intell Fuzzy Syst 36:2381–2386. https://doi.org/10.3233/JIFS-169949
64. Tian H, He J, Ding Y (2019) Medical data management on blockchain with privacy. J Med Syst 43(2):6. https://doi.org/10.1007/s10916-018-1144-x
65. Kuo T-T, Gabriel RA, Ohno-Machado L (2019) Fair compute loads enabled by blockchain: sharing models by alternating client and server roles. J Am Med Inform Assoc 26(5):392–403. https://doi.org/10.1093/jamia/ocy180
66. Kuo T-T, Ohno-Machado L, Zavaleta Rojas H (2019) Comparison of blockchain platforms: a systematic review and healthcare examples. J Am Med Inform Assoc 26(5):462–478. https://doi.org/10.1093/jamia/ocy185
67. Firdaus A, Anuar NB, Razak MFA, Hashem IAT, Bachok S, Sangaiah AK (2018) Root exploit detection and features optimization: mobile device and blockchain based medical data management. J Med Syst 42(6):112. https://doi.org/10.1007/s10916-018-0966-x
68. Mamoshina P, Ojomoko L, Yanovich Y, Ostrovski A, Botezatu A, Prikhodko P, Izumchenko E, Aliper A, Romantsov K, Zhebrak A, Ogu IO, Zhavoronkov A (2018) Converging blockchain and next-generation artificial intelligence technologies to decentralize and accelerate biomedical research and healthcare. Oncotarget 9(5):5665–5690. https://doi.org/10.18632/oncotarget.22345
69. Durneva P, Cousins K, Chen M (2020) The current state of research, challenges, and future research directions of blockchain technology in patient care: systematic review. J Med Internet Res 22(7):e18619. https://doi.org/10.2196/18619
70. Quaini T, Roehrs A, Da Costa CA, Da Rosa Righi R (2018) A model for blockchain-based distributed electronic health records. IADIS Int J WWW/Internet 16(2):66–79. https://doi.org/10.33965/ijwi_2018161205
71. Dwivedi AD, Srivastava G, Dhar S, Singh R (2019) A decentralized privacy-preserving healthcare blockchain for IoT. Sensors (Basel) 19(2):326. https://doi.org/10.3390/s19020326
72. Yang J, Onik MMH, Kim C-S (2020) Blockchain technology for protecting personal information privacy. In: Ahmed M (ed) Blockchain in data analytics. Cambridge Scholars Publisher, pp 122–144. https://books.google.com/books?id=z_zLDwAAQBAJ&dq
73. Al Omar A, Bhuiyan MZA, Basu A, Kiyomoto S, Rahman MS (2019) Privacy-friendly platform for healthcare data in cloud based on blockchain environment. Futur Gener Comput Syst 95:511–521. https://doi.org/10.1016/j.future.2018.12.044
74. Meyyan P (2018, January 16) Decrypting the utility of blockchain in clinical data management. VertMarkets. https://www.clinicalleader.com/doc/decrypting-the-utility-of-blockchain-in-clinical-data-management-0001. Accessed 23 Oct 2018
75. Goossens M (2018, June 6) Blockchain and how it can impact clinical trials. ICON. http://www2.iconplc.com/blog/blockchain. Accessed 18 Dec 2018

Dr. Wendy Charles has been involved in clinical trials from every perspective for 30 years, with a strong background in operations and regulatory compliance. She currently serves as Chief Scientific Officer for BurstIQ, a healthcare information technology company specializing in blockchain and AI, where she leads the Life Sciences division. Dr. Charles augments her blockchain healthcare experience by serving on the EU Blockchain Observatory and Forum Expert Panel, HIMSS Blockchain Task Force, Government Blockchain Association healthcare group, and IEEE Blockchain working groups. She is also involved as an assistant editor and reviewer for academic journals. Dr. Charles obtained her PhD in Clinical Science with a specialty in Health Information Technology from the University of Colorado, Anschutz Medical Campus. She is certified as an IRB Professional, Clinical Research Professional, and Blockchain Professional.

Printed in the United States
by Baker & Taylor Publisher Services